Recent Progress in Solid Dispersion Technology

Recent Progress in Solid Dispersion Technology

Special Issue Editor
Kohsaku Kawakami

MDPI • Basel • Beijing • Wuhan • Barcelona • Belgrade

Special Issue Editor
Kohsaku Kawakami
National Institute for Materials Science
Japan

Editorial Office
MDPI
St. Alban-Anlage 66
4052 Basel, Switzerland

This is a reprint of articles from the Special Issue published online in the open access journal *Pharmaceutics* (ISSN 1999-4923) in 2019 (available at: https://www.mdpi.com/journal/pharmaceutics/special_issues/solid_dispersion)

For citation purposes, cite each article independently as indicated on the article page online and as indicated below:

LastName, A.A.; LastName, B.B.; LastName, C.C. Article Title. *Journal Name* **Year**, *Article Number*, Page Range.

ISBN 978-3-03921-501-0 (Pbk)
ISBN 978-3-03921-502-7 (PDF)

© 2019 by the authors. Articles in this book are Open Access and distributed under the Creative Commons Attribution (CC BY) license, which allows users to download, copy and build upon published articles, as long as the author and publisher are properly credited, which ensures maximum dissemination and a wider impact of our publications.

The book as a whole is distributed by MDPI under the terms and conditions of the Creative Commons license CC BY-NC-ND.

Contents

About the Special Issue Editor vii

Preface to "Recent Progress in Solid Dispersion Technology" ix

Kohsaku Kawakami
Crystallization Tendency of Pharmaceutical Glasses: Relevance to Compound Properties, Impact of Formulation Process, and Implications for Design of Amorphous Solid Dispersions
Reprinted from: *Pharmaceutics* 2019, 11, 202, doi:10.3390/pharmaceutics11050202 1

Djordje Medarević, Jelena Djuriš, Panagiotis Barmpalexis, Kyriakos Kachrimanis and Svetlana Ibrić
Analytical and Computational Methods for the Estimation of Drug-Polymer Solubility and Miscibility in Solid Dispersions Development
Reprinted from: *Pharmaceutics* 2019, 11, 372, doi:10.3390/pharmaceutics11080372 18

Phuong Tran, Yong-Chul Pyo, Dong-Hyun Kim, Sang-Eun Lee, Jin-Ki Kim and Jeong-Sook Park
Overview of the Manufacturing Methods of Solid Dispersion Technology for Improving the Solubility of Poorly Water-Soluble Drugs and Application to Anticancer Drugs
Reprinted from: *Pharmaceutics* 2019, 11, 132, doi:10.3390/pharmaceutics11030132 51

Felix Ditzinger, Catherine Dejoie, Dubravka Sisak Jung and Martin Kuentz
Polyelectrolytes in Hot Melt Extrusion: A Combined Solvent-Based and Interacting Additive Technique for Solid Dispersions
Reprinted from: *Pharmaceutics* 2019, 11, 174, doi:10.3390/pharmaceutics11040174 77

Ryoma Tanaka, Yusuke Hattori, Yukun Horie, Hitoshi Kamada, Takuya Nagato and Makoto Otsuka
Characterization of Amorphous Solid Dispersion of Pharmaceutical Compound with pH-Dependent Solubility Prepared by Continuous-Spray Granulator
Reprinted from: *Pharmaceutics* 2019, 11, 159, doi:10.3390/pharmaceutics11040159 94

Joanna Szafraniec, Agata Antosik, Justyna Knapik-Kowalczuk, Krzysztof Chmiel, Mateusz Kurek, Karolina Gawlak, Joanna Odrobińska, Marian Paluch and Renata Jachowicz
The Self-Assembly Phenomenon of Poloxamers and Its Effect on the Dissolution of a Poorly Soluble Drug from Solid Dispersions Obtained by Solvent Methods
Reprinted from: *Pharmaceutics* 2019, 11, 130, doi:10.3390/pharmaceutics11030130 107

Hanah Mesallati, Anita Umerska and Lidia Tajber
Fluoroquinolone Amorphous Polymeric Salts and Dispersions for Veterinary Uses
Reprinted from: *Pharmaceutics* 2019, 11, 268, doi:10.3390/pharmaceutics11060268 129

Tuan-Tu Le, Abdul Khaliq Elzhry Elyafi, Afzal R. Mohammed and Ali Al-Khattawi
Delivery of Poorly Soluble Drugs via Mesoporous Silica: Impact of Drug Overloading on Release and Thermal Profiles
Reprinted from: *Pharmaceutics* 2019, 11, 269, doi:10.3390/pharmaceutics11060269 149

Tereza Školáková, Michaela Slámová, Andrea Školáková, Alena Kadeřábková, Jan Patera and Petr Zámostný
Investigation of Dissolution Mechanism and Release Kinetics of Poorly Water-Soluble Tadalafil from Amorphous Solid Dispersions Prepared by Various Methods
Reprinted from: *Pharmaceutics* 2019, 11, 383, doi:10.3390/pharmaceutics11080383 165

About the Special Issue Editor

Kohsaku Kawakami is currently working for International Center for Materials Nanoarchitectonics (MANA); National Institute for Materials Science (NIMS), where he is leading the Medical Soft Matter Group; and the Graduate School of Pure and Applied Sciences, University of Tsukuba, where he is serving as professor. His interests are in the basic science and development of amorphous dosage forms and on the development of a novel drug carrier using phospholipids. He has published more than 150 papers and book chapters and has given more than 150 invited lectures. He was working for pharmaceutical companies such as Merck & Co. and Shionogi & Co. for thirteen years as a senior scientist prior to joining NIMS, where he was responsible for the areas of physicochemical characterization, formulation studies, and DDS studies for new chemical entities. He was in University of Connecticut, School of Pharmacy from 2001 to 2002 as a visiting scholar. He was conferred a Ph.D. in chemical engineering from Kyoto University.

Preface to "Recent Progress in Solid Dispersion Technology"

Amorphous solid dispersion (ASD) has been recognized as a powerful formulation technology to improve oral absorption of poorly soluble drugs for more than half a century. Because ASD is in non-equilibrium state, it is sometimes challenging to control its stability and performance. Remarkable advances have recently been made in ASD technology and have led to the finding that supersaturation created after dissolution of ASDs may not be a simple solution but may involve colloidal structure. This knowledge can transform our ability to design superior ASDs capable of effectively maintain a supersaturated state. Another current hot topic in ASD is the crystallization behavior of active pharmaceutical ingredients. General understanding on the crystallization of small organic compounds is particularly challenging, as their dynamics are affected by both strong (covalent) and weak (noncovalent) interactions, unlike inorganic glasses. Industrial formulators who employ ASDs are therefore particularly concerned about the solid state stability of ASDs (especially their physical stability), which cannot be predicted from conventional accelerated testing protocols. Many studies using various experimental and computational methods are ongoing in hopes of deepening our understanding of the physical stability of ASDs.

Because of such technological innovations, the hurdles for the development of ASDs have been greatly reduced compared to a decade ago. This Special Issue therefore focuses on topics regarding recent progress in ASD technology.

Kohsaku Kawakami
Special Issue Editor

Review

Crystallization Tendency of Pharmaceutical Glasses: Relevance to Compound Properties, Impact of Formulation Process, and Implications for Design of Amorphous Solid Dispersions

Kohsaku Kawakami

World Premier International Research Center for Materials Nanoarchitectonics (WPI-MANA), National Institute for Materials Science, 1-1 Namiki, Tsukuba, Ibaraki 305-0044, Japan; kawakami.kohsaku@nims.go.jp; Tel.: +81-29-860-4424

Received: 2 April 2019; Accepted: 24 April 2019; Published: 1 May 2019

Abstract: Amorphous solid dispersions (ASDs) are important formulation strategies for improving the dissolution process and oral bioavailability of poorly soluble drugs. Physical stability of a candidate drug must be clearly understood to design ASDs with superior properties. The crystallization tendency of small organics is frequently estimated by applying rapid cooling or a cooling/reheating cycle to their melt using differential scanning calorimetry. The crystallization tendency determined in this way does not directly correlate with the physical stability during isothermal storage, which is of great interest to pharmaceutical researchers. Nevertheless, it provides important insights into strategy for the formulation design and the crystallization mechanism of the drug molecules. The initiation time for isothermal crystallization can be explained using the ratio of the glass transition and storage temperatures (T_g/T). Although some formulation processes such as milling and compaction can enhance nucleation, the T_g/T ratio still works for roughly predicting the crystallization behavior. Thus, design of accelerated physical stability test may be possible for ASDs. The crystallization tendency during the formulation process and the supersaturation ability of ASDs may also be related to the crystallization tendency determined by thermal analysis. In this review, the assessment of the crystallization tendency of pharmaceutical glasses and its relevance to developmental studies of ASDs are discussed.

Keywords: pharmaceutical glass; crystallization tendency; crystallization; nucleation; milling; accelerated stability test

1. Introduction

Amorphous solid dispersions (ASDs) are among of the most effective enabling formulations for improving the dissolution process and therefore the oral absorption of poorly soluble drugs [1–7]. Because of their high energy, amorphous solids can reach a supersaturated state during their dissolution process. Although solubilization techniques that increase the equilibrium solubility, including the use of micelles and organic solvents, can inhibit membrane permeation [8,9], it does not happen for supersaturated systems originated from ASDs [10]. It is now widely recognized that the supersaturation created by ASDs can cause phase separation into concentrated and diluted phases, based on the spinodal decomposition mechanism, followed by the formation of a quasi-equilibrium colloidal structure consisting of a concentrated dispersed phase suspended in a diluted continuum phase [11,12]. Although the role of the dispersed phase in the oral absorption is still under debate, this process can maintain high levels of supersaturation for the continuum phase, which are beneficial for oral absorption [13]. The stability of the colloidal phase is significantly influenced by the polymer species [13–15]. Since the supersaturation behavior of ASDs, including phase separation and its impact on membrane transport and oral absorption, are outside the scope of this review, readers interested in these aspects are referred to recent studies [11,13,16–18] for further details.

Drug molecules in ASDs must remain in the amorphous state to exert their beneficial effects during the dissolution process. Even a trace amount of crystals would undermine these favorable effects, because it induces crystallization after suspension of the ASD in aqueous media [19,20]. Polymeric excipients in ASDs serve not only for improving the supersaturation behavior as mentioned above, but also for inhibiting crystallization of the drug. Miscibility is an important factor for exploiting the stabilization effect by the polymer [21–23]. Obviously, the crystallization tendency of the drug molecule itself is another important factor affecting the storage stability.

Table 1 summarizes generally accepted ideas for good glass formers in the case of small organic compounds. Good glass formers tend to have a large molecular weight [24]; other chemical-structural properties of these compounds include a low number of benzene rings, a high degree of molecular asymmetry, as well as large numbers of rotatable bonds, branched carbon skeletons, and electronegative atoms [25–27]. Specific tendencies can be found for the physicochemical properties as well. Good glass formers should have a high melting temperature and enthalpy/entropy, as well as a large free energy difference between crystalline and amorphous states [26]. Fragility [28,29], which quantifies the degree of non-Arrhenius behavior of a glass, is another parameter that can correlate with the crystallization tendency [26,30,31]. However, it should be emphasized that the crystallization tendency of a certain compound is frequently determined by observing its crystallization during rapid cooling or cooling/reheating cycles using differential scanning calorimetry, which does not necessarily reflect easiness of the isothermal crystallization, which is of interest for pharmaceutical researchers. The difference between hot (non-isothermal) and isothermal crystallization is schematically illustrated in Figure 1. Hot crystallization proceeds upon a decrease in free volume, and each molecule has a relatively high conformational flexibility during the crystallization. On the other hand, isothermal crystallization occurs under almost constant volume, and the molecular motion is more restricted. Crystallization can only be achieved after overcoming the energetic barrier to structural transformation, in which noncovalent "weak" interactions play an important role, unlike in inorganic glasses.

Table 1. Features of good glass formers based on small organic molecules.

Chemical-Structural Features	Physicochemical Features
Large molecular weight	Large melting enthalpy/entropy
Low number of benzene rings	High melting temperature
Low symmetry	Large crystal/amorphous energy difference
Large number of rotatable bonds	Large fragility
High branching degree	Large T_g/T_m
Large number of electronegative atoms	Large viscosity above T_g

T_g, glass transition temperature; T_m, melting temperature.

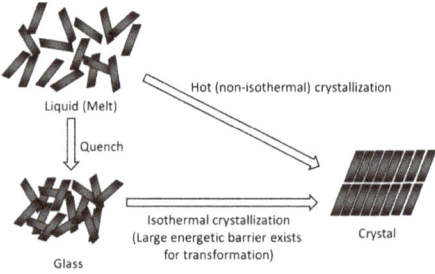

Figure 1. Schematic representation of hot (non-isothermal) and isothermal crystallization.

The following sections review the crystallization tendency of pharmaceutical glasses, with emphasis on relationship with their chemical structure, remark on its evaluation process, relevance for glass properties including the storage stability (i.e., isothermal crystallization), relevance to manufacture, and possible correlation with the supersaturation ability. In addition to discussion on ideal glasses that can be prepared by melt–quench procedure, the stability of real glasses, which are prepared through formulation process such as milling, is also discussed.

2. Classification of Crystallization Tendencies

In the field of pharmaceutical sciences, many research groups have evaluated the crystallization tendency of drug molecules by applying a cooling/reheating cycle to the melt in a differential scanning calorimetry (DSC) [26,32]. The following classification, as proposed by Taylor et al. [26], is widely recognized:

Class 1: Compounds that crystallize during cooling from the melt at 20 °C/min.
Class 2: Compounds that do not crystallize during cooling from the melt, but crystallize during subsequent reheating at 10 °C/min.
Class 3: Compounds that do not crystallize during the cooling/reheating cycle mentioned above.

Examples are shown in Figure 2. Haloperidol, a Class 1 compound, always crystallizes at 100 °C during cooling from the melt, regardless of the cooling rate achievable by conventional DSC (Figure 2a) [33], which means that crystallization is entirely governed by the temperature. It should be noted that the crystallization temperature of some Class 1 compounds such as tolbutamide depends on the cooling rate [33]. Class 1 compounds can be further divided into two groups according to their crystallization behavior during cooling in liquid nitrogen, whereby compounds that crystallize and remain amorphous are categorized as Class 1a and Class 1b, respectively [34]. This difference is likely to be analogous to the dependence of the crystallization temperature on the cooling rate mentioned above, that is, haloperidol and tolbutamide can be identified as Class 1a and Class 1b compounds, respectively. In the case of haloperidol, crystallization is inhibited when the melt is cooled at a rate faster than 100 °C/s to produce a mesophase [33]. Acetaminophen, a Class 2 compound, does not crystallize during cooling, but crystallizes during the subsequent reheating (Figure 2b). Fenofibrate, a Class 3 compound, does not crystallize during the cooling/reheating cycle (Figure 2c). Tables 2–4 summarizes examples of compounds belonging to each class.

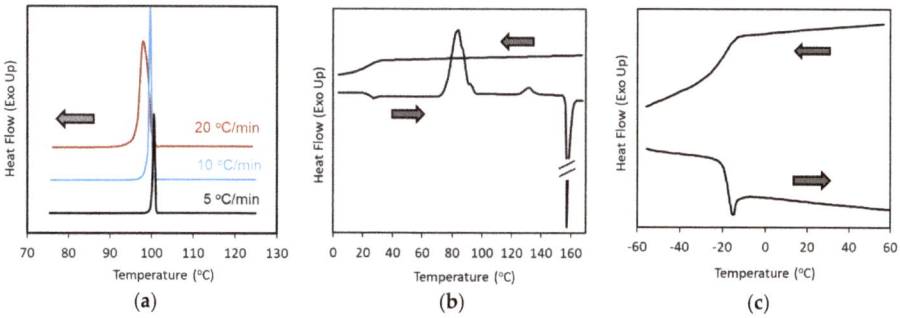

Figure 2. Examples of cooling/reheating differential scanning calorimetry (DSC) curves from the melt: (**a**) cooling curves of haloperidol (Class 1) at various cooling rates, as indicated in the figure; (**b**) cooling/reheating curves of acetaminophen (Class 2); and (**c**) cooling/reheating curves of fenofibrate (Class 3).

Table 2. Examples of Class 1 compounds.

Compounds	M_w (Da)	T_m (°C)	T_g (°C)	T_g/T_m	ΔH (kJ/mol)	m	Reference
Antipyrin	188	111	−25	0.65	25.2	81	[31]
Anthranilic acid	137	147	5	0.66	22.8	-	[26]
Atenolol	266	153	22	0.69	37.5	-	[26]
Atovaquone	367	219	-	-	33.5	-	[35]
Benzamide	121	127	−10	0.66	21.7	-	[26]
Benzocaine	165	89	−31	0.67	22.6	-	[26]
Caffeine	194	237	72	0.68	20.8	-	[26]
Carbamazepine	236	192	61	0.72	25.5	-	[26]
Chlorpropamide	277	118	17	0.74	27.4	219	[31]
Chlorzoxazone	170	191	38	0.67	25.6	-	[26]
Clofibric acid	215	121	-	-	29.0	-	[35]
Diflunisal	250	213	-	-	35.6	-	[35]
Felbinac	212	164	24	0.68	29.8	-	[26]
Flufenamic acid	281	135	17	0.71	27.1	78	[26,36]
Griseofulvin	353	218	89	0.74	39.1	74	[26,37]
Haloperidol	376	152	33	0.72	54.3	-	[26]
Indoprofen	281	212	50	0.67	36.0	-	[26]
Lidocaine	234	68	−39	0.69	16.7	-	[26]
Mefenamic acid	241	231	-	-	39.4	-	[35]
Naproxen	230	157	56	0.77	32.4	-	[35,38]
Nepafenac	254	183	-	-	42.8	-	[35]
Phenacetin	179	136	2	0.67	31.5	-	[26]
Piroxicam	331	201	-	-	35.6	-	[35]
Probenecid	285	199	-	-	40.4	-	[35]
Saccharin	183	228	-	-	29.5	-	[35]
Salicylic acid	138	159	-	-	24.9	-	[35]
Theophylline	180	272	94	0.67	29.6	-	[26]
Tolbutamide	270	128	5	0.69	26.2	122	[31]
Tolfenamic acid	262	213	63	0.69	38.8	-	[26]
Average	237	172	27	0.69	31.1	115	-

M_w, molecular weight; ΔH, melting enthalpy; m, fragility. Although the fragility can be determined by various methods, the evaluation based on the temperature dependence of T_g is preferentially employed because it exhibits the best correlation with the crystallization tendency [31].

Table 3. Examples of Class 2 compounds.

Compounds	M_w (Da)	T_m (°C)	T_g (°C)	T_g/T_m	ΔH (kJ/mol)	m	Reference
Acetaminophen	151	169	23	0.67	27.2	77	[31]
Bifonazole	310	149	16	0.68	39.2	76	[31]
Celecoxib	381	163	58	0.76	37.4	85	[26]
Cinnarizine	369	120	7	0.71	40.9	84	[31]
Clofoctol	365	88	−4	0.75	35.2	70	[26]
Dibucaine	343	65	−35	0.70	29.2	132	[26]
Droperidol	379	143	29	0.73	40.0	108	[26]
Flurbiprofen	244	115	−5	0.69	27.4	88	[31]
Nifedipine	346	172	46	0.72	38.2	112	[31]
Phenobarbital	233	174	42	0.70	28.7	96	[31]
Phenylbutazone	308	106	−6	0.70	27.6	79	[36]
Tolazamide	311	172	18	0.65	43.4	18	[26]
Average	312	136	16	0.71	34.5	85	-

Table 4. Examples of Class 3 compounds.

Compounds	M_w (Da)	T_m (°C)	T_g (°C)	T_g/T_m	ΔH (kJ/mol)	m	Reference
Aceclofenac	354	153	10	0.66	42.3	25	[26]
Clotrimazole	345	141	28	0.73	33.3	63	[31]
Curcumin	368	182	62	0.74	50.1	87	-
Felodipine	384	147	45	0.76	31.0	66	[26]
Fenofibrate	361	80	−19	0.72	33.0	82	[31]
Ibuprofen	206	76	−44	0.66	26.5	75	[31]
Indomethacin	358	161	45	0.73	37.6	85	[31]
Itraconazole	706	168	58	0.75	57.6	731	[26]
Ketoconazole	531	147	44	0.75	52.9	97	[31]
Ketoprofen	254	95	−3	0.73	28.3	67	[31]
Loratadine	383	134	35	0.76	27.3	72	[31]
Miconazole	417	86	1	0.76	32.8	61	[26]
Nilutamide	317	155	33	0.72	31.0	106	[26]
Nimesulide	308	150	21	0.70	33.4	103	[26]
Pimozide	462	219	54	0.66	42.7	170	[26]
Probucol	517	126	27	0.75	39.3	138	[39]
Procaine	236	61	−39	0.70	26.2	90	[31]
Ribavirin	244	168	56	0.75	45.7	70	[40]
Ritonavir	721	122	47	0.81	65.3	86	[31]
Average	393	135	24	0.73	38.8	120	-

Average parameters are also presented in the table for each class of compounds. The molecular weight shows an increase with increasing classification number, which reflects the importance of the complexity of the molecular structure. The melting enthalpy also increases with increasing classification number, which can be explained in terms of the strength of the molecular interactions. On the other hand, the effect of the melting temperature was opposite to the expectation, while the effect of the fragility was not clear. However, the effect of the fragility is difficult to evaluate, because this parameter could not be calculated for most Class 1 compounds. Moreover, the fragility obtained for chlorpropamide exhibited an unusual value, 219, which significantly influenced the overall average.

Figure 3 visualizes individual data of molecular weight and melting enthalpy of compounds in each class. Figure 3a clearly shows that all compounds with the molecular weight larger than 400 Da are involved in Class 3, whereas the molecules smaller than 200 Da are not included in Class 3 at all. However, molecular weight was found to be the only parameter that shows some extent of correlation with the crystallization tendency, if all the data are plotted, as presented in Figure 3. As an example, Figure 3b shows relationship between the melting enthalpy and crystallization tendency. Although the averaged values indicated correlation with the crystallization tendency, it is not obviously statistically meaningful. Other structural/thermodynamic parameters did not exhibit any correlations with the crystallization tendency, either. Special attention to molecular weight was also made by Mahlin et al. [24], who found the molecules larger than 300 Da to be good glass formers during formulation processes. Note that the structural feature of compounds that may be correlated with the crystallization tendency, as shown in Table 1, has been mainly concluded by observing series of compounds that have similarity in their chemical structure. When variety of compounds is collected for examination, focus on single parameter does not seem to be sufficient. The combination of molecular volume and melting enthalpy was reported to be an excellent predictor of the crystallization tendency by Wyttenbach et al., based on theoretical considerations centered on the so-called Prigogine–Defay ratio [35]. In their study, the trend of the T_g/T_m ratio also agreed with the expected trend; interestingly, the T_g/T_m parameter was also shown to be correlated with the Prigogine–Defay ratio [35,41].

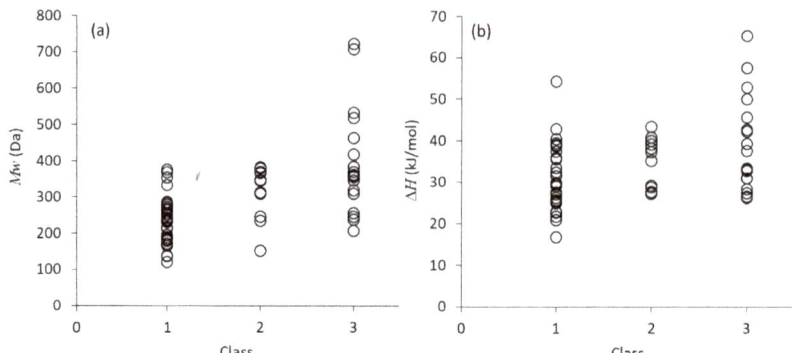

Figure 3. Visualization of: (**a**) molecular weight; and (**b**) melting enthalpy of compounds belonging to each class.

A common strategy to improve biopharmaceutical performance of poorly soluble candidates includes increase in hydrophilicity, which frequently has trade-off relationship with affinity to therapeutic targets. However, another approach may be suppression of crystallization tendency based on the information described in Table 1 to increase applicability of ASD. As noted below, suppression of crystallization tendency may also be related to increase in supersaturation ability after dissolution. Further understanding on relationship between chemical structure and crystallization tendency should increase options of chemical modification strategy of candidate compounds.

Alternatively, the critical cooling rate for achieving vitrification has also been employed for the classification; for example, compounds that crystallize even at 750 °C/min were classified as Class 1, those with moderate crystallization ability and that can be vitrified at ca. 10–20 °C/min were designated as Class 2, while Class 3 compounds only require a very slow cooling rate, below 2 °C/min, for vitrification [38,42]. Despite the different criteria employed, the classifications based on this methodology agreed well with those in Tables 2–4, except that tolbutamide and cinnarizine were placed in Classes 2 and 3, respectively [38].

The different behavior of Classes 1 and 2 compounds likely reflects differences in nucleation and crystal growth temperatures (Figure 4). For Class 1 compounds, the optimum nucleation and crystal growth temperatures should be close to each other; hence, after reaching an optimum temperature where both nucleation and crystal growth proceed, the melt can crystallize. This process is expected to be based on homogeneous nucleation. In contrast, the optimum nucleation temperature for Class 2 compounds should be located far below the optimum crystal growth temperature. Thus, the melt must be first cooled to the nucleation temperature range and then heated to the crystal growth temperature for crystallization to proceed. However, if the cooling rate is sufficiently slow, there is a finite chance for nucleation to occur at the optimum crystal growth temperature even though the nucleation rate is very low, which could explain the similar classifications produced by the two methods.

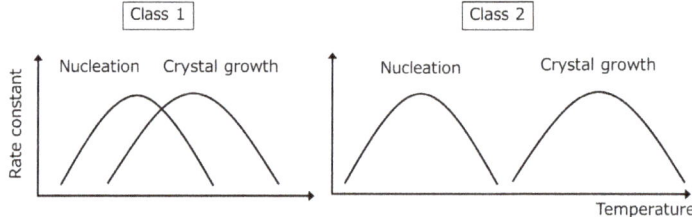

Figure 4. Schematic representation of the temperature dependence of the nucleation and crystal growth temperatures for Classes 1 and 2 compounds.

Figure 5 shows reheating DSC curves of celecoxib melt, illustrating the dependence of the cold crystallization on the target temperature of the cooling process [43]. When the melt was cooled down to −20 °C, a crystallization exotherm was observed during the subsequent heating process. However, no crystallization was observed when the melt was cooled down to 30 °C, although celecoxib is known as a Class 2 compound. Our investigation revealed that the optimum nucleation temperature of celecoxib was ca. −50 °C; thus, cooling to 30 °C was obviously not enough for inducing nucleation. In the classification criteria discussed above, the minimum temperature of the cooling process is not specified. However, a poor understanding of the nucleation process may result in the misclassification of a particular compound.

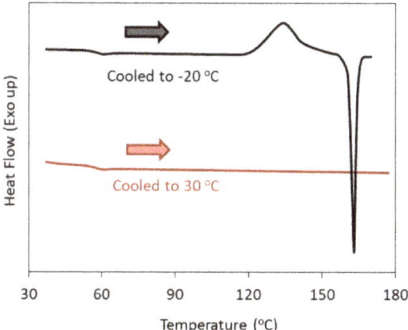

Figure 5. Reheating DSC curves of celecoxib melt, illustrating the dependence of the cold crystallization on the target temperature of the cooling process (shown in the figure).

The different behavior of Classes 2 and 3 compounds is likely due to the different strength of their molecular interactions. Thus, the presence of neighboring molecules during the crystallization cannot be ignored, and the crystallization is based on heterogeneous nucleation.

3. Relationship between Crystallization Tendency and Isothermal Crystallization

The crystallization tendency discussed above does not directly correlate with the physical stability under isothermal conditions. However, these two processes do have some indirect relationships. Figure 6 shows the time to reach 10% crystallinity (t_{10}, expressed in minutes) for pharmaceutical glasses as a function of T_g/T, where T is the storage temperature [44]. These data were acquired for quenched glass pellets under dry conditions. Crystallization has frequently been observed to start at the surface [45,46]. Since the pellets have a very small surface area, the surface effects on the crystallization were almost eliminated in this experiment. Clearly, the data corresponding to most compounds fell on a universal line; in particular, the compounds located on the line belonged to Classes 1 and 2. The other compounds, which exhibited better stability especially above T_g, belonged to Class 3.

The above data were obtained by fitting the crystallinity value at each time point to the Avrami–Erofeev equation. The obtained Avrami exponents are shown in Table 5 Smaller Avrami exponents were obtained for higher classification numbers, which indicates that the nucleation mechanism becomes more homogeneous with decreasing classification number. This hypothesis is also supported by a previous in-situ analysis of the isothermal crystallization process of tolbutamide and acetaminophen using synchrotron X-ray diffraction [44].

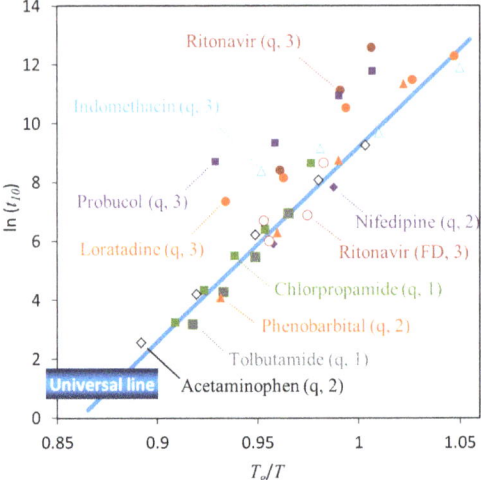

Figure 6. Initiation time of crystallization (t_{10}, min) as a function of T_g/T. The q and FD labels in the parentheses indicate that the glass was prepared by quenching and freeze-drying, respectively. The numbers in parentheses denote the crystallization tendency classification. The universal line is the best fit for Classes 1 and 2 compounds ($\ln(t_{10}) = 66.2 T_g/T - 57.0$).

Table 5. Ranges of Avrami exponents for isothermal crystallization.

Classification	Compound	Avrami Exponent
Class 1	Tolbutamide	3.7–4.6
	Chlorpropamide	3.0–4.2
Class 2	Acetaminophen	2.1–3.0
	Nifedipine	2.0
Class 3	Ritonavir	2.2–3.1
	Indomethacin	1.0–2.6
	Loratadine	1.1–1.5
	Probucol	1.2–1.3

The crystallization of some glasses was observed to start at the surface. In the case of indomethacin, crystallization is enhanced with decreasing particle size, which is most likely due to the increasing surface area [45]. Moreover, the crystallization of indomethacin glass particles is retarded by a polymer coating of the surface [46]. Quenched ritonavir glass exhibited higher stability relative to that of the compounds located on the universal line in Figure 6. However, the stability of freeze-dried ritonavir glass could be explained by the universal line, which is likely due to the increase in surface area [47]. The lower packing of the glass structure might also partially contribute to eliminate the effect of molecular interactions. The surface effects are usually explained in terms of the higher mobility of surface molecules [48], due to a decreased number of nearest neighbor molecules [49].

The results in Figure 6 suggest that the physical stability of Classes 1 and 2 compounds was strongly affected by the temperature. In these cases, physical stabilization of the glasses appears difficult to achieve without adding excipients. However, as the crystallization of Class 3 compounds is influenced by molecular interactions, physical stabilization of these systems may be achieved by manipulating these interactions. In fact, quenched ritonavir glass had higher stability compared to that of the freeze-dried glass, as discussed above.

Sub-T_g annealing based on this strategy was found to be an effective strategy for stabilizing ritonavir glass [50]. For example, ritonavir glass annealed at 40 °C for two days was much more stable compared to fresh glass. The fresh glass reached a crystallinity of 58% after annealing at 60 °C for six days, whereas the glass pre-annealed at 40 °C reached a crystallinity of only 8% after the same annealing procedure at 60 °C. Structural analysis revealed a change in the packing volume and hydrogen-bonding pattern during the pre-annealing at 40 °C, which was the most likely source of the stabilization. Such pre-annealing strategy did not work for Classes 1 and 2 compounds [50].

4. Non-Ideal Crystallization of Practical Glasses

The discussion presented above is based on observation under well-defined conditions, where effect of mechanical stress, moisture sorption, and surface area were minimized. Crystallization behavior of practical glasses, especially in the case of powder samples, may not be explained in such an ideal manner. Glasses prepared by grinding typically exhibit lower stability than the intact ones most likely because of remaining nuclei and/or small crystals that cannot be detected by X-ray powder diffraction. In the observation of Crowley et al. [51], crystallization behavior of indomethacin glasses prepared by cryogenic grinding of various crystal forms depended on the initial crystal form used, suggesting that the ground glasses remembered their original forms even after the grinding. In their study, they also observed significant differences in the crystallization rates of ground and quenched glasses. Thus, although grinding is a simple process to prepare amorphous form in a laboratory scale, it is not recommended because of difficulty in transformation into the amorphous state in a molecular level.

Even for melt–quenched glasses, application of subsequent grinding process can accelerate crystallization [52]. Moreover, very weak stresses such as crack formation [53] and transfer to different vessels [52] are also suspected as causes of nucleation. Figure 7 shows comparison of crystallization behavior of melt–quenched indomethacin glasses at 30 °C with or without grinding process before the storage. In the absence of the grinding process, the quenched glass remained completely in an amorphous state for more than one month. However, if the grinding process is applied for the melt–quenched glass, crystallization is initiated within one day. This comparison clearly indicates significant effect of the grinding process on the crystallization behavior, which appeared to be due to increase in the surface area and mechanical stress. It is also interesting to note that the crystal form obtained was not identical in these examples. Since no relevance between the preparation process and crystal form could be found, it might be because of difference in impurity profiles.

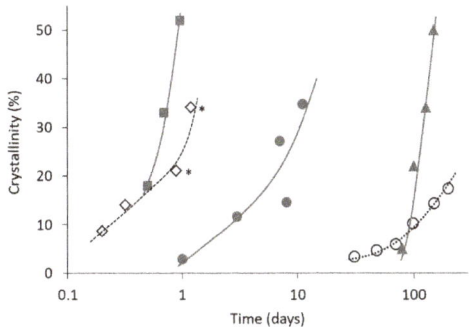

Figure 7. Isothermal crystallization of indomethacin glasses at 30 °C under dried condition. (■) Quenched and ground for 6 min. Crystallized to form γ [51]. (◇) Quenched and ground. Crystallized to form α except that symbols with asterisk involves small amount of form γ [54]. (●) Quenched and cryoground. Crystallized to mixture of form α and γ (our data). (▲) Quenched. Crystallized to form γ [55]. (○) Quenched and stored in DSC pan (our data). Crystallized to form α which contains small amount of form γ.

Crystallization of nifedipine is very sensitive to various factors including moisture sorption and mechanical stress. Thus, extensive care is required to investigate the ideal crystallization behavior as presented in Figure 6. In our experiments, crystalline powder was dried in a vacuum oven at 50 °C and stored in a desiccator with silica gel before use. Then, the dried powder was loaded in a hermetically sealed pan under flow of dried nitrogen air, and subjected to the melt–quench procedure to initiate the stability study. Only after such careful treatment, the data which could be explained by the universal line were obtained.

Therefore, the data for nifedipine crystallization found in literature are usually faster than the expectation from the universal line. Figure 8 shows onset crystallization time of nifedipine glasses extracted from various literature sources. As already presented in Figure 6, the nifedipine data obtained after the careful treatment mentioned above were explainable by the universal line. However, the crystallization was much faster for the glasses loaded in normal sealed pans without pretreatment. Observation using polarized light microscopy by Bhugra et al. was done in a very careful manner [56], where cracked glasses were eliminated from the analysis, because it can enhance the crystallization. However, the crystallization was much faster, presumably because the glasses could not be shielded from outer atmosphere completely.

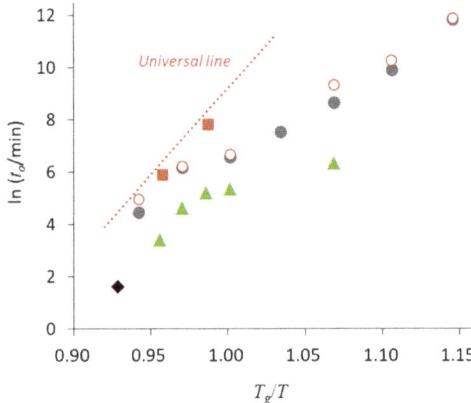

Figure 8. Onset crystallization time (t_o, min) of nifedipine glass as a function of T_g/T. (■) After the pretreatment (see text), quenched in hermetically sealed pan (our data) [44]. (○) Quenched in sealed DSC pan without pretreatment (our data). (●) Quenched in DSC pan [57]. (▲) Quenched on glass slides and crystallization was observed by polarized light microscopy [56]. Cracked glasses were excluded from the analysis. (◆) Quenched in DSC pan [58]. All the literature data were recalculated using the T_g value of 45.5 °C. Definition of onset crystallization time, which is analogous to t_{10}, is slightly different depending on literature, but its impact is ignorable in the analysis here.

Compression process is also recognized to affect the crystallization kinetics. Figure 9 shows effect of compression pressure on crystallization of sucrose glass investigated by isothermal microcalorimetry [59]. Initiation time for crystallization was rarely influenced below 0.5 MPa; however, crystal growth was enhanced with increasing pressure. It was shortened at 2.5 MPa, suggesting that condensation of glass structure can enhance nucleation after application of such relatively weak compression force. Similarly, Ayenew et al. reported that cold crystallization of indomethacin glass was enhanced by compression at ca. 43.7 MPa [60]. In their study, uncompressed glass, which was prepared by cooling the melt at 0.2 °C/min, was observed to crystallize at 121.4 °C during subsequent reheating at 5 °C/min. However, it decreased to 114.1, 113.1, and 112.7 °C, if the compression was applied for 1 s, 2.5 min, or 5 min, respectively. Rams-Baron et al. observed that isothermal crystallization of etoricoxib was significantly enhanced after compression at 300 MPa; however, it could be prevented by mixing with polyvinylpyrroridone (PVP), where investigations were done under the identical relaxation time conditions [61]. This result indicated that physical barrier by excipients were very effective for inhibiting pressure-induced nucleation.

Based on the universal line, the only requirement for assuring three-year stability of pharmaceutical glasses at 25 °C is the T_g higher than 48 °C [44]. Its applicability to practical glasses which are produced under various mechanical stresses without protection from outer atmosphere is discussed next.

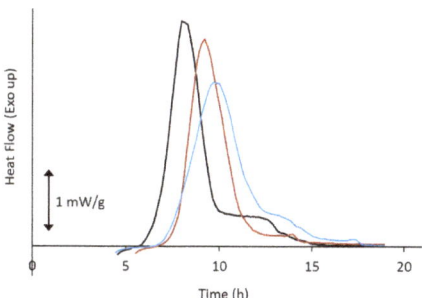

Figure 9. Effect of compression pressure on crystallization heat flow curves of sucrose glasses investigated by isothermal microcalorimetry (30 °C). Freeze-dried sucrose was compressed at pressure of ca. 2.5 MPa (black), 0.5 MPa (red), or 0.1 MPa (blue) for 10 s and subjected to the measurement.

5. Relevance to Formulation Research

Practical ASDs cannot be manufactured by the melt–quenching procedure. The crystallization tendency during practical formulation processes is also of great interest to formulators. This is similar but different phenomena from the crystallization from the melt; therefore, some attempts have been made to find relevance between them. The tendency to crystallize from solution after drying is of great importance for evaluating applicability of spray-drying. Eerdenbrugh et al. investigated the crystallization of 51 compounds after removal of the solvent by spin coating, in order to identify possible correlations between the crystallization tendencies evaluated by thermal analysis upon cooling/reheating and during drying from solutions [62]. In their analysis, the compounds that exhibited a tendency to crystallize immediately after spin coating were denoted as Class 1, those that crystallized within one week were categorized as Class 2, and the remaining compounds were regarded as Class 3. Approximately 76% of the compounds classified as Class 1 by DSC were also assigned to Class 1 by the spin coating method, whereas 76% of the compounds classified as Class 3 by DSC were again classified in the same group by the spin coating approach. The T_g values seemed to break the correlation between the two classification methods.

The relevance to vitrification during milling has also been studied. Blaabjerg et al. reported minimum milling times to achieve vitrification as 90 and 270 min for Classes 3 and 2 compounds, respectively, whereas no vitrification was achieved for any of the Class 1 compounds [42]. It should be noted that some Classes 2 and 3 compounds failed to form amorphous systems, most likely due to their low T_g. This observation suggests that the crystallization tendency from the melt can be used to guide the design of hot-melt extrusion processes, along with additional information on the T_g values.

Thus, applicability of ASD technology to poorly soluble candidates may be judged from the crystallization tendency determined by DSC with the information on T_g. In the formulations, polymeric excipients are used for two purposes: physical stabilization and improvement of dissolution/supersaturation behaviors [12]. Class 3 compounds can be expected to be transformed into the amorphous state using typical formulation processes for ASDs even without excipients. Main purposes of addition of polymeric excipients in the ASD design are to raise T_g for ensuring storage stability and to improve dissolution behavior. Amount of excipient may be kept small for these compounds. Even if amount of the drug exceeds solid solubility limit, the drug is expected to remain in the amorphous state [63]. In contrast, Class 1 compounds must be completely mixed with excipients in a molecular level for the successful transformation to the amorphous state. The amounts of excipients are expected to be larger compared to that for Class 3 compounds. Typically, solid solubility of drug in polymeric matrix under ambient temperature is below 30%, sometimes below 10%, depending on combination of drug and excipient [21–23]. Moreover, it is frequently observed that the effective polymer for physical stabilization and dissolution improvement is different. Hydroxypropyl methylcellulose acetate succinate frequently offers great effect for maintaining high level of supersaturation; however, its miscibility with drug is typically low. In contrast, PVP and its derivatives have relatively high miscibility with drug, but its supersaturation effect cannot be maintained for long duration in many cases. It must be recognized as well that prepared ASDs are not necessarily in the equilibrium state. If ASDs are prepared under an elevated temperature condition, as in case of hot-melt extrusion, the mixing state at this temperature may be kinetically frozen even after cooling to ambient temperature. In spray-drying, the drug and excipient molecules may be separated based on the difference in their molecular weights, because diffusion rate during evaporation process is different [64,65]. This kinetically-separated structure may also be frozen after the drying [65,66]. If solvents are used during the preparation, as in the cases of spray-drying and coprecipitation, the mixing state of the ASDs is affected by the solvent species [67]. In such cases, the mixing state may change with time [23].

How the universal line in Figure 6 is applicable to multi-component ASDs is of great interest. Figure 10 shows comparison of the onset crystallization time of single phase ASDs appearing in the literature. As an overall trend, the universal line seems to work even for the multicomponent systems. Comparison of nifedipine/PVP ASDs from three different stuides implies that milling enhances the nucleation. However, presence of polymeric excipients appears to stabilize the ASDs more than expected from change in T_g (i.e., molecular mobility), most likely because of dilution effect and interaction with drug. The result for Sanofi–Aventis compounds is the most informative from a practical point of view, because the ASD was prepared by spray-drying. Stability of this ASD is a little lower but roughly agrees with the universal line regardless of absence/presence of the moisture. When each dataset is fitted with a regression line, their slopes are almost the same, suggesting that activation energy for nucleation does not significantly depend on the type of ASDs. Design of accelerated physical stability test may be possible for ASDs based on this information.

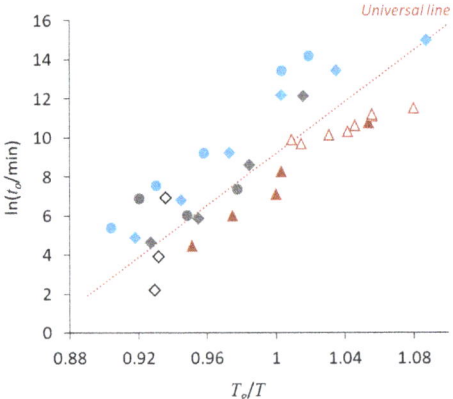

Figure 10. Onset crystallization time (t_o, min) of various ASDs as a function of T_g/T. (◆,♦,◇) Nifedipine/PVP ASDs prepared by melt–quench [68,69] followed by milling [58], respectively. (●, ●) Phenobarbital/PVP ASDs prepared by melt–quench [68,69]. (▲,△) Sanofi–Aventis compound/HPMCP ASDs prepared by spray-drying stored under dried and humid conditions, respectively [70]. Definition of onset crystallization time, which is analogous to t_{10}, is slightly different depending on the study, but its impact is ignorable in the analysis here. HPMCP, Hydroxypropyl methylcellulose phthalate.

6. Relevance to the Dissolution Benefits of ASDs

The greatest advantage of ASDs is that they can achieve supersaturation of poorly soluble drugs. Supersaturation can be maintained unless crystallization occurs in aqueous environments [71]. Thus, it is important to understand the controlling factors that cause crystallization of pharmaceutical glasses in aqueous media. Our preliminary investigation revealed that crystallization proceeds immediately above T_g of the solid [33]. Blaabjerg et al. reported that the degree of supersaturation tended to be high for good glass formers, but no correlation was found between crystallization tendency and supersaturation lifetime [72]. We have shown that amount of orally absorbed fenofibrate was correlated with liquid–liquid phase separation concentration, which is analogous to degree of supersaturation, if the oral absorption was limited by solubility [13]. Thus, suppression of the crystallization tendency can be an option of chemical modification of the poorly soluble candidates instead of increasing aqueous solubility. Alhalaweh analyzed the relationship between the crystallization tendencies from the melt and those in solution, highlighting the need to consider structural factors to improve the correlation between these tendencies [73]. From the viewpoints of physical stability and supersaturation ability, Class 3 compounds seem to be suitable candidates for ASDs. In fact, most of the marketed ASDs consist of Class 3 compounds [74].

7. Summary

This review provides the classification of the crystallization tendencies of pharmaceutical compounds, focusing on its relevance for the glass properties. Possible relationships discussed in this review are summarized in Table 6. In addition to its effectiveness for describing the physical stabilities of ASDs, this classification provides important insights into the glass properties. The investigation of the crystallization mechanism of small organic compounds is an attractive subject because of their structural diversity and complicated molecular interactions, in contrast to the inorganic compounds that have dominated the field of glass science so far. Further progress in this field can make a significant contribution to both basic glass science and practical developmental studies of pharmaceutical products.

Table 6. Relevance of crystallization tendency classification for glass properties.

With increasing classification number:
- Nucleation becomes more heterogeneous
- The nucleation barrier becomes larger
- Surface effects become more important
- Vitrification during the formulation processes may become easier
- The supersaturation ability may increase
- The universal line in Figure 6 is applicable for Class 1 and 2 compounds, whereas the stability of Class 3 compounds may be better
- Stabilization may be achieved via thermal treatment

Preparation of practical formulations involves some procedure to enhance nucleation. Ideality of the nucleation/crystallization behavior is destroyed by application of some activation processes such as milling. However, presence of polymeric excipients can contribute to stabilization, presumably due to dilution effect and its interaction with drug molecules. As a result, deviation from the ideal behavior due to formulation processes is suppressed for enabling rough prediction of the crystallization time. Design of accelerated physical stability test may be possible for ASDs based on this observation.

Although compounds in any classes can be formulated as ASDs, Class 3 compounds obviously have the highest applicability. In addition to their high physical stability, they may have an advantage in supersaturation ability that has great contribution to enhanced absorption. Therefore, chemical modification to decrease crystallization tendency may be considered as an option for drug design instead of increasing solubility.

Funding: This research received no external funding.

Conflicts of Interest: The author declares no conflicts of interest.

References

1. Brouwers, J.; Brewster, M.E.; Augustijns, P. Supersaturating drug delivery systems: The answer to solubility-limited oral bioavailability? *J. Pharm. Sci.* **2009**, *98*, 2549–2572. [CrossRef]
2. Kawakami, K. Modification of physicochemical characteristics of active pharmaceutical ingredients and application of supersaturatable dosage forms for improving bioavailability of poorly absorbed drugs. *Adv. Drug Deliv. Rev.* **2012**, *64*, 480–495. [CrossRef] [PubMed]
3. Paudel, A.; Worku, Z.A.; Meeus, J.; Guns, S.; Van den Mooter, G. Manufacturing of solid dispersions of poorly soluble drugs by spray drying: Formulation and process considerations. *Int. J. Pharm.* **2013**, *453*, 253–284. [CrossRef] [PubMed]
4. Kawakami, K. Theory and practice of supersaturatable formulations for poorly soluble drugs. *Ther. Deliv.* **2015**, *6*, 339–352. [CrossRef] [PubMed]
5. Singh, A. Spray drying formulation of amorphous solid dispersions. *Adv. Drug Deliv. Rev.* **2016**, *100*, 27–50. [CrossRef] [PubMed]
6. Repka, M.A.; Bandari, S.; Kallakunta, V.R.; Vo, A.Q.; McFall, H.; Pimparade, M.B.; Bhagurkar, A.M. Melt extrusion with poorly solubloe drugs—An integrated review. *Int. J. Pharm.* **2018**, *535*, 68–85. [CrossRef]
7. Jermain, S.V.; Brough, C.; Williams, R.O., III. Amorphous solid dispersions and nanocrystal technologies for poorly water-soluble drug delivery—An update. *Int. J. Pharm.* **2018**, *535*, 379–392. [CrossRef] [PubMed]
8. Twist, J.N.; Zatz, J.L. Influence of solvents on paraben permeation through idealized skin model membranes. *J. Soc. Cosmet. Chem.* **1986**, *37*, 429–444.
9. Hens, B.; Brouwers, J.; Corsetti, M.; Augustijns, P. Gastrointestinal behavior of nano- and microsized fenofibrate: In vivo evaluation in man and in vitro simulation by assessment of the permeation potential. *Eur. J. Pharm. Sci.* **2015**, *77*, 40–47. [CrossRef]
10. Porat, D.; Dahan, A. Active intestinal drug absorption and the solubility-permeability interplay. *Int. J. Pharm.* **2018**, *537*, 84–93. [CrossRef] [PubMed]
11. Taylor, L.S.; Zhang, G.G.Z. Physical chemistry of supersaturated solutions and implications for oral absorption. *Adv. Drug Deliv. Rev.* **2016**, *101*, 122–142. [CrossRef] [PubMed]
12. Kawakami, K. Supersaturation and Crystallization: Non-equilibrium dynamics of amorphous solid dispersions for oral drug delivery. *Exp. Opin. Drug Deliv.* **2017**, *14*, 735–743. [CrossRef] [PubMed]
13. Kawakami, K.; Sato, K.; Fukushima, M.; Miyazaki, A.; Yamamura, Y.; Sakuma, S. Phase separation of supersaturated solution created from amorphous solid dispersions: Relevance to oral absorption. *Eur. J. Pharm. Biopharm.* **2018**, *132*, 146–156. [CrossRef] [PubMed]
14. Ilevbare, G.A.; Liu, H.; Edgar, K.J.; Taylor, L.S. Maintaining supersaturation in aqueous drug solutions: Impact of different polymers on induction times. *Cryst. Growth Des.* **2013**, *13*, 740–751. [CrossRef]
15. Jackson, M.J.; Toth, S.J.; Kestur, U.S.; Huang, J.; Qian, F.; Hussain, M.A.; Simpson, G.J.; Taylor, L.S. Impact of polymers on the precipitation behavior of highly supersaturated aqueous danazol solutions. *Mol. Pharm.* **2014**, *11*, 3027–3038. [CrossRef] [PubMed]
16. Raina, S.A.; Zhang, G.G.Z.; Alonzo, D.E.; Wu, J.; Zhu, D.; Catron, N.D.; Gao, Y.; Taylor, L.S. Impact of solubilizing additives on supersaturation and membrane transport of drugs. *Pharm. Res.* **2015**, *32*, 3350–3364. [CrossRef]
17. Indulkar, A.S.; Gao, Y.; Raina, S.A.; Zhang, G.G.Z.; Taylor, L.S. Exploiting the phenomenon of liquid-liquid phase separation for enhanced and sustained membrane transport of a poorly soluble drug. *Mol. Pharm.* **2016**, *13*, 2059–2069. [CrossRef]
18. Stewart, A.M.; Grass, M.E.; Brodeur, T.J.; Goodwin, A.K.; Morgen, M.M.; Friesen, D.T.; Vodak, D.T. Impact of drug-rich colloids of itraconazole and HPMCAS on membrane flux in vitro and oral bioavailability in rats. *Mol. Pharm.* **2017**, *14*, 2437–2449. [CrossRef]
19. Liu, H.; Wang, P.; Zhang, X.; Shen, F.; Gogos, C.G. Effects of extrusion process parameters on the dissolution behavior of indomethacin in Eudragit E PO solid dispersions. *Int. J. Pharm.* **2010**, *383*, 161–169. [CrossRef]

20. Zhang, S.; Kawakami, K.; Yamamoto, M.; Masaoka, Y.; Kataoka, M.; Yamashita, S.; Sakuma, S. Coaxial electrospray formulations for improving oral absorption of a poorly water-soluble drug. *Mol. Pharm.* **2011**, *8*, 807–813. [CrossRef]
21. Xie, T.; Taylor, L.S. Effect of temperature and moisture on the physical stability of binary and ternary amorphous solid dispersion of celecoxib. *J. Pharm. Sci.* **2017**, *106*, 100–110. [CrossRef]
22. Lehmkemper, K.; Kyeremateng, S.O.; Heinzerling, O.; Dagenhardt, M.; Sadowski, G. Impact of polymer type and relative humidity on the long-term physical stability of amorphous solid dispersions. *Mol. Pharm.* **2017**, *14*, 4374–4386. [CrossRef]
23. Kawakami, K.; Bi, Y.; Yoshihashi, Y.; Sugano, K.; Terada, K. Time-dependent phase separation of amorphous solid dispersions: Implications for accelerated stability studies. *J. Drug Deliv. Sci. Technol.* **2018**, *46*, 197–206. [CrossRef]
24. Mahlin, D.; Bergström, C.A.S. Early drug development predictions of glass-forming ability and physical stability of drugs. *Eur. J. Pharm. Sci.* **2013**, *49*, 323–332. [CrossRef]
25. Miyazaki, T.; Yoshioka, S.; Aso, Y.; Kawanishi, T. Crystallization rate of amorphous nifedipine analogues unrelated to the glass transition temperature. *Int. J. Pharm.* **2007**, *336*, 191–195. [CrossRef]
26. Baird, J.A.; van Eerdenbrugh, B.; Taylor, L.S. A classification system to assess the crystallization tendency of organic molecules from undercooled melts. *J. Pharm. Sci.* **2010**, *99*, 3787–3806. [CrossRef]
27. Mahlin, D.; Ponnambalam, S.; Hockerfelt, M.H.; Bergström, C.A.S. Toward in silico prediction of glass-forming ability from molecular structure alone: A screening tool in early drug development. *Mol. Pharm.* **2011**, *8*, 498–506. [CrossRef]
28. Angell, C.A. Relaxation in liquids, polymers and plastic crystals—Strong/fragile patterns and problems. *J. Non-Cryst. Solids* **1991**, *131–133*, 13–31. [CrossRef]
29. Crowley, K.J.; Zografi, G. The use of thermal methods for predicting glass-former ability. *Thermochim. Acta* **2001**, *380*, 79–93. [CrossRef]
30. Senkov, O.N. Correlation between fragility and glass-forming ability of metallic alloys. *Phys. Rev. B* **2007**, *76*, 104202. [CrossRef]
31. Kawakami, K.; Harada, T.; Yoshihashi, Y.; Yonemochi, E.; Terada, K.; Moriyama, H. Correlation between glass forming ability and fragility of pharmaceutical compounds. *J. Phys. Chem. B* **2015**, *119*, 4873–4780. [CrossRef]
32. Pajula, K.; Taskinen, M.; Lehto, V.P.; Ketolainen, J.; Korhonen, O. Predicting the formation and stability of amorphous small molecule binary mixtures from computationally determined Flory-Huggins interaction parameter and phase diagram. *Mol. Pharm.* **2010**, *7*, 795–804. [CrossRef] [PubMed]
33. Kawakami, K.; Usui, T.; Hattori, M. Understanding the glass-forming ability of active pharmaceutical ingredients for designing supersaturating dosage forms. *J. Pharm. Sci.* **2012**, *101*, 3239–3248. [CrossRef] [PubMed]
34. Baird, J.A.; Thomas, L.C.; Aubuchon, S.R.; Taylor, L.S. Evaluating the non-isothermal crystallization behavior of organic molecules from the undercooled melt state during rapid heat/cool calorimetry. *CrystEngComm* **2013**, *15*, 111–119. [CrossRef]
35. Wyttenback, N.; Kirchmeyer, W.; Alsenz, J.; Kuentz, M. Theoretical considerations of the prigogine-defay ratio with regard to the glass-forming ability of drugs from undercooled melts. *Mol. Pharm.* **2016**, *13*, 241–250. [CrossRef] [PubMed]
36. Tu, W.; Li, X.; Chen, Z.; Liu, Y.D.; Labardi, M.; Capaccioli, S.; Paluch, M.; Wang, L.M. Glass formability in medium-sized molecular system/pharmaceuticals. I. Thermodynamics vs. kinetics. *J. Chem. Phys.* **2016**, *144*, 174502. [CrossRef] [PubMed]
37. Mao, C.; Chamarthy, S.P.; Byrn, S.R.; Pinal, R. A calorimetric method to estimate molecular mobility of amorphous solids at relatively low temperatures. *Pharm. Res.* **2006**, *23*, 2269–2276. [CrossRef] [PubMed]
38. Blaabjerg, L.I.; Lindenberg, E.; Löbmann, K.; Grohganz, H.; Rades, T. Glass forming ability of amorphous drugs investigated by continuous cooling and isothermal transformation. *Mol. Pharm.* **2016**, *13*, 3318–3325. [CrossRef]
39. Kawakami, K.; Ohba, C. Crystallization of probucol from solution and the glassy state. *Int. J. Pharm.* **2017**, *517*, 322–328. [CrossRef]
40. Kawakami, K. Dynamics of ribavirin glass in sub-T_g temperature region. *J. Phys. Chem. B* **2011**, *115*, 11375–11381. [CrossRef]

41. Schmelzer, J.W.P.; Gutzow, I. Structural order parameters, the Prigogine-Defay ratio and the behavior of the entropy in vitrification. *J. Non-Cryst. Solids* **2009**, *355*, 653–662. [CrossRef]
42. Blaabjerg, L.I.; Lindenberg, E.; Rades, T.; Grohganz, H.; Löbmann, K. Influence of preparation pathway on the glass forming ability. *Int. J. Pharm.* **2017**, *521*, 232–238. [CrossRef]
43. Kawakami, K. Nucleation and crystallization of celecoxib glass: Impact of experience of low temperature on physical stability. *Thermochim. Acta* **2019**, *671*, 43–47. [CrossRef]
44. Kawakami, K.; Harada, T.; Miura, K.; Yoshihashi, Y.; Yonemochi, E.; Terada, K.; Moriyama, H. Relationship between crystallization tendencies during cooling from melt and isothermal storage: Toward a general understanding of physical stability of pharmaceutical glasses. *Mol. Pharm.* **2014**, *11*, 1835–1843. [CrossRef]
45. Wu, T.; Yu, L. Surface crystallization of indomethacin below T_g. *Pharm. Res.* **2006**, *23*, 2350–2355. [CrossRef]
46. Wu, T.; Sun, Y.; Li, N.; de Villiers, M.M.; Yu, L. Inhibiting surface crystallization of amorphous indomethacin by nanocoating. *Langmuir* **2007**, *23*, 5148–5153. [CrossRef]
47. Kawakami, K. Surface effects on the crystallization of ritonavir glass. *J. Pharm. Sci.* **2015**, *104*, 276–279. [CrossRef]
48. Bell, R.C.; Wang, H.; Iedema, M.J.; Cowin, J.P. Nanometer-resolved interfacial fluidity. *J. Am. Chem. Soc.* **2003**, *125*, 5176–5185. [CrossRef]
49. Mirigian, S.; Schweizer, K.S. Slow relaxation, spatial mobility gradients, and vitrification in confined films. *J. Chem. Phys.* **2014**, *141*, 161103. [CrossRef]
50. Tominaka, S.; Kawakami, K.; Fukushima, M.; Miyazaki, A. Physical stabilization of pharmaceutical glasses based on hydrogen bond reorganization under sub-T_g temperature. *Mol. Pharm.* **2017**, *14*, 264–273. [CrossRef]
51. Crowley, K.J.; Zografi, G. Cryogenic grinding of indomethacin polymorphs and solvates: Assessment of amorphous phase formation and amorphous phase physical stability. *J. Pharm. Sci.* **2002**, *91*, 492–507. [CrossRef] [PubMed]
52. Bhugra, C.; Shmeis, R.; Pikal, M.J. Role of mechanical stress in crystallization and relaxation behavior of amorphous indomethacin. *J. Pharm. Sci.* **2008**, *97*, 4446–4458. [CrossRef] [PubMed]
53. Descamps, M.; Dudognon, E. Crystallization from the amorphous state: Nucleation-growth decoupling, polymorphism interplay, and the role of interfaces. *J. Pharm. Sci.* **2014**, *103*, 2615–2628. [CrossRef] [PubMed]
54. Otsuka, M.; Kaneniwa, N. A kinetic study of the crystallization process of noncrystalline indomethacin under isothermal conditions. *Chem. Pharm. Bull.* **1988**, *36*, 4026–4032. [CrossRef]
55. Andronis, V.; Zografi, G. Crystal nucleation and growth of indomethacin polymorphs from the amorphous state. *J. Non-cryst. Solids* **2000**, *271*, 236–248. [CrossRef]
56. Bhugra, C.; Shmeis, R.; Krill, S.L.; Pikal, M.J. Prediction of onset of crystallization from experimental relaxation times. II. Comparison between predicted and experimental onset times. *J. Pharm. Sci.* **2008**, *97*, 455–472. [CrossRef]
57. Aso, Y.; Yoshioka, S.; Kojima, S. Relationship between the crystallization rates of amorphous nifedipine, phenobarbital, and flopropione, and their molecular mobility as measured by their enthalpy relaxation and ^{1}H NMR relaxation times. *J. Pharm. Sci.* **2000**, *89*, 408–416. [CrossRef]
58. Kothari, K.; Ragoonanan, V.; Suryanarayanan, R. The role of polymer concentration on the molecular mobility and physical stability of nifedipine solid dispersions. *Mol. Pharm.* **2015**, *12*, 1477–1484. [CrossRef]
59. Kawakami, K.; Miyoshi, K.; Tamura, N.; Yamaguchi, T.; Ida, Y. Crystallization of sucrose glass under ambient conditions: Evaluation of crystallization rate and unusual melting behavior of resultant crystals. *J. Pharm. Sci.* **2006**, *95*, 1354–1363. [CrossRef]
60. Ayenew, Z.; Paudel, A.; Rombaut, P.; Van den Mooter, G. Effect of compression on non-isothermal crystallization behaviour of amorphous indomethacin. *Pharm. Res.* **2012**, *29*, 2489–2498. [CrossRef]
61. Rams-Baron, M.; Pacult, J.; Jedrzejowska, A.; Knapik-Kowalczuk, J.; Paluch, M. Changes in physical stability of supercooled etoricoxib after compression. *Mol. Pharm.* **2018**, *15*, 2969–3978. [CrossRef] [PubMed]
62. van Eerdenbrugh, B.; Baird, J.A.; Taylor, L.S. Crystallization tendency of active pharmaceutical ingredients following rapid solvent evaporation—Classification and comparison with crystallization tendency from undercooled melts. *J. Pharm. Sci.* **2010**, *99*, 3826–3838. [CrossRef]
63. Kawakami, K. Miscibility analysis of particulate solid dispersions prepared by electrospray deposition. *Int. J. Pharm.* **2012**, *433*, 71–78. [CrossRef] [PubMed]
64. Vehring, R. Phamaceutical particle engineering via spray drying. *Pharm. Res.* **2008**, *25*, 999–1022. [CrossRef] [PubMed]

65. Kawakami, K.; Hasegawa, Y.; Deguchi, K.; Ohki, S.; Shimizu, T.; Yoshihashi, Y.; Yonemochi, E.; Terada, K. Competition of thermodynamic and dynamic factors during formation of multicomponent particles via spray drying. *J. Pharm. Sci.* **2013**, *102*, 518–529. [CrossRef]
66. Calahan, J.L.; Azali, S.C.; Munson, E.J.; Nagapudi, K. Investigation of phase mixing in amorphous solid dispersions of AMG 517 in HPMC-AS using DSC, solid-state NMR, and solution calorimetry. *Mol. Pharm.* **2015**, *12*, 4115–4123. [CrossRef] [PubMed]
67. Bank, M.; Leffingwell, J.; Thies, C. The influence of solvent upon the compatibility of polystyrene and poly(vinyl methyl ether). *Macromolecules* **1971**, *4*, 43–46. [CrossRef]
68. Aso, Y.; Yoshioka, S.; Kojima, S. Molecular mobility-based estimation of the crystallization rates of amorphous nidedipine and phenobarbital in poly(vinylpyrrolidone) solid dispersions. *J. Pharm. Sci.* **2004**, *93*, 384–391. [CrossRef]
69. Caron, V.; Bhugra, C.; Pikal, M.J. Prediction of onset of crystallization in amorphous pharmaceutical systems: Phenobarbital, nifedipine/PVP, and phenobarbital/PVP. *J. Pharm. Sci.* **2010**, *99*, 3887–3900. [CrossRef]
70. Greco, S.; Authelin, J.R.; Leveder, C.; Segalini, A. A practical method to predict physical stability of amorphous solid dispersions. *Mol. Pharm.* **2012**, *29*, 2792–2805. [CrossRef] [PubMed]
71. Greco, K.; Bogner, R. Solution-mediated phase transformation: Significance during dissolution and implications for bioavailability. *J. Pharm. Sci.* **2012**, *101*, 2996–3018. [CrossRef] [PubMed]
72. Blaabjerg, L.I.; Lindenberg, E.; Löbmann, K.; Grohganz, H.; Rades, T. Is there a correlation between the glass forming ability of a drug and its supersaturation propensity? *Int. J. Pharm.* **2018**, *538*, 243–249. [CrossRef] [PubMed]
73. Alhalaweh, A.; Alzghoul, A.; Bergström, C.A.S. Molecular divers of crystallization kinetics for drugs in supersaturated aqueous solutions. *J. Pharm. Sci.* **2019**, *108*, 252–259. [CrossRef] [PubMed]
74. Wyttenbach, N.; Kuentz, M. Glass-forming ability of compounds in marketed amorphous drug products. *Eur. J. Pharm. Biopharm.* **2017**, *112*, 204–208. [CrossRef]

© 2019 by the author. Licensee MDPI, Basel, Switzerland. This article is an open access article distributed under the terms and conditions of the Creative Commons Attribution (CC BY) license (http://creativecommons.org/licenses/by/4.0/).

Review

Analytical and Computational Methods for the Estimation of Drug-Polymer Solubility and Miscibility in Solid Dispersions Development

Djordje Medarević [1,*], Jelena Djuriš [1], Panagiotis Barmpalexis [2], Kyriakos Kachrimanis [2] and Svetlana Ibrić [1]

1. Department of Pharmaceutical Technology and Cosmetology, Faculty of Pharmacy, University of Belgrade, Vojvode Stepe 450, 11221 Belgrade, Serbia
2. Department of Pharmaceutical Technology, Faculty of Pharmacy, Aristotle University of Thessaloniki, 54124 Thessaloniki, Greece
* Correspondence: djordje.medarevic@pharmacy.bg.ac.rs; Tel.: +381-11-395-1356

Received: 28 June 2019; Accepted: 22 July 2019; Published: 1 August 2019

Abstract: The development of stable solid dispersion formulations that maintain desired improvement of drug dissolution rate during the entire shelf life requires the analysis of drug-polymer solubility and miscibility. Only if the drug concentration is below the solubility limit in the polymer, the physical stability of solid dispersions is guaranteed without risk for drug (re)crystallization. If the drug concentration is above the solubility, but below the miscibility limit, the system is stabilized through intimate drug-polymer mixing, with additional kinetic stabilization if stored sufficiently below the mixture glass transition temperature. Therefore, it is of particular importance to assess the drug-polymer solubility and miscibility, to select suitable formulation (a type of polymer and drug loading), manufacturing process, and storage conditions, with the aim to ensure physical stability during the product shelf life. Drug-polymer solubility and miscibility can be assessed using analytical methods, which can detect whether the system is single-phase or not. Thermodynamic modeling enables a mechanistic understanding of drug-polymer solubility and miscibility and identification of formulation compositions with the expected formation of the stable single-phase system. Advance molecular modeling and simulation techniques enable getting insight into interactions between the drug and polymer at the molecular level, which determine whether the single-phase system formation will occur or not.

Keywords: solid dispersions; miscibility; solubility; thermodynamic modeling; phase diagram; molecular dynamics simulation; thermal analysis; spectroscopic techniques

1. Introduction

Rise in the number of poorly soluble drugs is accompanied by simultaneous progress in the development of techniques for improving solubility and bioavailability of these drugs, which include but are not limited to: formation of salts and soluble prodrugs, particle size reduction up to nano-size range, using of cosolvents or surfactants in the formulation, complexation with cyclodextrins, formulation of micro or nanoemulsions, and solid dispersions. Solid dispersions, as systems where the drug is dispersed within the polymeric matrix in the crystalline or amorphous state, or dissolved in the polymeric matrix, have been proved to be one of the most successful approaches for overcoming drugs' poor solubility and bioavailability [1–6]. Even though since the introduction of solid dispersions in 1961, by Sekiguchi and Obi [7], thousands of studies have proved their benefits, both in vitro and in vivo, only a few such formulations have appeared on the market up to date. One of the main reasons for this is certainly the difficulty of ensuring long-term product stability due to phase separation between the drug and the polymer and/or drug recrystallization from the initial amorphous form that can cause unacceptable variations in drug dissolution rate and oral bioavailability. It has been well established that apart from differentiation whether the drug is present in the crystalline or amorphous form within the polymeric matrix, it should be determined whether the drug forms single-phase system with amorphous polymer or system separates into drug-rich and polymer-rich phases. It has been shown that the maximum improvement of drug dissolution rate and maintenance of long-term formulation stability are achieved if the formation of a single-phase system occurs [8,9]. Otherwise, if the phase separation occurs, properties of pure components will dominate in the respective phases, and polymer effect on inhibition of drug molecular mobility and reduction of driving force for crystallization will be diminished. Additionally, separation into drug-rich and polymer-rich phases will lead to the fast dissolution of the polymer phase leaving undissolved drug phase [10,11]. Currently available analytical techniques can distinguish between drug and polymer domains of different size, but only at the moment of analysis, which does not guarantee that initially single-phase system will maintain this structure during the whole storage period. Also, direct measurement of drug-polymer miscibility or solubility of the drug in the polymer is challenging due to the high viscosity of polymers below glass transition temperature (T_g), which makes it difficult to achieve equilibrium in the drug-polymer system in the glassy state [12,13]. Only in the last 10 years, it has been recognized that the evaluation of thermodynamics of drug-polymer mixing should be included in the rational design of solid dispersion formulations. Thermodynamic models, initially developed for polymer-polymer and polymer-solvent systems, have been successfully adapted to drug-polymer systems and showed the good prediction of drug-polymer miscibility and solubility of the drug in the polymeric matrix [8,11,14–20]. Although the terms solubility and miscibility are sometimes used with confusion, they can be distinguished, since the term solubility describes the ability of a polymer to dissolve a crystalline drug, while miscibility describes the ability of an amorphous drug to mix with an amorphous polymer giving a single-phase system [11,12]. Although only below the drug solubility limit in the polymer, solid dispersion systems are stable without any concern for drug crystallization, the low solubility of most drugs in the common polymers limits formulation of solid solutions only to very low dose drugs. Therefore, particular efforts are invested to estimate the miscibility of the drug with the polymer, which is always greater than the drug solubility in particular polymer, and below miscibility limit, only large temperature and/or composition fluctuations can destabilize system toward drug crystallization. Thermodynamic modeling allows estimation of the free energy of mixing between the drug and the polymer, with regards to formulation composition and temperature, i.e., whether mixing between the drug and polymer is spontaneous or not at a particular drug:polymer ratio and temperature. This approach, based on well-known Florry-Huggins theory [21], allows construction of temperature-composition phase diagrams, which separates stable, metastable, and unstable regions and helps formulation scientists to choose appropriate formulation compositions and processing conditions during solid dispersions preparation. Apart from the estimation of formulation stability, usage of thermodynamic modeling is particularly beneficial in the early stages of formulation development, when a limited amount of material is available, since totally immiscible formulations can be rejected at this stage, saving both materials and time. In this review, we have given an overview of the currently available methodology for the estimation of miscibility between the drug and polymer as well as solubility of the drug in the polymer. Although in the text, methods are separated into analytical methods and computational methods (based on thermodynamic modeling and molecular modeling and simulations), one should be aware that estimation of the drug-polymer miscibility and the solubility of the drug in the polymer is a complex problem, requiring multi-methodological approach.

2. Analytical Techniques for the Assessment of Drug-Polymer Solubility/Miscibility

Once a solid dispersion system is formed, the solid-state characterization is performed using various techniques to estimate the physical state of the drug and/or potential interactions, which can be attributed to the miscibility, with the selected polymers. In the case of the well-mixed system, only one phase exists since the system components (i.e., drug and the polymer) are intimately mixed at the molecular level. On the other hand, the presence of at least two different phases shows that components are immiscible. These differences are reflected in the physical properties and can be analyzed using a variety of analytical techniques for solid-state characterization.

2.1. Thermal Techniques

Thermal characterization, using Differential Scanning Calorimetry (DSC) or Modulated-temperature Differential Scanning Calorimetry (M-DSC), is used to determine the solid-state of the drug and possible drug-polymer interactions in the prepared solid dispersions. M-DSC enables determination of both the specific heat capacity and the heat flow data from a kinetically controlled process [22]. Polymers used for the preparation of the solid dispersions are usually amorphous and thermoplastic with specific glass transition temperatures (T_g). If the selected drug is crystalline, it usually preserves this state in physical mixtures with the polymer, which is evident on the DSC thermogram of the physical mixture as sharp endothermic peak(s), corresponding to the drug melting point. However, the absence of drug-specific endothermic melting peak(s) may suggest either that the drug is present in the solid dispersion in its amorphous state (drug forms single-phase or multi-phase system with polymer), or it is solubilized during DSC analysis by the polymer (or other excipients) used for the preparation of solid dispersions. Furthermore, shifts in the polymer T_g may also occur, which is also indicative of molecular interactions between the drug and the polymer [23]. Since solid dispersions, with miscible drug and polymer, create a single-phase amorphous system, single T_g peak is considered as the marker of the miscible drug-polymer system [24]. When the two components are miscible, the single T_g of the formed solid dispersion lies between the T_gs of the individual components [25].

Melting point depression is one of the most common analytical methods for the assessment of the drug and polymer miscibility. The changes in the onset of the melting endotherm or the drug melting enthalpy are monitored as a function of the polymer amount in the prepared solid dispersions. In the case of the miscible system, a decrease in the drug melting point(s) and/or enthalpies is expected with the increase of the polymer amount. If the DSC scan represents separate glass transition points, T_g, specific of the drug and the polymer, it is an indication that the prepared solid dispersions do not constitute a miscible system, i.e., two individual phases are present [23]. Therefore, immiscibility is usually manifested as the phase separation, i.e., the existence of crystalline or amorphous domains within the polymer or two separated amorphous domains. There may also be a gradient of drug concentrations in different regions of the dispersion [26].

Several different approaches have been established to estimate T_g of drug-polymer mixtures based on the known mixture composition. Certainly, the most widely used equation for the estimation of mixture T_g (T_g^{mix}) is the Gordon-Taylor equation [27]:

$$T_g^{mix} = \frac{w_1 T_{g1} + K w_2 T_{g2}}{w_1 + K w_2} \quad (1)$$

where w and T_g are weight fraction and glass transition temperature of each component, respectively, while subscripts 1 and 2 represent components with the lowest and the highest T_g, respectively. Constant K is originally defined as a parameter whose value depends on the change of thermal expansion coefficient of the components upon their transformation from glassy to the rubbery state during glass transition. This constant is usually calculated using Simha-Boyer rule [28]:

$$K = \frac{\rho_1 T_{g1}}{\rho_2 T_{g2}} \quad (2)$$

where ρ_1 and T_{g1} are the density and the glass transition temperature of the amorphous component with the lowest T_g, respectively, and ρ_2 and T_{g2} are the density and glass transition temperature of the amorphous component with the highest T_g, respectively. Couchman and Karasz [29] proposed a thermodynamic model to predict the T_g of mixtures:

$$\ln T_g^{mix} = \frac{w_1 \Delta C_{p1} \ln T_{g1} + w_2 \Delta C_{p2} \ln T_{g2}}{w_1 \Delta C_{p1} + w_2 \Delta C_{p2}} \qquad (3)$$

where ΔC_p is a change in heat capacity of the component between liquid-like and glassy state. Another approach for the prediction of mixture T_g represents Fox equation [30]:

$$\frac{1}{T_g^{mix}} = \frac{w_1}{T_{g1}} + \frac{w_2}{T_{g2}} \qquad (4)$$

When using these theoretical approaches to predict mixture T_g, one should be aware of some inherent limitations of these methods. These approaches assume the absence of specific interactions between components (i.e., ideal mixing behavior is assumed), ideal volume additivity of the components at T_g, and linear change in volume with temperature [31,32]. Therefore, the presence of interactions between components will result in deviations between experimentally observed T_g of the mixture and those predicted by previously described models. Negative deviation of experimental T_g from predicted one can indicate that cohesive interactions between individual components are more pronounced than adhesive drug-polymer interactions, as observed for indomethacin-polyvinylpyrrolidone (PVP) system [33,34]. Positive deviation of experimental T_g from predicted one indicates that drug-polymer interactions are stronger than drug-drug and polymer-polymer interactions. This effect has been observed for numerous solid dispersion systems, such as indomethacin-Eudragit® E [34,35], lapatinib-hydroxypropylmethylcellulose phthalate (HPMCP) [36], nimodipine-PVP [37]. However, the presence of positive or negative deviations of experimental from predicted T_g is not a reliable indicator whether adhesive or cohesive interactions are predominant. This is demonstrated for curcumin-hydroxypropylmethylcellulose (HPMC) solid dispersions, where the negative deviation of experimental T_g from predicted one is observed, even though the presence of drug-polymer intermolecular interactions is proved by FT-IR spectroscopy [38]. An additional limitation of the presented models is that they do not take into account entropic contribution to the drug-polymer mixing. It should be also noted that the chosen experimental conditions can influence the measured values of T_g. Gordon-Taylor equation has been adapted for ternary solid dispersions (Equation (5)); however, above-mentioned basic assumptions of this model significantly limit its application for ternary systems, making difficult to draw any conclusions from the obtained results:

$$T_g^{mix} = \frac{w_1 T_{g1} + K_1 w_2 T_{g2} + K_2 w_3 T_{g3}}{w_1 + K_1 w_2 + K_2 w_3} \qquad (5)$$

$$K_1 = \frac{\rho_1 T_{g1}}{\rho_2 T_{g2}} \qquad (6)$$

$$K_2 = \frac{\rho_1 T_{g1}}{\rho_3 T_{g3}} \qquad (7)$$

It has been reported that some microstructural phase separations could not be detected by the DSC method due to its resolution limitation (~30 nm) [39]. If the drug and the polymer have similar T_g values, then it is also difficult to estimate their miscibility using DSC studies [26]. Another limitation of the conventional DSC analysis is the fact that the drug may dissolve in the molten polymeric material below its melting point, which may be mistakenly considered as solubility/miscibility [40]. If the crystalline drug dissolves in the molten polymer during heating, it is better to use fast DSC analysis (such as M-DSC) because higher heating rates may hinder the drug dissolution process [26]. Fule and Amin [41] used M-DSC studies to investigate whether the absence of drug melting endotherm in the DSC scan is a consequence of the presence of drug amorphous form, or drug dissolution within the molten excipients during DSC scan. They demonstrate that the endothermic peak, corresponding to the melting of crystalline drug, broadens during the first heating cycle and disappears in the second heating cycle of M-DSC analysis. On the other hand, melting peak of the drug is absent on the DSC thermogram since the crystalline drug gradually dissolves in the molten polymers during conventional DSC heating process, giving false evidence of the presence of amorphous drug [41]. Tao et al. [13] have used slow heating rates for DSC measurements and extrapolated the temperature of the final dissolution of the crystalline drug to zero heating rate to determine the solubility of the small molecule crystalline drugs in the polymer. If mixture containing known composition of the crystalline drug (x) is heated, the broad endothermic peak occurs due to the drug dissolution, and the drug solubility in the polymer is defined as x at the end temperature of drug dissolution (T_{end}). Specifically, cryogenic milling is used for sample preparation to ensure uniform mixing and facilitate determination of dissolution endpoint. However, even at a low heating rate (0.1 °C/min), the available time during DSC analysis may not be sufficient to reach equilibrium (i.e., T_{end} is higher than equilibrium solution temperature), and solubility of the drug in the polymeric matrix may be underestimated. This problem is particularly pronounced at temperatures close to T_g, due to high polymer viscosity, which causes the time for reaching equilibrium to be much higher, compared to the timescale of the DSC scan. Therefore, this method was further refined by Sun et al. [42] who proposed a method where the drug-polymer mixture is annealed during 4–10 h near an equilibrium solution temperature followed by the scan at standard scanning speed (10 °C/min) to detect the presence of undissolved drug crystals. If the annealing temperature is lower than the equilibrium solution temperature, the melting endotherm will appear in the heating scan due to the presence of undissolved crystals. By annealing at different temperatures, boundaries of equilibrium solution temperature can be determined. Although this method enables determination of solubility at a temperature closer to T_g and improves sensitivity to detect residual drug crystals, it is still considerably time-consuming and requires several experiments for only one point in the solubility plot. The method proposed by Mahieu et al. [43] is based on the generation of the supersaturated solid solution of drug in polymer and further induction of demixing by annealing above T_g. The equilibrium concentration of a dissolved drug is subsequently determined by measuring T_g of the annealed mixture and further calculation from Gordon-Taylor equation (Equation (1)), which gives the relationship between T_g of the mixture and mixture composition. This method is much faster, since the demixing process is faster than dissolution, due to enhanced mobility in the supersaturated system caused by plasticization effect of drug molecules, and only one experiment is required to generate one point in the drug-polymer solubility plot. Schematic drawing of methods for the determination of drug solubility in the polymer, described in the references [13], [42], and [43] is given in Figure 1. Tian et al. [44] recently proposed an improved method for the determination of equilibrium drug solubility within the polymeric matrix and the determination of the solid-liquid transition curve. In this method, a mixture of drug and polymer is firstly undergone to hot-melt extrusion and then subject to isothermal annealing at elevated temperatures (above T_g of polymer and below the melting temperature (T_m) of the drug) during 24 h. High-speed DSC (Hyper DSC) analysis with a heating rate of 200 °C/min is used to detect the presence of undissolved drug crystals remained after sample annealing. This method should provide a more reliable determination of drug solubility within the polymer as long annealing process provides sufficient time for dissolution of drug crystals and overcoming of high polymer viscosity, which can delay completion of drug dissolution. High heating rate after sample annealing provides greater sensitivity to detect melting endotherm of remaining drug crystals, compared to usual DSC heating rates (1–10 °C/min), with the lower possibility that drug crystal will dissolve during DSC scan, leading to overestimation of drug equilibrium solubility in the polymer [44].

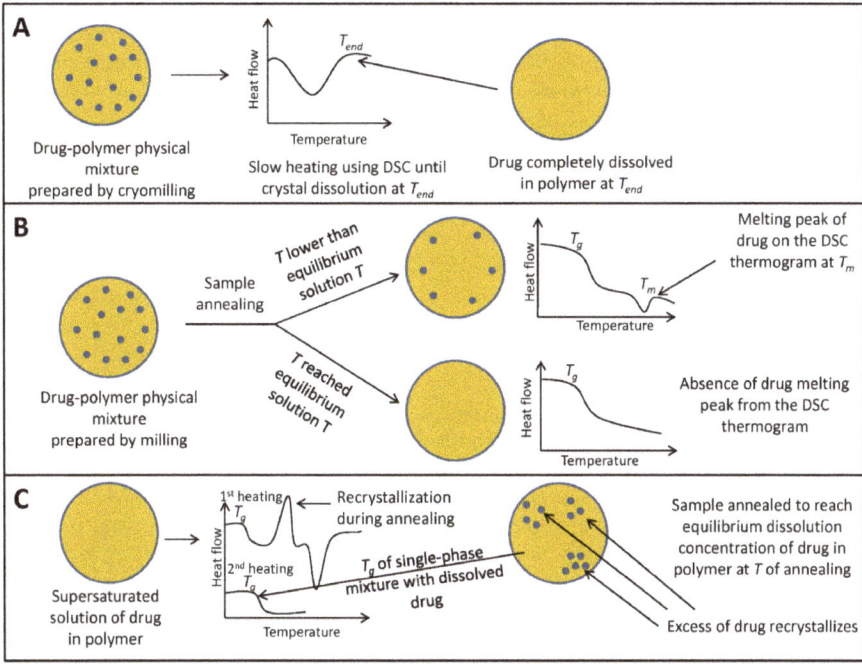

Figure 1. Schematic drawing of methods for the determination of drug solubility in polymer described by (**A**) Tao et al. [13], (**B**) Sun et al. [42], and (**C**) Mahieu et al. [43] (*T*—temperature, T_g—glass transition temperature, T_{end}—end temperature of drug dissolution, T_m—melting temperature).

Thermally stimulated depolarization current (TSDC) is a method for measurement of dielectric properties through thermally stimulated depolarization of materials molecules. This technique can also be used to study the miscibility of the drug and the selected polymer [45]. Shmeis et al. [46] have compared TSDC and DSC for the assessment of the miscibility of the novel drug substance and PVP and demonstrated the superiority of the TSDC method. At higher drug loads, the saturation level of the drug within the polymer has been only possible to be detected by the TSDC method of analysis.

2.2. Spectroscopic Techniques

Apart from the thermal methods, miscibility within the solid dispersions is often analyzed by spectroscopic techniques. Fourier Transform Infrared (FT-IR) Spectroscopy can be used to study specific interactions between the polymer and the drug functional groups. Hydrogen bonding is a predominant mechanism for the stabilization of the miscible drug-polymer systems. When such bonds are formed, subtle changes in the FT-IR spectra are visible [23]. IR spectroscopy with principal component analysis can be utilized to verify drug-polymer mixing at the molecular level [47].

Taylor and Zografi [33] have criticized the traditional approach where the spectra of the pure crystalline drug and solid dispersion are compared to assess changes in the crystallinity of the drug and potential for miscibility with the polymer. More detailed studies on the amorphous structures should be performed, using the spectra of the amorphous form of the drug as the reference [33].

Solid-State Nuclear Magnetic Resonance (SSNMR) spectroscopy is often employed for structural analysis and the assessment of interactions between the molecules. ^{1}H-NMR spin-lattice relaxation measurements can be used for the assessment of the drug-polymer miscibility [48]. The main benefit of this technique is that it can detect phase-separated domains with size below the detection limit of DSC. The miscibility of the drug-polymer and size of the phase-separated domains are estimated based on the spin-lattice relaxation times in the laboratory (^{1}H T$_1$) and rotating frames (^{1}H T$_{1\rho}$). In the phase-separated system, protons in each phase will have their relaxation times, so individual relaxation times of drug and polymer will be observed. On the other hand, in a single-phase system, spin diffusion should average individual relaxations, resulting in uniform average relaxation time, different from relaxation time measured for pure components [36,49,50]. Single ^{1}H T$_1$ in the drug-polymer system indicates miscibility down to 20–50 nm domain size, while single ^{1}H T$_{1\rho}$ indicates miscibility with domain size below 5 nm [51]. Formation of single-phase solid dispersions with indomethacin and Eudragit® E, with drug loading up to ~50%, has been shown based on similar ^{1}H T$_1$ and ^{1}H T$_{1\rho}$ values for the drug and polymer [50]. Miscibility of the selected drug (nifedipine) and polymers (PVP and HPMC) is appointed by mono-exponential spin-lattice relaxation decay for measurements of solid dispersions in the rotating frame [48]. Geppi et al. [52] have used several high-resolution solid-state NMR techniques to confirm the miscibility of ibuprofen and Eudragit® L100 and the chemical nature of their interactions. Phase separation, which is indicative of the immiscible system, has been also assessed by NMR relaxometry study [26].

X-ray Photon Spectroscopy (XPS) is an advanced surface analysis technique, which can also be used to assess the magnitude of the intermolecular interactions between the drug and polymer, which are indicative of the miscibility within the system [24]. Due to the drug-polymer interactions in the obtained spectra, new peaks are formed, and several bond peaks are shifted. It has been demonstrated that drug-polymer interactions observed through XPS analysis are directly related to their miscibility [24].

2.3. X-ray Powder Diffraction (XRPD)

X-ray Powder Diffraction (XRPD) technique, in general, can be used to assess the crystalline state of the material. It is to be expected that solid dispersions prepared from miscible drug-polymer systems are amorphous and lack typical crystalline patterns in an X-ray diffractogram. Newman et al. [39] have developed the XRPD method coupled with the computation of pair distribution functions (PDF) to analyze miscibility in the drug-polymer systems. A lack of agreement of the PDF profiles of the solid dispersions and individual components indicates that the mixture with a unique PDF is miscible [39]. The method proposed by Newman et al. [39] has revealed the superiority of XRPD studies over the DSC analysis for the assessment of the drug-polymer miscibility due to an inability of the DSC technique to detect T_g values for amorphous domains smaller than 30 nm. XRPD with PDF and pure curve resolution method (PCRM) analysis may be used to verify drug-polymer mixing for both completely and partially miscible systems. These techniques are especially useful for the examination of miscibility when DSC measurements are inconclusive or yield variable results [47].

2.4. Microscopic and Imaging Techniques

Methods for the visual analysis of solid dispersion samples are also of great importance to study the solubility/miscibility within the solid dispersion systems. Scanning electron microscopy (SEM) or transmission electron microscopy (TEM) studies of prepared solid dispersions may be indicative of intrinsic miscibility of the drug and polymer. The surface and cross-section morphological features of prepared solid dispersions are studied to analyze whether the miscible system is formed, usually in addition to DSC and XRPD analyses. An example of the assessment of the drug and polymer miscibility using SEM demonstrates that solid dispersions appear to be agglomerated with a rough surface, which is attributed to the miscibility of the drug and polymer [41].

3D surface Atomic Force Microscopy (AFM) imaging analysis of solid dispersions is used to further elucidate drug-polymer surface morphology and interactions [41]. Although AFM provides nanoscale resolution, which is desirable in miscibility evaluation, it cannot provide information regarding chemical composition in different regions of analyzed samples. To overcome this drawback, AFM imaging is coupled with the source of IR radiation (nanoscale infrared spectroscopy—nano IR, AFM-IR) or heating source (nanoscale thermal analysis—nanoTA). In the AFM-IR technique, light from IR source is focused on the contact area between the sample and AFM tip. Absorption of IR radiation causes thermal expansion of the sample, which induces cantilever oscillation. The spectrum of the amplitude of cantilever oscillations as a function of IR wavelength is unique for each sample and provides information about chemical composition in the analyzed sample. In the nanoscale thermal analysis, AFM probe is heated and moved along the region of interest and, when the thermal event occurs, a region of the sample softens and AFM probe penetrates the sample. By measuring thermal properties across the sample surface, nanoscale thermal analysis enables evaluation of whether a system is single-phased or phase separation occurred. Nanoscale infrared spectroscopy and nanoscale thermal analysis have been successfully used to evaluate miscibility between telaprevir and three different polymers [53]. Crystalline structures may also be visually observed by polarized light microscopy (PLM) due to the characteristic appearance of birefringence of crystalline structures, which are formed in immiscible systems [54]. Miscibility within the drug-polymer system can be further confirmed by the hot stage microscopy (HSM) [55], which is often coupled with polarized light microscopy.

Raman mapping (or Raman spectral imaging) is a method whereby detailed chemical images are generated from samples' Raman spectra. Raman mapping may also be used to investigate the drug crystallinity [56]. Analysis of small-size samples through micro-Raman mapping can be used to detect the phase separation in systems in which multiple glass transition events cannot be resolved by DSC [57]. The main benefit of using Raman mapping in the evaluation of drug-polymer miscibility is providing information regarding the chemical composition of phase-separated domains. Qian et al. [58] have demonstrated the superiority of a Raman mapping method over conventional DSC and XRPD studies for analysis of the drug-polymer miscibility within the two batches of amorphous solid dispersion systems that exhibit different physical stability against crystallization over time. They demonstrated that single distinctive T_g might not always be a reliable indicator of homogeneity and optimal stability since Raman maps of the less stable systems are indicative of wide distribution ranges of the drug concentration [58].

Thermal analysis by structural characterization (TASC) is a novel thermal analysis technique that combines image analysis with hot stage microscopy. TASC technique is based on the algorithm that converts any change in the sample appearance during heating into a quantified signal, i.e., TASC curve [59]. Suitability of TASC as a technique for fast screening of the drug-polymer miscibility by evaluating the melting point depression has been demonstrated for felodipine and ten commonly used polymers in solid dispersion formulations [60]. Fast analysis, high sensitivity, and the requirement of a small amount of sample make this technique very attractive in the early stage of solid dispersion formulation development; however, further studies are necessary to confirm the usefulness of this technique.

Martinez-Marcos et al. [55] have highlighted the potential of the novel technique, micro-computed tomography (μ-CT), to be used for the characterization of internal materials properties. This technique enables X-ray imaging in three dimensions. X-ray micro-computed tomography and TASC technique are used in conjunction with conventional thermal, microscopic, and spectroscopic techniques to analyze the miscibility of felodipine with several excipients [61].

2.5. Other Techniques

Crowley and Zografi [62] have estimated the miscibility of PVP and three hydrophobic drugs through water vapor absorption studies. They demonstrated that interactions in amorphous dispersions affect the water uptake properties of the individual components [62]. Liu et al. [35] have reported on the potential for application of rheological measurements to accompany thermal and spectroscopic analysis for the assessment of the drug and polymer miscibility. Gupta et al. [63] have proposed that if the viscosity versus temperature plots for different drug concentrations are parallel to each other (without observable drug melting transition), it is indicative of complete drug-polymer miscibility.

2.6. Techniques Used in Combination

It might be of great interest to use several analytical techniques for investigation of the miscibility of the selected drug-polymer system. Rumondor et al. [47] have demonstrated that DSC, FT-IR spectroscopy, and XRPD analysis provide complementary results to each other. Marsac et al. [64] have used DSC, AFM, and TEM techniques to study the effect of temperature and moisture on the miscibility within the drug-polymer solid dispersions. It is also of great interest to study ternary systems, i.e., to estimate the solubility/miscibility of the drug within the polymer mixtures. Janssens et al. [65] have analyzed the miscibility in ternary systems of itraconazole with polyethylene glycol (PEG) and HPMC polymer blends, of different molecular weights, using M-DSC and XRPD. Miscibility in ternary polymer-drug-surfactant systems hydroxypropylmethylcellulose acetate succinate (HPMCAS)-itraconazole-Soluplus® has been also analyzed using DSC, XRPD, and PLM [66]. Parikh et al. [67] have proposed the preparation of the film-casted samples to investigate miscibility of the drug and the polymer using techniques, such as DSC, XRPD, and PLM. Several analytical techniques are also used to investigate miscibility of felodipine with polymer blends used for the fused deposition modeling 3D printing [68]. Examples of usage of different analytical methods for the estimation of drug-polymer solubility/miscibility are given in Table 1.

Table 1. Examples of analytical methods used for the characterization of solid dispersions.

Drug	Polymer(s) and Other Excipients	Method for Preparation of Solid Dispersions	Analytical Methods Used to Study Solubility/Miscibility	References
Albendazole	PVP	Hot-melt extrusion	DSC, XRPD, HSM, μ-CT SEM	[55]
Carbamazepine	Soluplus®	Hot-melt extrusion	Rheological properties, DSC, XRPD	[63]
Carbamazepine, Prednisolone	PVP, Eudragit® E 100	Electrospray deposition	DSC, XRPD	[69]
Chloramphenicol	Poly(ε-caprolactone)	Film casting	DSC, XRPD, FT-IR, AFM	[70]
Diphenhydramine, Propranolol	Eudragit® L 100, Eudragit® L 100-55	Hot-melt extrusion	XPS, DSC, XRPD, SEM	[24]
	PVP	Solvent evaporation	DSC, FT-IR, XRPD	[23]
Felodipine	Eudragit® E PO	Hot-melt extrusion	SEM, DSC, M-DSC, NMR	[26]
	Soluplus®, HPMCAS, PVP, Eudragit® E PO, PVPVA, HPC, PAA, Na CMC, PVA, HEC	Spin coating	TASC, IR imaging	[60]
Felodipine, nifedipine, ketoconazole	PVP, PAA	Solvent evaporation	DSC, FT-IR, XRPD	[47]
Griseofulvin	HPMCAS	Co-milling	FT-IR, XRPD, DSC	[71]
Ibuprofen	Eudragit® L 100	Solvent evaporation	NMR	[52]
Indomethacin	PVP	Solvent evaporation	FT-IR, FT-Raman	[33]
	Eudragit® E PO	Melting or compression methods	M-DSC, rheological properties, FT-IR	[35]
Indomethacin, dextran	PVP	Solvent evaporation	XRPD, DSC	[39]
Indomethacin, nifedipine, D-mannitol	PVP, PVA	Co-milling	DSC, XRPD	[13]
Indomethacin, ursodeoxycholic acid, indapamide	PVP	Solvent evaporation	Water vapor absorption studies	[62]

Table 1. Cont.

Drug	Polymer(s) and Other Excipients	Method for Preparation of Solid Dispersions	Analytical Methods Used to Study Solubility/Miscibility	References
Itraconazole	PEG and HPMC	Solvent evaporation (spray drying)	M-DSC, XRPD	[65]
	HPMCAS and Soluplus®	Film casting	XRPD, DSC, PLM	[66]
	HPMCP, Soluplus®, PVPVA 64, Eudragit® E PO			[67]
	HPMCAS with the addition of Poloxamer 188, Poloxamer 407, or TPGS	Film casting and hot-melt extrusion	DSC, XRPD	[72]
Lacidipine	PVP K30, PVP VA64, Soluplus®	Hot-melt extrusion	XRPD, DSC, PLM, FT-IR	[73]
Lapatinib ditosylate	Soluplus®	Hot-melt extrusion and solvent evaporation	DSC, XRPD, SEM	[74]
n.a. (new chemical entity)	PVP	Solvent evaporation	TSDC, DSC	[46]
Naproxen	PVP	Solvent evaporation (spray drying)	M-DSC, FT-IR, XRPD	[75]
Nifedipine	PVP, HPMC, PHPA	Solvent evaporation (spray drying)	NMR, DSC	[48]
Posaconazole	Soluplus®, with the addition of PEG 4000, Poloxamer 188, Poloxamer 407 or TPGS	Hot-melt extrusion	DSC, M-DSC, SEM, AFM	[41]
Telaprevir	HPMC, HPMCAS, PVPVA	Solvent evaporation	AFM, AFM-IR, nanoTA, Fluorescence microscopy	[53]

AFM—Atomic Force Microscopy; AFM-IR—Nanoscale Infrared Spectroscopy; DSC—Differential Scanning Calorimetry; FT-IR—Fourier Transform Infrared Spectroscopy; FT-Raman—Fourier Transform Raman Spectroscopy; HEC—Hydroxyethyl Cellulose; HPC—Hydroxypropyl Cellulose; HPMC—Hydroxypropylmethyl Cellulose; HPMCAS—Hydroxypropylmethyl Cellulose Acetate Succinate; HPMCP—Hydroxypropylmethyl Cellulose Phthalate; HSM—Hot Stage Microscopy; IR imaging—Infrared imaging; M-DSC—Modulated-temperature Differential Scanning Calorimetry; µCT—Micro-computed Tomography; Na CMC—Sodium Carboxymethylcellulose; nanoTA—Nanoscale Thermal Analysis; NMR—Nuclear Magnetic Resonance; PAA—Polyacrylic Acid; PEG—Polyethylene Glycol; PHPA—α,β-poly(N-5-hydroxypentyl)-L-aspartamide; PLM—Polarized Light Microscopy; PVA—Poly(vinyl alcohol); PVP—Polyvinylpyrrolidone; PVPVA—Polyvinylpyrrolidone Vinyl Acetate; SEM—Scanning Electron Microscopy; TASC—Thermal Analysis by Structural Characterization; TPGS—D-α-Tocopheryl Polyethylene Glycol 1000 Succinate; XRPD—X-ray Powder Diffraction.

3. Computational Methods for the Assessment of Drug-Polymer Solubility/Miscibility

3.1. Solubility Parameters

The early phase of formulation development requires fast screening methods capable of making rough differentiation between the drug-polymer immiscible systems and the drug-polymer systems that are likely to be miscible. The usage of solubility parameters, as a purely theoretical-based approach, perfectly fits this description since it enables evaluation of the drug-polymer miscibility based on only drug's and polymer's chemical structures and without the need for performing even single experiment. The concept of solubility parameters has been introduced by Hildebrand [76,77] and Scatchard [78] who have linked solubility of the solute in different solvents with cohesion energy density.

Cohesion energy is defined as the increase in internal energy per mole of a substance if all of the intermolecular forces are eliminated, i.e., cohesion energy represents the strength of attractive forces of constituent molecules in the substance. For low molecular weight molecules, cohesion energy (E_{coh}) can be calculated from experimentally determined heat of vaporization using the following equation [79]:

$$E_{coh} = \Delta H_{vap} - p\Delta V \approx \Delta H_{vap} - RT \tag{8}$$

where ΔH_{vap} is molar heat of evaporation, p is pressure, ΔV is the volume change, R is the universal gas constant, and T is temperature.

In the concept proposed by Hildebrand, solubility parameter (δ) is calculated as a square root of cohesive energy density:

$$\delta = \sqrt{\frac{E_{coh}}{V}} \tag{9}$$

Since it is not possible to determine the heat of vaporization for polymers, δ for these high molecular weight molecules can be determined by indirect methods, such as dissolving or swelling of polymers in series of solvents of known δ [80,81], measurements of polymers viscosity in the solvents of known δ [82], or using inverse gas chromatography [83]. Because these methods are time and material consuming, several group contribution methods are developed, which enable calculation of solubility parameters from the knowledge of the molecule's chemical structure. The basic postulate of group contribution methods is that properties of the polymer can be estimated by summation of the contributions of its structural fragments. Although there are several group contribution methods for the estimation of the cohesion energy of the polymers, in the following text we have given an overview of the few of them, which are the most commonly used. One of the earliest group contribution methods for the estimation of Hildebrand solubility parameters has been proposed by Small [84] who defined parameter $F = (E_{coh}V)^{1/2}$, called molecular attraction constant, and provided values of this parameter for numerous structural groups. According to Small's group contribution method, the Hildebrand solubility parameter can be estimated by summation of F for structural fragments of the molecule, using the following equation:

$$\delta = \frac{\sum F}{V} \tag{10}$$

This system has been further refined by Hoy [85] and Hoftyzer and Van Krevelen [86] who provided refined and updated tables for group contributions to the overall F value of the molecule. Fedors [87] proposed a slightly different concept that provides contributions of a much larger number of structural groups to both E_{coh} and volume of the molecule. According to Fedors' method, the Hildebrand solubility parameter can be estimated according to the equation:

$$\delta = \left(\frac{\sum_i \Delta e_i}{\sum_i \Delta v_i}\right)^{1/2} \tag{11}$$

where Δe_i and Δv_i are the additive atomic and group contribution for the energy of vaporization and molar volume, respectively.

The specificity of this approach is that it offers the way for estimating not only E_{coh} but also volume from group contributions and allows estimation of the temperature dependence of δ from the knowledge of density-temperature dependence.

Although the introduction of the Hildebrand solubility parameter resulted in huge progress in studying solute-solvent interactions, its application is limited to the systems with predominant dispersion force between molecules (non-polar molecules). Therefore, this concept was further extended by Hansen [88], who defined three-dimensional Hansen solubility parameter, which applies to substances that, in addition to dispersion forces, also interact by hydrogen bonding and polar forces. Total cohesion energy (E_{coh}) is redefined as the sum of the contributions from dispersion forces (E_d), polar forces (E_p), and hydrogen bonding (E_h):

$$E_{coh} = E_d + E_p + E_h \tag{12}$$

Accordingly, the three-dimensional solubility parameter can be expressed as follows:

$$\delta^2 = \delta_d^2 + \delta_p^2 + \delta_h^2 \tag{13}$$

where δ_d, δ_p, and δ_h are dispersion, polar, and hydrogen bonding partial solubility parameters, respectively.

Several group contribution methods have been developed to estimate the Hansen solubility parameter. Two most widely used group contribution methods have been developed by Hoftyzer and Van Krevelen [86] and Hoy [89]. According to the method proposed by Hoftyzer and Van Krevelen [86], partial solubility parameters are estimated using the following equations, and the total solubility parameter is calculated according to the Equation (13).

$$\delta_d = \frac{\sum F_{di}}{V} \tag{14}$$

$$\delta_p = \frac{\sqrt{\sum F_{pi}^2}}{V} \tag{15}$$

$$\delta_h = \sqrt{\frac{\sum E_{hi}}{V}} \tag{16}$$

where F_{di} is molar attraction constant due to dispersion component, F_{pi} is molar attraction constant due to the polar component, E_{hi} is hydrogen bonding energy, and V is molar volume.

If the two identical polar groups are presented in the symmetrical positions, δ_p is reduced by multiplying value obtained using Equation (15) with one of the following correction factors: 0.5 for one plane of symmetry, 0.25 for two planes of symmetry, or 0 for more than two planes of symmetry. Besides, for molecules with several planes of symmetry, δ_h is 0. The method developed by Hoy [89,90] is more complex and requires the usage of four additive functions and several auxiliary equations to estimate values of partial solubility parameters and the total solubility parameter. Equations that are used for calculations in Hoy's method are given in Table 2 [79]. Although presented group contribution methods up to now have been enriched and provide a huge collection of group contributions for solubility parameters estimation with reasonable accuracy, further research in this field pointed out some drawbacks of this concept and provided its further refinement. It has been recognized that presented group contribution methods are quite simplified and neglected how groups are connected as well as interactions between adjacent groups and electron delocalization. Therefore, these methods are unable to distinguish between different isomers and estimate the same δ values for those molecules [91]. To overcome these drawbacks, Stefanis and co-workers developed a new group contribution system for the estimation of solubility parameters, through the division of molecules into the first order and second order groups. The first order groups describe the basic structure of the molecule, similar to the previously described group contribution methods. Additional second-order groups consist of two or three adjacent first-order groups, and they are based on the conjugation theory [92–94]. Conjugation is considered as one of the important stabilization mechanisms, whereby compounds with a higher number of conjugates are considered as more energetically stable. This theory, known as ABC approach (ABC-contribution of Atoms and Bonds to the properties of Conjugates), considers chemical compounds not as single structures but as hybrids of different conjugates formed by a different arrangement of valence electrons. Therefore, this method can capture intramolecular interactions between adjacent atoms as well as atoms separated by several bonds [91]. Second-order groups are formed from the two or three adjacent first-order groups and ranked based on their contribution to the standard enthalpy of formation. Structures that exhibit the highest contribution to the enthalpy of formation are considered as second-order groups for further calculations [92,93]. The basic equation for the estimation of Hansen solubility parameters according to Stefanis and Panayiotou group contribution method is given as follows [94]:

$$\delta = \sum_i N_i C_i + W \sum_j M_j D_j \quad (17)$$

where C_i is the contribution of the first-order group of type i that appears N_i times in the compound and D_j is the contribution of the second-order group of type j that appears M_j times in the compound. The constant W is equal to 0 for compounds without second-order groups and equal to one for compounds, which have second-order groups.

Partial solubility parameters can be estimated by the following equations [94]:

$$\delta_d = \left(\sum_i N_i C_i + \sum_j M_j D_j + 959.11 \right)^{0.4126} [MPa]^{1/2} \quad (18)$$

$$\delta_p = \left(\sum_i N_i C_i + \sum_j M_j D_j + 7.6134 \right) [MPa]^{1/2} \quad (19)$$

$$\delta_h = \left(\sum_i N_i C_i + \sum_j M_j D_j + 7.7003 \right) [MPa]^{1/2} \quad (20)$$

whereby different equations are used for the cases where values of δ_p and δ_h are lower than 3 MPa$^{1/2}$:

$$\delta_p = \left(\sum_i N_i C_i + \sum_j M_j D_j + 2.6560 \right) [MPa]^{1/2} \quad (21)$$

$$\delta_h = \left(\sum_i N_i C_i + \sum_j M_j D_j + 1.3720 \right) [MPa]^{1/2} \quad (22)$$

Stefanis and Panayiotou further proposed subdivision of δ_h into acidic and basic components δ_a and δ_b, respectively, and extended the three-parameter Hansen solubility parameter to the four-parameter solubility parameter. The donor and acceptor parameters of the hydrogen bonds have been obtained by evaluation of third moments of sigma profiles of charge density distribution on the surface of molecules. These profiles for a large number of compounds are available in the software databases (COSMObase or VT Sigma Profile Databases) or can be calculated using suitable software (Dmol^3 or TURBOMOLE). Therefore, calculations do not require so many computational resources. The main benefit of this concept is that it takes into account acid-base interactions that favor solubility and miscibility [94]. Just et al. [95] for the first time developed group contribution set based exclusively only on pharmaceutical relevant solids to predict the solubility of the drug in polymer for solid dispersion systems prepared by hot-melt extrusion. Although in the initial study, this group contribution system showed improved prediction ability of drug solubility compared to existing methods, further experiments are necessary to enrich group contribution tables with more data as well as to validate obtained methods on additional experimental data.

Table 2. The equations used for the estimation of the solubility parameter and its components in Hoy's (1985) group contribution system [79].

Equations Used in the Calculation	Low Molecular Weight Substances	Amorphous Polymers
Additive molar functions	$Ft = \sum N_i F_{t,i}$ $Fp = \sum N_i F_{p,i}$ $V = \sum N_i V_i$ $\Delta_T = \sum N_i \Delta_{T,i}$	$\Delta_T^{(P)} = \sum N_i \Delta_{T,i}^{(P)}$
Auxiliary equations	$\log \alpha =$ $3.39 \log(T_b/T_{cr}) - 0.1585 - \log V$ $T_b/T_{cr} = 0.567 + \Delta T - (\Delta T)^2$	$\alpha^{(P)} = 777 \Delta_T^{(P)}/V$ $n = 0.5/\Delta_T^{(P)}$
Calculation of total and partial solubility parameters	$\delta_t = (F_i + B)/V$ $\delta_p = \delta_t \left(\frac{1}{\alpha} \frac{F_p}{F_i+B}\right)^{1/2}$ $\delta_h = \delta_t[(\alpha-1)/\alpha]^{1/2}$ $\delta_d = (\delta_t^2 - \delta_p^2 - \delta_h^2)^{1/2}$	$\delta_t = (F_i + B/n)/V$ $\delta_p = \delta_t \left(\frac{1}{\alpha^{(P)}} \frac{F_p}{F_i+B/n}\right)^{1/2}$ $\delta_h = \delta_t \left[(\alpha^{(P)}-1)/\alpha^{(P)}\right]^{1/2}$

F_t—total molar attraction constant for each group; F_p—polar molar attraction constant; F_i—sum of molar attraction constants of constituent groups; V—molar volume of the molecule or repeated unit in the polymer; Δ_T—Lydersen correction for non-ideality (values for low molecular substances have been provided by Lydersen [96], while values for polymers $\Delta_T^{(P)}$ have been derived by Hoy; T_b—boiling point; T_{cr}—critical temperature; B—base value ($B = 277$).

Solubility parameters are used as a simple screening tool for miscibility evaluation in the early solid dispersion formulations development. Simply, values of the solubility parameters of drug and polymer should be close to each other, if the drug and polymer are miscible. This means that energy released due to cohesive interactions between like molecules is counterbalanced by adhesive interactions between unlike molecules [97,98]. While the difference in the solubility parameters ($\Delta\delta$) in the case of Hildebrand solubility parameters is easily calculated by subtracting δ of the drug from δ of polymer, for Hansen solubility parameters difference is calculated as the Euclidean distance according to the following equation:

$$\Delta\delta = \sqrt{(\delta_{d1} - \delta_{d2})^2 + (\delta_{p1} - \delta_{p2})^2 + (\delta_{h1} - \delta_{h2})^2} \quad (23)$$

where subscripts 1 and 2 denote drug and polymer, respectively.

Bagley et al. [99] considered the effects of δ_d and δ_p as similar and combined them into single parameter δ_v, while the effect of δ_h is considered as different. After applying this transformation, the difference between Hansen solubility parameters of drug and polymer can be evaluated on a two-dimensional plot with δ_v and δ_h on the axes. Since Bagley's plot shows superior performances over three-dimensional Hansen plot in locating regions where the polymer is soluble in solvents, this plot should be preferably used in evaluating miscibility between drug and polymer. Although the difference between solubility parameters of drug and polymer should be small if components are miscible, it is difficult to establish a threshold for $\Delta\delta$ value below which components are considered as miscible. According to Greenhalgh et al. [100], if $\Delta\delta < 7.0$ MPa$^{1/2}$, components are likely to be miscible, while if $\Delta\delta > 10.0$ MPa$^{1/2}$, components are like to be immiscible. In their study, Forster et al. [101] suggested more rigorous criteria, which predict the formation of solid solution in the cases where $\Delta\delta < 2.0$ MPa$^{1/2}$, while immiscibility is predicted for systems with $\Delta\delta > 10.0$ MPa$^{1/2}$. For drug-polymer systems with Δδ between 5.0 and 10.0 MPa$^{1/2}$, it is difficult to make a reliable conclusion of whether the system is miscible or immiscible. Further melt extrusion experiments showed that this system successfully predicts whether a system is miscible or not in the cases where Δδ < 2.0 MPa$^{1/2}$ and Δδ > 10.0 MPa$^{1/2}$.

Solubility parameters have been extensively used as the screening tool to get some information regarding drug-polymer miscibility, alone or more commonly in conjunction with thermodynamic modeling [20,38,98,102–105]. Although this method is more or less successful to distinguish between drug-polymer miscible pairs from those pairs where miscibility problems may occur, the application of this method alone is not highly reliable and can give misleading results. When applying solubility parameters for the evaluation of drug-polymer miscibility, one should be aware of some limitations of this concept. This is a purely theoretical concept, wherein drug-polymer interactions are based on chemical similarity, and it takes into account only enthalpic contribution to drug-polymer mixing. For further thermodynamic interpretation of the drug-polymer mixing, this method is used in conjunction with Florry-Huggins thermodynamic modeling, which has been explained in the further text. One of the main limitations of this method is that it is qualitative and does not provide any quantitative information regarding drug-polymer miscibility as well as the physical state of the API (active pharmaceutical ingredient) after mixing with the polymer [98]. It should also bear in mind that the application of different group contribution methods will inevitably give different values of solubility parameters and even the same group contribution method will give a different result if the structure of the molecule is divided in different ways. However, solubility parameters can still be considered as a useful screening tool in the early formulation development, wherein bringing of any conclusions regarding drug-polymer miscibility/immiscibility requires further application of experimental techniques and thermodynamic modeling.

3.2. Thermodynamic Modeling

Although analytical methods, described in Section 2, are capable of more or less accurate determination whether the drug and polymer form single-phase system or not, obtained results are valid only at the moment of analysis. It is more important to get insight into thermodynamics of the drug-polymer mixing since this will enable identification of drug and polymer composition ranges, where the formation of single-phase system is more likely to occur as well as identification of potential destabilization mechanisms. Additionally, miscibility or solubility of the drug in the polymeric matrix, below polymer's T_g, can be only estimated by model prediction due to very slow system equilibration, which gives more significance to this approach.

It has been recognized that thermodynamic models describing the mixing of small molecules with a solvent are not suitable to describe mixing between the drug and polymer since they do not take into account large volume differences between polymer and drug molecules [11]. The suitable model should relate free energy change upon mixing to volume fractions, rather than mole fractions of the components since entropic contribution to the free energy change is significantly reduced by mixing a large molecular weight molecule with a small molecular weight molecule. Flory-Huggins lattice theory that has been originally developed to describe mixing in the polymer-polymer and polymer-solvent blends is further applied to describe mixing between the drug and polymer by considering drug molecule analogous to the solvent molecule. According to this theory, free energy change upon mixing (ΔG_{mix}) of the drug and polymer is given by the following equation [11]:

$$\frac{\Delta G_{mix}}{RT} = n_{drug} ln \Phi_{drug} + n_{polymer} ln \Phi_{polymer} + n_{drug} \Phi_{polymer} \chi \qquad (24)$$

where n_{drug} is the number of moles of the drug, $n_{polymer}$ is the number of moles of polymer, Φ_{drug} is the volume fraction of the drug, $\Phi_{polymer}$ is the volume fraction of the polymer, R is the gas constant, T is the absolute temperature, and χ is Flory-Huggins interaction parameter.

The first two terms on the right side of the Equation (24) describe entropic contribution to the free energy of mixing, which always favors mixing, since mixing of two components increases system disorder. Since entropic contribution to the free energy of mixing in drug-polymer systems is much lower compared to the small molecule-solvent system, enthalpy of mixing will mainly determine whether drug-polymer mixing will occur spontaneously ($\Delta G_{mix} < 0$) or not ($\Delta G_{mix} > 0$). The contribution of the enthalpy of mixing to the overall free energy of mixing is determined by the sign and magnitude of the Flory-Huggins interaction parameter χ, which reflects the strength of drug-polymer adhesive interactions relative to cohesive interactions between drug-drug and polymer-polymer pairs. Negative values of χ indicate stronger adhesive interactions, which facilitate drug-polymer mixing, as a result of negative ΔG_{mix}. In the case of positive χ, which indicates stronger cohesive interactions, drug-polymer mixing will be thermodynamically favored only if the entropic contribution to ΔG_{mix} overcomes unfavorable enthalpy of mixing and gives negative overall free energy of mixing [11,106,107]. Since the strength of adhesive interactions between the drug and polymer is determined by their chemical structures, shifting from one to other chemical class of polymers is a better approach to achieve miscibility than shifting from higher to lower molecular weight grade of the same polymer [11]. Although values of χ can be determined by different approaches, the most common way to determine χ in drug-polymer systems is a melting point depression method. In this method, physical mixtures of drug and polymer of various compositions are subjected to DSC scan and, if the system is miscible, drug melting point should be reduced in the mixture compared to the melting point of the pure drug. Melting of the drug occurs at a temperature where the chemical potential of the crystalline drug becomes equal to the chemical potential of the molten drug. Mixing of drug with polymer will reduce drug chemical potential and, therefore, the melting of the drug will occur at a lower temperature, compared to the pure drug. If the drug and polymer are immiscible, the melting of the drug will be unaltered in the presence of polymer [14,15,20]. The relationship between the melting point depression upon drug-polymer mixing and χ is given by the following equation [11]:

$$\left(\frac{1}{T_M^{mix}} - \frac{1}{T_M^{pure}}\right) = \frac{-R}{\Delta H_{fus}} \left[\ln \Phi_{drug} + \left(1 - \frac{1}{m}\right)\Phi_{polymer} + \chi \Phi_{polymer}^2\right] \quad (25)$$

where T_M^{mix} is the melting temperature of the drug in the presence of the polymer, T_M^{pure} is the melting temperature of the pure drug, ΔH_{fus} is the heat of fusion of the pure drug, and m is the ratio of the volume of the polymer to that of the lattice site (defined as the volume of the drug molecule). This equation is further rearranged to give the plot of

$$\left(\frac{1}{T_M^{mix}} - \frac{1}{T_M^{pure}}\right) \times \left(\frac{\Delta H_{fus}}{-R}\right) - \ln(\Phi_{drug}) - \left(1 - \frac{1}{m}\right)\Phi_{polymer} \quad vs. \quad \Phi_{polymer}^2 \quad (26)$$

which exhibits linear relationship within the low polymer concentrations with the slope equal to χ. When using melting point depression method to estimate χ, one should be aware of some limitations of this approach. Firstly, T_g of polymer should be sufficiently below drug melting temperature because the crystalline drug should interact with the polymer in a supercooled liquid state sufficiently long before it starts to melt. Additionally, linearity in the plot used to calculate χ is limited to low polymer concentrations. Although markedly different values of χ are obtained if used onset, midpoint, or offset of drug melting peak in the DSC scan as a drug melting temperature [14], there is no consensus in the literature regarding which value should be used. Marsac et al. [14] and Paudel et al. [15] proposed that offset of melting endotherm should be used since it represents the melting of the final composition after the occurrence of mixing. However, in most of the studies, onset values have been used [17,19,20,104], where the low heating rate is used to facilitate mixing within the experimental time scale. Calculated χ enables estimation of the free energy changes upon drug-polymer mixing (ΔG_{mix}) as a function of the drug weight fraction, according to the Equation (24). However, values of χ obtained by the melting point depression method are valid only at drug melting temperature and cannot be used to predict drug-polymer miscibility at lower temperatures. It has been shown that χ varies with temperatures and composition. Since the effect of the composition on χ is considered as negligible, compared to the effect of temperature, the temperature dependence of χ can be expressed using the following simplified equation [12,18]:

$$\chi = A + \frac{B}{T} \qquad (27)$$

where A is referred to as the non-combinatorial entropic contribution to χ, while B/T is the enthalpic contribution [108]. By measuring melting points of the drug in mixtures of different compositions, different values of χ can be obtained, and by plotting these values as a function of corresponding temperatures, parameters A and B can be obtained. This enables calculation of χ at any temperature and prediction of ΔG_{mix} as a function of temperature and composition. Solubility parameters can be also used to calculate χ, according to Equation (28):

$$\chi = \frac{V_{site}}{RT}(\delta_{drug} - \delta_{polymer})^2 \qquad (28)$$

where V_{site} is the volume of the hypothetical lattice (approximated as the volume of the drug).

Since χ calculated in this way reflects drug-polymer interactions at 25 °C, this value can be additionally used to estimate temperature dependence of χ, according to Equation (27) [17,20]. Although the calculation of χ using solubility parameters is the simplest approach and does not require any experiment, it has been shown that obtained χ can fail to predict drug-polymer miscibility [11]. This probably comes from the inherent limitation of solubility parameters approach in systems with pronounced specific intermolecular interactions. An additional limitation of this approach is that it does not take into account possible exothermic mixing since calculated χ is always positive [15].

Besides the estimation of the drug-polymer miscibility, thermodynamic modeling is used to estimate drug solubility in polymers used for solid dispersions preparation. Only if drug concentration in solid dispersion is below the solubility limit, physical stability of this system is guaranteed without the tendency of drug toward crystallization. Therefore, estimation of the drug solubility in the polymeric matrix is of particular importance since it is an indicator of the degree of supersaturation, which determines driving force for drug crystallization. If the amount of drug in solid dispersion is above its solubility in the polymer, but below the miscibility limit, the system is considered as metastable and is stabilized against crystallization through intimate mixing with polymer, unless large fluctuations of temperature and/or composition occur, which makes favorable conditions for crystallization [12]. Marsac et al. [14] developed a model for the estimation of the solubility of the drug in polymers by using the measured drug solubility in low molecular weight analog of the polymer. The solubility of the drug in the low molecular weight analog is given by the following equation [14]:

$$\ln x_{drug}\gamma_{drug} = \frac{-\Delta G_{fus}}{RT} = -\frac{\Delta H_{fus} T_m}{RT}\left[1 - \frac{T}{T_m}\right] - \frac{1}{RT}\int_{T_m}^{T}\Delta C_p dT + \frac{1}{R}\int_{T_m}^{T}\frac{\Delta C_p}{T}dt \qquad (29)$$

where γ_{drug} is the activity coefficient of the drug in the mixture at the solubility limit, x_{drug} is the mole fraction of dissolved drug, ΔG_{fus} is the free energy difference between supercooled liquid and crystal, T is the temperature of interest, T_m is melting temperature, R is universal gas constant, ΔH_{fus} is heat of fusion, and ΔC_p is the capacity difference between the liquid and crystal.

By considering that ΔC_p does not change significantly in the temperature range of interest, Equation (29) can be rewritten into the following form [109]:

$$\ln x_{drug}\gamma_{drug} = -\frac{\Delta H_{fus}}{RT}\left[1 - \frac{T}{T_m}\right] - \frac{\Delta C_p}{R}\left[1 - \frac{T_m}{T} + \ln\left(\frac{T_m}{T}\right)\right] \qquad (30)$$

Since all parameters in the Equations (29) and (30) can be easily experimentally determined, except activity coefficient (γ_{drug}), calculation of γ_{drug} is a necessary prerequisite to determine drug solubility. By considering ideal mixing ($\gamma_{drug} = 1$), ideal solubility of the drug in the low molecular weight analog can be calculated from Equation (30). The ratio of ideal solubility to experimentally determined the solubility of the drug in the low molecular weight analog of polymer gives γ_{drug} in low molecular weight analog of the polymer (γ_{drug}^{LMW}). The activity coefficient of the drug in the polymer ($\gamma_{drug}^{polymer}$) is considered equal to γ_{drug}^{LMW} after the addition of correction factor to reduce the entropy of mixing of the drug in polymer compared to the low molecular weight analog:

$$\ln \gamma_{drug}^{polymer} = \frac{MV_{drug}}{MV_{lattice}} \left[\frac{1}{m_{drug}} \ln \frac{\Phi_{drug}}{x_{drug}} + \left(\frac{1}{m_{drug}} - \frac{1}{m_{polymer}} \right) \Phi_{polymer} \right] + \ln \gamma_{drug}^{LMW} \qquad (31)$$

where $MV_{lattice}$ is lattice molecular volume (in this case, defined as volume of low molecular weight analog of polymer), MV_{drug} is drug molecular volume, $m_{polymer}$ is the ratio of the volume of the polymer to that of the lattice site, m_{drug} is the ratio of the volume of the drug to the lattice site [14].

Calculated $\gamma_{drug}^{polymer}$ can be further used to calculate χ, as an alternative approach compared to commonly used melting point depression method [14]:

$$\ln \gamma_{drug}^{polymer} = \frac{MV_{drug}}{MV_{lattice}} \left[\frac{1}{m_{drug}} \ln \frac{\Phi_{drug}}{x_{drug}} + \left(\frac{1}{m_{drug}} - \frac{1}{m_{polymer}} \right) \Phi_{polymer} + \chi \Phi_{polymer}^2 \right] \qquad (32)$$

Marsac et al. [14] calculated the solubility of several drugs in different grades of PVP using solubility data in the 1-ethyl-2-pyrrolidone as the low molecular weight analog of PVP. Obtained results show significantly reduced solubility in polymer due to reduced entropy of mixing up to the certain molecular weight of the polymer after which solubility is only slightly changed. Additionally, χ calculated using this approach is in agreement with those obtained by melting point depression method. The same method has been successfully used by Paudel et al. [15] to predict the solubility of naproxen in different grades of PVP using measured solubility of naproxen in N-methylpyrrolidone as a low molecular weight analog of PVP. Although this approach is quite simple and does not require so many experimental resources, it assumes that interactions between the drug and polymer are the same as between the drug and the low molecular weight analog (i.e., χ is the same in both cases) and is applicable only for polymers with available low molecular weight analog in the liquid state. Djuris et al. [19], in their study, used Hansen solubility parameters to calculate the activity coefficient of carabamazepine in polyethyleneglycol-polyvinylcaprolactam-polyvinyl acetate grafted copolymer (Soluplus®) according to the following equation:

$$\ln \gamma_{drug} = \frac{V_{drug}}{RT} \left[\left(\delta_d^{drug} - \overline{\delta}_d \right)^2 + 0.25 \left(\left(\delta_p^{drug} - \overline{\delta}_p \right)^2 + \left(\delta_h^{drug} - \overline{\delta}_h \right)^2 \right) \right] + \ln \frac{V_{drug}}{\overline{V}} + 1 - \frac{V_{drug}}{\overline{V}} \qquad (33)$$

$$\overline{\delta} = \sum_{k=1}^{n} \Phi_k \delta_k \qquad (34)$$

where $\overline{\delta}$ is the molar volume-weighted solubility parameter, and $\overline{V}$ is the mixture volume, where the subscript k denotes the different components of the mixture.

Obtained activity coefficient shows strong composition dependence and can be used to estimate mole fraction of dissolved drug within the polymeric matrix using either Equation (30) or Equation (32). Results obtained via both ways are in close agreement and show that the amount of carbamazepine that can be molecularly dispersed in the Soluplus® matrix is limited to below 5% (w/w) [19].

Prudic et al. used thermodynamic modeling based on perturbed-chain statistical associating fluid theory (PC-SAFT) to estimate drug solubility in polymer. According to this theory, each molecule is described as a chain composed of spherical segments that can interact with segments of other molecules through different types of interactions. In PC-SAFT model, the residual Helmholtz energy a^{res} of a system containing drug and polymer is calculated as the sum of reference hard-chain contribution accounting for repulsive interactions between molecules (a^{hc}), a dispersion contribution accounting for van der Waals attraction forces (a^{disp}), and a contribution caused by association via hydrogen bonds (a^{assoc}) [110]:

$$a^{res} = a^{hc} + a^{disp} + a^{assoc} \tag{35}$$

Segment number m_i^{seg} and segment diameter σ_i are used to calculate hard-chain contribution, while the contribution from the van der Waals attraction forces between segments (a^{disp}) is calculated using dispersion-energy parameter u_i/k_b (u_i—dispersion energy, k_b—the Boltzmann constant). Additionally, for drugs and polymers capable of hydrogen bonds formation, it is necessary to calculate a contribution caused by association via hydrogen bonds (a^{assoc}). The calculation of this term requires definition of the number of association sites (electron acceptors and donors) N_i^{assoc}, defined based on the molecule's chemical structure, the association-energy parameter $\varepsilon^{A_iB_i}/k_B$ (related to the strength of association), and the association-volume parameter $\kappa^{A_iB_i}$ (related to the distance between two molecules necessary to form a hydrogen bond). These pure-component parameters of drugs and polymers, required for the calculation of the residual Helmholtz energy, are usually determined by fitting to experimental solubility data of these components in organic solvents. The calculation of the residual Helmholtz energy of the system is described in detail elsewhere [110].

The activity coefficient of the drug in the liquid drug/polymer phase (γ_i^L), required for the calculation of solid-liquid equilibrium curve (Equation (30)), can be calculated by PC-SAFT method using the following equations [111]:

$$\gamma_i^L = \frac{\phi_i^L}{\phi_{0i}^L} \tag{36}$$

$$\ln \phi_i^L = \frac{\mu_i^{res}}{k_B T} - \ln Z \tag{37}$$

$$Z = \frac{pV}{k_B N_A T} \tag{38}$$

$$\frac{\mu_i^{res}}{k_B T} = \frac{a^{res}}{k_B T} + Z - 1 + \left(\frac{\partial(a^{res}/k_B T)}{\partial x_i}\right) - \sum_{j=1}^{N} x_j \left(\frac{\partial(a^{res}/k_B T)}{\partial x_j}\right) \tag{39}$$

where ϕ_i^L is fugacity coefficient of component i in the mixture, ϕ_{0i}^L is fugacity coefficient of the pure component, μ_i^{res} is residual chemical potential, Z is compressibility factor, p is system pressure, and N_A is the Avogadro number.

PC-SAFT is successfully used to predict the solubility of artemisinin and indomethacin in PEGs of different molecular weights as a function of temperature and predicted results are in close agreement with experimentally obtained data [110]. This approach is also used to predict long-term stability of the drug in both binary and ternary solid dispersions and evaluate the impact of relative humidity on drug recrystallization and amorphous-amorphous phase separation [110–114]. The solubility of acetaminophen in PVP K25 and PVP VA64 and the impact of relative humidity on the solubility are predicted by PC-SAFT and Flory-Huggins modeling and further used as an indicator of long-term stability of acetaminophen solid dispersions in these two polymers. Obtained results show that the PC-SAFT method gives more accurate prediction and can better differentiate whether solid dispersions remain stable or undergo recrystallization under elevated humidity [111]. Advantage of PC-SAFT method is that each component is characterized with parameters that are physically meaningful and do not depend on the temperature, component molecular weight, concentration, etc. Additionally, this method takes into account different types of interactions in the system, such as association (hydrogen bonding) and ionic and polar interactions between the compounds [115]. It has been also demonstrated that the PC-SAFT method enables accurate prediction of the drug solubility in copolymers if the drug solubility in the respective homopolymers is known [115]. Although the PC-SAFT method requires less experimental work than Flory-Huggins method, it requires more complicated calculations. However, once determined, parameters of the pure component can be further used for other systems, which contains that component, so it is expected that this method will be more frequently used in the future upon an increase in the availability of necessary component parameters in the literature.

Construction of the Phase Diagram

As stated above, the calculation of χ at different temperatures, using Equation (27), enables prediction of the free energy change upon drug-polymer mixing as a function of both temperature and composition. As long as $\Delta G_{mix} < 0$ and ΔG_{mix} vs. composition curve is concave up, the formation of the single-phase system occurs since free energy of the mixture is lower than the free energy of the two-phase system. Phase separation can occur only if the single-phase system can lower its free energy by separating into two phases [104]. Determination of ΔG_{mix} vs. composition curve at different temperatures is a necessary prerequisite to constructing temperature vs. composition phase diagram, which shows phase behavior of drug-polymer mixture, and differentiate regions of stability, metastability, and instability. Phase behavior has not been studied for so many drug-polymer systems, and available studies describe the different methodology to construct a drug-polymer phase diagram. Phase diagrams have been described for solid dispersions of dipyridamole and cinnarizine in PVP and polyacrylic acid (PAA) [20], cinarizine in Soluplus® [8], indomethacin in polyvinylpyrrolidone-vinyl acetate (PVP VA) copolymer [17], itraconazole in HPMC [106], felodipine in HPMCAS HF grade and Soluplus® [18], PAA [16] and different grades of PVP [103], aceclofenac in Soluplus® [104], naproxen and acetaminophen in HPMCAS, PVP K25 and PVP VA64 [112], and binary polymeric blends containing HPMCAS and either PVP K25 or PVP VA64 [113]. Typical phase diagram includes a solid-liquid phase transition curve, amorphous phase separation curve, and glass transition curve. In their work, Tian et al. [8] calculated drug solubility curve for cinarizine in Soluplus® using solid-liquid equilibrium equation, which considers that the polymer behaves as a solvent for a crystalline drug:

$$\ln x_{drug} = \frac{\Delta H_{fus}}{RT_m}\left(1 - \frac{T_m}{T}\right) - \ln \gamma_{drug} \tag{40}$$

Lehmkemper et al. suggested that solubility of the drug in polymer should be assessed using Equation (30), which includes ΔC_p term and, therefore, should give more accurate results [111]. Activity coefficient, required for the calculation of the solubility curve, can be calculated using Hansen solubility parameters, according to above-mentioned Equation (33). Solid-liquid phase transition curve can be also calculated using melting point depression approach (Equation (25)), considering Flory-Huggins interaction parameter calculated at different temperatures, as described by Lin et al. [16] and Tian et al. [18]. The PC-SAFT method also enables the prediction of the solid-liquid phase transition curve in the drug-polymer systems and, in some cases, gives more accurate results compared to Flory-Huggins modeling [110,111].

While the solid-liquid phase transition curve describes the solubility limit of the drug in the polymer, miscibility limit of two phases, i.e., the tendency towards amorphous-amorphous phase separation is described by binodal and spinodal curves. The binodal curve is determined by the common tangent rule to free energy vs. composition curve, where the first derivative of this curve is set to zero [17,108]. Above this curve, single-phase amorphous system is formed, while in the region below binodal and above the spinodal curve, the system is metastable, i.e., large composition fluctuation is necessary to induce phase separation [8]. Phase separation process between binodal and spinodal curves can occur via nucleation and growth mechanisms, only if the significant energetic barrier is overcome [18]. The spinodal curve is obtained by setting the second derivative of free energy vs. composition curve to zero according to the following equation [17,18,20]:

$$\frac{1}{\Phi_{drug}} + \frac{1}{m_{polymer}(1-\Phi_{drug})} - 2\chi_{drug-polymer} = 0 \tag{41}$$

After the determination of the temperature dependence of χ, Equation (41) can be transformed in the following form, which enables the construction of the spinodal curve on the temperature vs. composition phase diagram [16,104]:

$$T_s = \frac{2B}{\frac{1}{\Phi_{drug}} + \frac{1}{m\,(1-\Phi_{drug})} - 2A} \qquad (42)$$

The spinodal curve represents phase boundary between metastable and unstable region, and, below this curve, spontaneous (barrier-free) phase separation into drug-rich and polymer-rich regions occurs, which is often denoted as spinodal decomposition. The glass transition curve is an important part of the temperature composition phase diagram as an indicator of system kinetic stabilization. Although solubility and miscibility limits can be exceeded, phase separation and crystallization can be avoided, through kinetic stabilization of the system below T_g of the mixture. Polymers with high T_g are preferred in the formulation of solid dispersions due to increasing T_g of the mixture, which is denoted as an antiplasticization effect. Below T_g, viscosity drastically increases and molecular mobility decreases, which altogether hinders crystallization of drug molecules. It is often reported that molecular mobility can be neglected at temperatures more than 50 °C below the mixture T_g [116]. Therefore, although the system is thermodynamically not stable, it can be kinetically stabilized during product shelf life. Kinetic stabilization is particularly important when using techniques for solid dispersions preparation where materials are processed under non-ambient conditions, such as hot-melt extrusion. In this technique, mixing of the molten drug and polymer above polymer T_g and/or dissolving of the crystalline drug within the polymer above its T_g is facilitated by high processing temperature and high shear stress applied by mixing elements. During cooling to room temperature, homogeneously mixed or dissolved drug can be kinetically frozen in that state for a sufficiently long time, although above solubility and miscibility limit for a particular temperature. After construction, the phase diagram should be validated. This is commonly performed by preparing the solid dispersions of different composition and evaluating drug crystallinity by XRPD as well as the presence of phase separation and/or drug recrystallization by DSC or similar thermal analysis techniques [17,18,104].

Due to simplicity and not a straightforward calculation of binodal curve, the phase diagram is often represented with solid-liquid phase transition curve (solubility curve), spinodal (miscibility) curve, and glass transition curve, as shown in Figure 2 [18]. Above the solubility curve (Zone A and B), the drug is dissolved in the polymer and is stable to temperature and concentration fluctuations. Even if crystallization starts, the thermodynamic driving force in this region is to dissolve the crystalline drug in the polymer. Although this should be the most desired region in the development of solid dispersions formulations, the solubility of the drug is usually too low and limits practical formulation development only to very low dose drugs. Since the amorphous drug has a higher chemical potential compared to the crystalline drug, the miscibility of the amorphous drug with the corresponding polymer is much higher than the solubility of the crystalline drug in the polymer. Therefore, the miscibility boundary is more relevant to the solid dispersion formulation development. Below this boundary line (zones E and F), the system is thermodynamically unstable, and spontaneous phase separation will occur. Only within the zone F, the system can be stabilized kinetically, if stored at temperatures sufficiently below T_g. Above miscibility and below solubility curve (zones C and D), the system is supersaturated with respect to drug solubility but is stabilized through mixing with polymer, and phase separation requires certain activation energy. In the zone D, the system is additionally kinetically stabilized due to reduced molecular mobility below T_g.

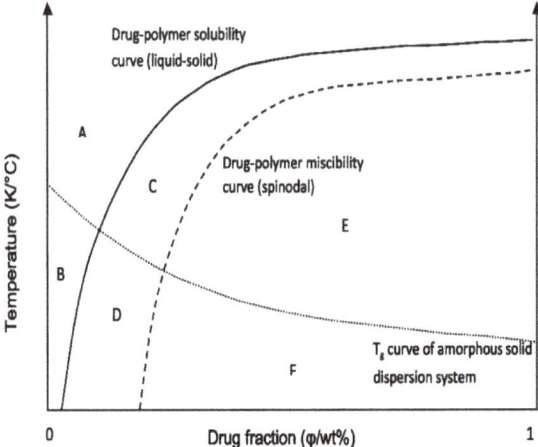

Figure 2. Typical temperature vs. composition phase diagram for the binary drug-polymer system (Reproduced with permission from Tian et al., 2013). [18]. Copyright (2013) American Chemical Society.

3.3. Computational Modeling and Simulations

Following the advances in informatics technologies, which led to increased computing power and speed, together with the high availability of reliable free or affordable proprietary molecular modeling software capable of handling large systems, the simulation of the solid-state has become possible to a very satisfactory degree of precision and time scale. Recent applications of molecular modeling relevant to the formulation of poorly soluble drugs are numerous and focusing on various aspects of the solid-state.

Specifically, molecular dynamics (MD) simulations, a highly powerful molecular modeling technique for the study of physical movement of atoms and molecules by numerically solving Newton's equation of motion, is gaining increased attention in the recent years. In MD simulations, interatomic potentials or molecular mechanics force fields are used to calculate the potential energies and forces occurring between the simulated atom particles [117,118]. In brief, during an MD simulation, the components are initially identified (molecules and concentrations), and the interaction functions (or else "force fields") are set. Then, after setting the desired thermodynamic conditions (i.e., density, pressure, and temperature), the initial positions of molecules are defined, and the velocities of atoms are randomly assigned. Finally, the simulation starts, and, depending on the property under investigation, the thermodynamic parameters may change or fluctuate until the whole system equilibrates in a given set of conditions, i.e., mimicking the procedure of macroscopic equilibration process in a real laboratory experiment [119].

Based on the satisfactory degree of precision and the rather simple and easy to interpret theoretical background, MD simulations have gained increased attention regarding the in-depth evaluation of pharmaceutical solid-state processes. In recent years, such attempts include the simulation of API amorphous state [120], API crystallization processes from supersaturated solutions [121], API—water interactions [122], and API—matrix carrier interactions [123]. Based on the required level of detail, simulations may be performed from picoseconds up to several hundred nanoseconds.

As noted from a recent expert review published by Edueng et al. [124], although the use of MD simulations exhibits extremely promising results in characterizing both pure API amorphous state and API—carrier molecular interactions, this methodology is still only used relatively sporadically. This may be attributed probably to the improper realization or training of the scientists working in the field. Therefore, to alleviate the poor perception of scientists on the subject, it is attempted in the following section, to present a detailed overview of the currently available advanced computational models, used specifically for the estimation of miscibility and molecular interactions (an indirect indication of miscibility) occurring between solid dispersion components.

In this direction, Gupta et al. have performed MD simulations to predict the miscibility of pharmaceutical compounds [120]. Specifically, the authors developed a computational model (verified experimentally via thermoanalytical techniques), which can predict the miscibility of indomethacin in several carriers (polyethylene oxide, glucose, and sucrose). In all applied MD simulations, the COMPASS (Condensed-phase Optimized Molecular Potentials for Atomistic Simulation Studies) force field is used, which is parameterized based on ab initio quantum mechanics calculations. According to the authors, after an initial energy minimization step, the MD simulations are carried out in two phases: (1) the equilibration run, where the amorphous cells are allowed to relax for 2 ns under isothermal (NVT) or isobaric-isothermal conditions (NPT-NVT) at 298 K and (2) the production run where the equilibrated structure is processed via the NVT ensemble for 200 to 500 ps at 298 K with a time step of 1 fs. The Andersen thermostat and barostat are used to maintain the temperature and pressure stable, respectively. The non-bonded van der Waals and electrostatic interactions are truncated using the group-based cut-off distance of 1.25 nm. Trajectory frames are captured during the production run, and the data from the final 50 ps are used for computing the Cohesive Energy Density (CED) and solubility parameter (δ). Results show that the employed MD simulations can predict successfully indomethacin miscibility with polyethylene oxide and immiscibility with sucrose and glucose.

In another paper, the same group of authors used MD simulations for predicting glass transition temperature and plasticization effect in amorphous pharmaceuticals [122]. Amorphous sucrose (widely used as a carrier in the preparation of solid dispersions) and water are selected as model compound and plasticizer, respectively. As in their previous work, MD simulations are performed using the COMPASS force field and isothermal-isobaric ensembles in two steps (equilibration and production phase). In this study, to predict T_g, the authors allowed the system to stepwise cooling from 440 K to 265 K at 5 K intervals by using the final structure from each MD run as the starting structure for the subsequent run. The density is measured at every picosecond interval during the last 50 ps run at each temperature step of the production run, and the average density values are used to calculate the specific volume. Specific volume vs. temperature plots is used to estimate MD-based T_g value for amorphous sucrose containing 0%, 3%, and 5% w/w water, respectively, which are in reasonable agreement with the experimental values reported in the literature. Additionally, radial distribution function analysis of the MD trajectories reveals strong hydrogen bond interactions between sucrose hydroxyl oxygen and water oxygen.

In another study, Maus et al. used MD simulations to predict miscibility and T_g for pharmaceutical solid dispersion systems prepared by a melt-based method, such as hot-melt extrusion [125]. Different mixtures containing theophylline or ibuprofen and water-soluble (triethyl citrate) or water-insoluble (acetyl tributyl citrate or dibutyl sebacate) plasticizers dissolved or dispersed in a cationic polymethacrylate matrix carrier have been evaluated. Initially, for the MD simulations, cubic simulation boxes (with periodic boundary conditions in all directions) are constructed (side length of ca. 4 nm). Then, after appropriate energy minimization, the structures are left to relax for 2 ns under NPT conditions at ambient conditions, to obtain a well-relaxed start structure with the correct density using the Andersen thermostat and barostat, at a time step of 1 fs. Afterward, a 200 ps run at constant volume and temperature (NVT) is carried out (100 ps for equilibration and 100 ps for data sampling). The cohesive energy (E_{coh}) is averaged over this latter period, and the corresponding cohesive energy density is calculated by dividing it through the volume (V) of the simulation cell (E_{coh}/V). In all MD simulations, a cut-off distance of 1.25 nm with a spline switching function is applied for the Coulomb and van der Waals interactions using charge groups to prevent dipoles from being artificially split. Atomic charges and interactions between atoms and molecules are accounted for by the use of the COMPASS force field. For T_g evaluation, the specific volume vs temperature diagrams are constructed by relaxing the systems for 2 ns under NPT conditions at a temperature of approximately 100 K above the supposed T_g, followed by a cooling process with a stepwise of 10 K until the temperature is

~100 K below T_g. Results show that the use of Hilderbrand's solubility parameter estimated via MD calculations leads to an incomplete picture of the system's miscibility, while better results are obtained when MD-based Gibbs free energy is used. Additionally, the correlation of the simulated T_g with the experimentally determined values reflects the different solubility behaviors of the plasticizers studied (less miscible plasticizers show a higher deviation from the experimental T_g).

In a similar work, published by Macháčková et al. [126], MD simulations are employed evaluate miscibility of cyclosporine-A in six biodegradable polymers, namely L-polylactide, D-polylactide, chitosan, polyglycolic acid, PEG, and cellulose [126]. All prepared models are optimized using PCFF (Polymer Consistent Force Field) force field, while smart algorithm (a cascade of steepest descent, conjugate gradient, and quasi-Newton methods) with 50,000 steps is used for geometry optimization, while atomic charges are assigned by a PCFF force field. For MD simulations, periodic boundary conditions are employed under NPT dynamics with Nose thermostat and Berendsen barostat for 1.5 ns. The Flory-Huggins interaction parameter, χ, describing API-polymer miscibility is calculated based on the mixing energy (E_{mix}) representing the difference in free energy between the mixture and the sum of pure state energy of both components (API and polymer). With the present work, the author revealed that MD-simulations could be a powerful tool for predicting component miscibility. Specifically, results show that miscibility is dependent on chain length and this dependence is more noticeable for flexible chains, while the best miscibility is strongly correlated with the polymer-drug interaction energy and with the number of hydrogen bonds between polymer and drug molecule. Additionally, MD simulations can show that the two polymers (polycellulose and polychitosan), with the best miscibility and the highest polymer-drug adhesion, exhibit surprisingly higher rigidity.

Barmpalexis et al. [127] have also used MD simulations to study the miscibility of three commonly used plasticizers (namely, citric acid, triethyl citrate, and PEG) with Soluplus®, a widely used polymeric matrix in hot-melt extrusion processes [127]. MD simulations are performed using pcff_d force field under NPT with a cut-off radius of 7 Å, spline distance of 1 Å, Berendsen thermostat, variable volume and shape option, and 1 fs time step. The cohesive energy (E_{coh}, i.e., the measure of the intermolecular forces acting between molecules) is calculated after 2 ns structure relaxation process and another 400 ps run under NVT, to calculate the Hildebrand solubility parameter by dividing the square root of E_{coh} with square root of simulation volume (V) (Equation (9)). Simulations show miscibility only in the case of Soluplus® and PEG, a result that is verified experimentally by the presence of significant molecular interactions between the two components.

In another paper, the same group of authors tried to expand their previously developed molecular model (polymer-plasticizer matrix system) by including two BCS (Biopharmaceutics Classification System) class II model drugs (namely ibuprofen and carbamazepine) having substantially different thermal properties and glass-forming ability [123]. The same set of MD simulation parameters, as the once selected previously, are used in this attempt. Simulations results suggest that both APIs are miscible in the selected solid dispersion matrix (Soluplus®-PEG) verified experimentally by thermoanalytical analysis (DSC).

MD simulations (using AMBER 11 force field) are also utilized to evaluate the molecular structures of solid dispersions by the simulated annealing method, mimicking the hot-melt preparation method [128]. During the minimization procedure, the structures are subjected to 1000 steps of steepest descent minimization followed by 1000 steps of conjugate gradient minimization. After minimization, a 1 ns simulated annealing simulation, with a Langevin dynamics is used in a 2 fs time step and a cut-off of 12 Å for non-bonded interactions. During the simulated annealing procedure, the system is initially heated from 0 to 500 K in 200 ps and then is kept at temperature for 300 ps to equilibrate. Next, the system is quickly cooled down from 500 to 300 K in 100 ps and is kept at that temperature for 400 ps. The simulated annealing procedure is repeated 10 times (10 ns) for complete convergence of the systems. Based on the presented results, the authors have succeeded in developing an all-atomic MD model for the formation of solid dispersions prepared by hot-melt method and the molecular mechanisms involved during such preparations.

Finally, in a similar attempt, Xiang and Anderson [129] have also used a simulated annealing method, mimicking the hot-melt preparation in an attempt to investigate the molecular interactions occurring between indomethacin and PVP. All MD simulations are performed via Amber-ff02 force field. The prepared initial structures after energy minimization are left to equilibrate at 600 K for approximately 10 ns and then subjected to cooling dynamic runs to a final temperature of 200 K at a cooling rate of 0.03 K/ps. The newly formed glasses are used as starting configurations for prolonged aging dynamic runs (~100 ns) at 298 K and 1 bar. MD simulations suggest that the two components are miscible, a result that is verified by the formation of strong specific interactions (hydrogen bonds).

4. Conclusions

Evaluation of the drug-polymer miscibility and solubility of the drug in the selected polymer has become an unavoidable part in the rational design of solid dispersion formulations. There are numerous approaches to estimate the drug-polymer solubility/miscibility, which we grouped into analytical and computational methods, although this classification is rather arbitrary since computational methods in most cases use data obtained by analytical methods. However, one should be aware that each analytical technique alone has its limitations to differentiate between the drug and polymer domains, whereas computational methods have some inherent assumptions since they have not been developed specifically for the drug-polymer systems. Since the application of any of either computational or analytical methods alone provides only one part of the complete picture, the evaluation of the drug-polymer miscibility and solubility of the drug in polymer requires a multimethodological approach. Future research in this field should enable the development of a standardized methodology for the estimation of drug-polymer solubility/miscibility, which is of the utmost importance for the development of solid dispersion formulations, as well as the final dosage forms, in the pharmaceutical industry. We propose that standardized methodology should be based on the thermodynamic modeling, a coupled computational-analytical method, as this approach enables straightforward construction of temperature vs. composition phase diagram, which further serves as a guidance for formulation scientists to choose suitable polymer, drug loading, processing conditions, and storage conditions to ensure long-term stability of solid dispersion systems. Certainly, analytical methods are an inevitable part of this methodology, as it is necessary to provide experimental data for thermodynamic modeling and validation of the phase diagrams. The time-consuming calculations required for the thermodynamic modeling hinder its implementation in the pharmaceutical industry, so the development of user-friendly software solutions for thermodynamic modeling should certainly facilitate the wider application of this approach in the formulation development. Additionally, it is of particular importance to refine currently used thermodynamic models to adjust them to drug-polymer systems and to include specific drug-polymer interactions with the overall aim to improve the prediction accuracy of the models. Thorough implementation of a methodology for the assessment of drug-polymer solubility/miscibility in the development of solid dispersion formulations should bridge the gap between the great success in solid dispersion technology on the laboratory scale and difficulties for such products to reach the market.

Funding: This research was funded by the Ministry of Education, Science and Technological Development, Republic of Serbia, grant number TR 34007.

Conflicts of Interest: The authors declare no conflict of interest.

Abbreviations

AFM	Atomic Force Microscopy;
AFM-IR (nano IR)	Nanoscale Infrared Spectroscopy;
BCS	Biopharmaceutics Classification System
COMPASS	Condensed-phase Optimized Molecular Potentials for Atomistic Simulation Studies
DSC	Differential Scanning Calorimetry;
HEC	Hydroxyethyl Cellulose;
HPC	Hydroxypropyl Cellulose;
HPMC	Hydroxypropylmethyl Cellulose;
HPMCAS	Hydroxypropylmethyl Cellulose Acetate Succinate;
HPMCP	Hydroxypropylmethyl Cellulose Phthalate;
HSM	Hot Stage Microscopy;
MD	Molecular Dynamics;
M-DSC	Modulated-temperature Differential Scanning Calorimetry;
μ-CT	Micro-computed Tomography;
Na CMC	Sodium Carboxymethyl Cellulose;
nanoTA	Nanoscale Thermal Analysis;
NMR	Nuclear Magnetic Resonance;
PAA	Polyacrylic Acid;
PCFF	Polymer Consistent Force Field
PCRM	Pure Curve Resolution Method;
PC-SAFT	Perturbed-Chain Statistical Associating Fluid Theory;
PDF	Pair Distribution Functions;
PEG	Polyethylene Glycol;
PHPA	α,β-poly(N-5-hydroxypentyl)-L-aspartamide;
PLM	Polarized Light Microscopy;
PVA	Poly(vinyl alcohol);
PVP	Polyvinylpyrrolidone;
PVPVA	Polyvinylpyrrolidone Vinyl Acetate;
SEM	Scanning Electron Microscopy;
SSNMR	Solid State Nuclear Magnetic Resonance;
TASC	Thermal Analysis by Structural Characterization;
TEM	Transmission Electron Microscopy;
Tg	Glass Transition Temperature;
TPGS	D-α-Tocopheryl Polyethylene Glycol 1000 Succinate;
TSDC	Thermally stimulated depolarization current

References

1. Vo, C.L.; Park, C.; Lee, B.J. Current trends and future perspectives of solid dispersions containing poorly water-soluble drugs. *Eur. J. Pharm. Biopharm.* **2013**, *85*, 799–813. [CrossRef] [PubMed]
2. Kanaujia, P.; Poovizhi, P.; Ng, W.K.; Tan, R.B.H. Amorphous formulations for dissolution and bioavailability enhancement of poorly soluble APIs. *Powder Technol.* **2015**, *285*, 2–15. [CrossRef]
3. Lee, T.W.; Boersen, N.A.; Hui, H.W.; Chow, S.F.; Wan, K.Y.; Chow, A.H. Delivery of poorly soluble compounds by amorphous solid dispersions. *Curr. Pharm. Des.* **2014**, *20*, 303–324. [CrossRef] [PubMed]
4. Medarević, D.P.; Kachrimanis, K.; Mitrić, M.; Djuriš, J.; Djurić, Z.; Ibrić, S. Dissolution rate enhancement and physicochemical characterization of carbamazepine-poloxamer solid dispersions. *Pharm. Dev. Technol.* **2016**, *21*, 268–276. [CrossRef] [PubMed]
5. Medarević, D.P.; Kleinebudde, P.; Djuriš, J.; Djurić, Z.; Ibrić, S. Combined application of mixture experimental design and artificial neural networks in the solid dispersion development. *Drug Dev. Ind. Pharm.* **2016**, *42*, 389–402. [CrossRef] [PubMed]
6. Djuris, J.; Ioannis, N.; Ibric, S.; Djuric, Z.; Kachrimanis, K. Effect of composition in the development of carbamazepine hot-melt extruded solid dispersions by application of mixture experimental design. *J. Pharm. Pharmacol.* **2014**, *66*, 232–243. [CrossRef]

7. Sekiguchi, K.; Obi, N. Studies on absorption of eutectic mixtures. I. A comparison of the behavior of eutectic mixtures of sulphathiazole and that of ordinary sulphathiazole in man. *Chem. Pharm. Bull.* **1961**, *9*, 866–872. [CrossRef]
8. Tian, B.; Wang, X.; Zhang, Y.; Zhang, K.; Zhang, Y.; Tang, X. Theoretical prediction of a phase diagram for solid dispersions. *Pharm. Res.* **2015**, *32*, 840–851. [CrossRef]
9. Higashi, K.; Hayashi, H.; Yamamoto, K.; Moribe, K. The effect of drug and EUDRAGIT® S 100 miscibility in solid dispersions on the drug and polymer dissolution rate. *Int. J. Pharm.* **2015**, *494*, 9–16. [CrossRef]
10. Six, K.; Leuner, C.; Dressman, J.; Verreck, G.; Peeters, J.; Blaton, N.; Augustijns, P.; Kinget, R.; Van den Mooter, G. Thermal properties of hot-stage extrudates of itraconazole and eudragit E100. Phase separation and polymorphism. *J. Therm. Anal. Calor.* **2002**, *68*, 591–601. [CrossRef]
11. Marsac, P.J.; Shamblin, S.L.; Taylor, L.S. Theoretical and practical approaches for prediction of drug-polymer miscibility and solubility. *Pharm. Res.* **2006**, *23*, 2417–2426. [CrossRef] [PubMed]
12. Qian, F.; Huang, J.; Hussain, M.A. Drug-polymer solubility and miscibility: Stability consideration and practical challenges in amorphous solid dispersion development. *J. Pharm. Sci.* **2010**, *99*, 2941–2947. [CrossRef] [PubMed]
13. Tao, J.; Sun, Y.; Zhang, G.G.; Yu, L. Solubility of small-molecule crystals in polymers: D-mannitol in PVP, indomethacin in PVP/VA, and nifedipine in PVP/VA. *Pharm. Res.* **2009**, *26*, 855–864. [CrossRef] [PubMed]
14. Marsac, P.J.; Li, T.; Taylor, L.S. Estimation of drug-polymer miscibility and solubility in amorphous solid dispersions using experimentally determined interaction parameters. *Pharm. Res.* **2009**, *26*, 139–151. [CrossRef] [PubMed]
15. Paudel, A.; Van Humbeeck, J.; Van den Mooter, G. Theoretical and experimental investigation on the solid solubility and miscibility of naproxen in poly(vinylpyrrolidone). *Mol. Pharm.* **2010**, *7*, 1133–1148. [CrossRef] [PubMed]
16. Lin, D.; Huang, Y. A thermal analysis method to predict the complete phase diagram of drug-polymer solid dispersions. *Int. J. Pharm.* **2010**, *399*, 109–115. [CrossRef] [PubMed]
17. Zhao, Y.; Inbar, P.; Chokshi, H.P.; Malick, A.W.; Choi, D.S. Prediction of the thermal phase diagram of amorphous solid dispersions by Flory-Huggins theory. *J. Pharm. Sci.* **2011**, *100*, 3196–3207. [CrossRef]
18. Tian, Y.; Booth, J.; Meehan, E.; Jones, D.S.; Li, S.; Andrews, G.P. Construction of drug-polymer thermodynamic phase diagrams using Flory-Huggins interaction theory: Identifying the relevance of temperature and drug weight fraction to phase separation within solid dispersions. *Mol. Pharm.* **2013**, *10*, 236–248. [CrossRef]
19. Djuris, J.; Nikolakakis, I.; Ibric, S.; Djuric, Z.; Kachrimanis, K. Preparation of carbamazepine-Soluplus solid dispersions by hot-melt extrusion, and prediction of drug-polymer miscibility by thermodynamic model fitting. *Eur. J. Pharm. Biopharm.* **2013**, *84*, 228–237. [CrossRef]
20. Baghel, S.; Cathcart, H.; O'Reilly, N.J. Theoretical and experimental investigation of drug-polymer interaction and miscibility and its impact on drug supersaturation in aqueous medium. *Eur. J. Pharm. Biopharm.* **2016**, *107*, 16–31. [CrossRef]
21. Flory, P.J. *Principles of Polymer Chemistry Ithaca*; Cornell University: New York, NY, USA, 1953.
22. Cassel, B.; Packer, R. *Modulated Temperature DSC and the DSC 8500: A Step Up in Performance*; Technical Note; PerkinElmer, Inc.: Waltham, MA, USA, 2010.
23. Karavas, E.; Ktistis, G.; Xenakis, A.; Georgarakis, E. Miscibility behavior and formation mechanism of stabilized felodipine-polyvinylpyrrolidone amorphous solid dispersions. *Drug. Dev. Ind. Pharm.* **2005**, *31*, 473–489. [CrossRef]
24. Maniruzzaman, M.; Morgan, D.J.; Mendham, A.P.; Pang, J.; Snowden, M.J.; Douroumis, D. Drug-polymer intermolecular interactions in hot-melt extruded solid dispersions. *Int. J. Pharm.* **2013**, *443*, 199–208. [CrossRef]
25. Zheng, X.; Yang, R.; Tang, X.; Zheng, L. Part I: Characterization of solid dispersions of nimodipine prepared by hot-melt extrusion. *Drug. Dev. Ind. Pharm.* **2007**, *33*, 791–802. [CrossRef]
26. Qi, S.; Belton, P.; Nollenberger, K.; Clayden, N.; Reading, M.; Craig, D.Q. Characterisation and prediction of phase separation in hot-melt extruded solid dispersions: A thermal, microscopic and NMR relaxometry study. *Pharm. Res.* **2010**, *27*, 1869–1883. [CrossRef]
27. Gordon, M.; Taylor, J.S. Ideal copolymers and the second-order transitions of synthetic rubbers. I. Non-crystalline copolymers. *J. Appl. Chem.* **1952**, *2*, 493–500. [CrossRef]

28. Simha, R.; Boyer, R.F. On a general relation involving the glass temperature and coefficients of expansion of polymers. *J. Chem. Phys.* **1962**, *37*, 1003–1007. [CrossRef]
29. Couchman, P.R.; Karasz, F.E. A Classical Thermodynamic Discussion of the Effect of Composition on Glass-Transition Temperatures. *Macromolecules* **1978**, *11*, 117–119. [CrossRef]
30. Fox, T.G. Influence of Diluent and of Copolymer Composition on the Glass Temperature of a Polymer System. *Bull. Am. Phys. Soc.* **1956**, *1*, 123.
31. Kalogeras, I.M. A novel approach for analyzing glass-transition temperature vs. composition patterns: Application to pharmaceutical compound + polymer systems. *Eur. J. Pharm. Sci.* **2011**, *42*, 470–483. [CrossRef]
32. Baird, J.A.; Taylor, L.S. Evaluation of amorphous solid dispersion properties using thermal analysis techniques. *Adv. Drug. Deliv. Rev.* **2012**, *64*, 396–421. [CrossRef]
33. Taylor, L.S.; Zografi, G. Spectroscopic characterization of interactions between PVP and indomethacin in amorphous molecular dispersions. *Pharm. Res.* **1997**, *14*, 1691–1698. [CrossRef]
34. Prasad, D.; Chauhan, H.; Atef, E. Amorphous stabilization and dissolution enhancement of amorphous ternary solid dispersions: Combination of polymers showing drug-polymer interaction for synergistic effects. *J. Pharm. Sci.* **2014**, *103*, 3511–3523. [CrossRef]
35. Liu, H.; Zhang, X.; Suwardie, H.; Wang, P.; Gogos, C.G. Miscibility studies of indomethacin and Eudragit®E PO by thermal, rheological, and spectroscopic analysis. *J. Pharm. Sci.* **2012**, *101*, 2204–2212. [CrossRef]
36. Song, Y.; Yang, X.; Chen, X.; Nie, H.; Byrn, S.; Lubach, J.W. Investigation of drug-excipient interactions in lapatinib amorphous solid dispersions using solid-state NMR spectroscopy. *Mol. Pharm.* **2015**, *12*, 857–866. [CrossRef]
37. Papageorgiou, G.Z.; Papadimitriou, S.; Karavas, E.; Georgarakis, E.; Docoslis, A.; Bikiaris, D. Improvement in chemical and physical stability of fluvastatin drug through hydrogen bonding interactions with different polymer matrices. *Curr. Drug Deliv.* **2009**, *6*, 101–112. [CrossRef]
38. Meng, F.; Trivino, A.; Prasad, D.; Chauhan, H. Investigation and correlation of drug polymer miscibility and molecular interactions by various approaches for the preparation of amorphous solid dispersions. *Eur. J. Pharm. Sci.* **2015**, *71*, 12–24. [CrossRef]
39. Newman, A.; Engers, D.; Bates, S.; Ivanisevic, I.; Kelly, R.C.; Zografi, G. Characterization of amorphous API: Polymer mixtures using x-ray powder diffraction. *J. Pharm. Sci.* **2008**, *97*, 4840–4856. [CrossRef]
40. Bikiaris, D.; Papageorgiou, G.Z.; Stergiou, A.; Pavlidou, E.; Karavas, E.; Kanaze, F.; Georgarakis, M. Physicochemical studies on solid dispersions of poorly water-soluble drugs: Evaluation of capabilities and limitations of thermal analysis techniques. *Thermochim. Acta* **2005**, *439*, 58–67. [CrossRef]
41. Fule, R.; Amin, P. Hot melt extruded amorphous solid dispersion of posaconazole with improved bioavailability: Investigating drug-polymer miscibility with advanced characterisation. *Biomed. Res. Int.* **2014**, *2014*, 1–16. [CrossRef]
42. Sun, Y.; Tao, J.; Zhang, G.G.; Yu, L. Solubilities of crystalline drugs in polymers: An improved analytical method and comparison of solubilities of indomethacin and nifedipine in PVP, PVP/VA, and PVAc. *J. Pharm. Sci.* **2010**, *99*, 4023–4031. [CrossRef]
43. Mahieu, A.; Willart, J.F.; Dudognon, E.; Danède, F.; Descamps, M. A new protocol to determine the solubility of drugs into polymer matrices. *Mol. Pharm.* **2013**, *10*, 560–566. [CrossRef]
44. Tian, Y.; Jones, D.S.; Donnelly, C.; Brannigan, T.; Li, S.; Andrews, G.P. A New Method of Constructing a Drug-Polymer Temperature-Composition Phase Diagram Using Hot-Melt Extrusion. *Mol. Pharm.* **2018**, *15*, 1379–1391. [CrossRef]
45. Shimizu, H.; Horiuchi, S.; Nakayama, K. Structural analysis of miscible polymer blends using thermally stimulated depolarization current method. *Proc. Int. Symp. Electrets.* **1999**, 553–556.
46. Shmeis, R.A.; Wang, Z.; Krill, S.L. A mechanistic investigation of an amorphous pharmaceutical and its solid dispersions, part I: A comparative analysis by thermally stimulated depolarization current and differential scanning calorimetry. *Pharm. Res.* **2004**, *21*, 2025–2030. [CrossRef]
47. Rumondor, A.C.; Ivanisevic, I.; Bates, S.; Alonzo, D.E.; Taylor, L.S. Evaluation of drug-polymer miscibility in amorphous solid dispersion systems. *Pharm. Res.* **2009**, *26*, 2523–2534. [CrossRef]
48. Aso, Y.; Yoshioka, S.; Miyazaki, T.; Kawanishi, T.; Tanaka, K.; Kitamura, S.; Takakura, A.; Hayashi, T.; Muranushi, N. Miscibility of nifedipine and hydrophilic polymers as measured by (1)H-NMR spin-lattice relaxation. *Chem. Pharm. Bull.* **2007**, *55*, 1227–1231. [CrossRef]

49. Calahan, J.L.; Azali, S.C.; Munson, E.J.; Nagapudi, K. Investigation of Phase Mixing in Amorphous Solid Dispersions of AMG 517 in HPMC-AS Using DSC, Solid-State NMR, and Solution Calorimetry. *Mol. Pharm.* **2015**, *12*, 4115–4123. [CrossRef]
50. Lubach, J.W.; Hau, J. Solid-State NMR Investigation of Drug-Excipient Interactions and Phase Behavior in Indomethacin-Eudragit E Amorphous Solid Dispersions. *Pharm. Res.* **2018**, *35*, 65. [CrossRef]
51. Yuan, X.; Sperger, D.; Munson, E.J. Investigating miscibility and molecular mobility of nifedipine-PVP amorphous solid dispersions using solid-state NMR spectroscopy. *Mol. Pharm.* **2014**, *11*, 329–337. [CrossRef]
52. Geppi, M.; Guccione, S.; Mollica, G.; Pignatello, R.; Veracini, C.A. Molecular properties of ibuprofen and its solid dispersions with Eudragit RL100 studied by solid-state nuclear magnetic resonance. *Pharm. Res.* **2005**, *22*, 1544–1555. [CrossRef]
53. Li, N.; Taylor, L.S. Nanoscale Infrared, Thermal, and Mechanical Characterization of Telaprevir-Polymer Miscibility in Amorphous Solid Dispersions Prepared by Solvent Evaporation. *Mol. Pharm.* **2016**, *13*, 1123–1136. [CrossRef]
54. Yoo, S.U.; Krill, S.L.; Wang, Z.; Telang, C. Miscibility/Stability Considerations in Binary Solid Dispersion Systems Composed of Functional Excipients towards the Design of Multi-Component Amorphous Systems. *J. Pharm. Sci.* **2009**, *98*, 4711–4723. [CrossRef]
55. Martinez-Marcos, L.; Lamprou, D.A.; McBurney, R.T.; Halbert, G.W. A novel hot-melt extrusion formulation of albendazole for increasing dissolution properties. *Int. J. Pharm.* **2016**, *499*, 175–185. [CrossRef]
56. Démuth, B.; Farkas, A.; Pataki, H.; Balogh, A.; Szabó, B.; Borbás, E.; Sóti, P.L.; Vigh, T.; Kiserdei, É.; Farkas, B.; et al. Detailed stability investigation of amorphous solid dispersions prepared by single-needle and high speed electrospinning. *Int. J. Pharm.* **2016**, *498*, 234–244. [CrossRef]
57. Padilla, A.M.; Ivanisevic, I.; Yang, Y.; Engers, D.; Bogner, R.H.; Pikal, M.J. The study of phase separation in amorphous freeze-dried systems. Part I: Raman mapping and computational analysis of XRPD data in model polymer systems. *J. Pharm. Sci.* **2011**, *100*, 206–222. [CrossRef]
58. Qian, F.; Huang, J.; Zhu, Q.; Haddadin, R.; Gawel, J.; Garmise, R.; Hussain, M. Is a distinctive single T_g a reliable indicator for the homogeneity of amorphous solid dispersion? *Int. J. Pharm.* **2010**, *395*, 232–235. [CrossRef]
59. Reading, M.; Qi, S.; Alhijjaj, M. Local Thermal Analysis by Structural Characterization (TASC). In *Thermal Physics and Thermal Analysis. Hot Topics in Thermal Analysis and Calorimetry*; Šesták, J., Hubík, P., Mareš, J., Eds.; Springer: Berlin/Heidelberg, Germany, 2017; pp. 1–10.
60. Alhijjaj, M.; Belton, P.; Fábián, L.; Wellner, N.; Reading, M.; Qi, S. Novel Thermal Imaging Method for Rapid Screening of Drug-Polymer Miscibility for Solid Dispersion Based Formulation Development. *Mol. Pharm.* **2018**, *15*, 5625–5636. [CrossRef]
61. Alhijjaj, M.; Yassin, S.; Reading, M.; Zeitler, J.A.; Belton, P.; Qi, S. Characterization of Heterogeneity and Spatial Distribution of Phases in Complex Solid Dispersions by Thermal Analysis by Structural Characterization and X-ray Micro Computed Tomography. *Pharm. Res.* **2017**, *34*, 971–989. [CrossRef]
62. Crowley, K.J.; Zografi, G. Water Vapor Absorption into Amorphous Hydrophobic Drug/Poly(vinylpyrrolidone) Dispersions. *J. Pharm. Sci.* **2002**, *91*, 2150–2165. [CrossRef]
63. Gupta, S.S.; Parikh, T.; Meena, A.K.; Mahajan, N.; Vitez, I.; Serajuddin, A.T.M. Effect of carbamazepine on viscoelastic properties and hot melt extrudability of Soluplus®. *Int. J. Pharm.* **2015**, *478*, 232–239. [CrossRef]
64. Marsac, P.J.; Rumondor, A.C.; Nivens, D.E.; Kestur, U.S.; Stanciu, L.; Taylor, L.S. Effect of temperature and moisture on the miscibility of amorphous dispersions of felodipine and poly(vinyl pyrrolidone). *J. Pharm. Sci.* **2010**, *99*, 169–185. [CrossRef]
65. Janssens, S.; Denivelle, S.; Rombaut, P.; Van den Mooter, G. Influence of polyethylene glycol chain length on compatibility and release characteristics of ternary solid dispersions of itraconazole in polyethylene glycol/hydroxypropylmethylcellulose 2910 E5 blends. *Eur. J. Pharm. Sci.* **2008**, *35*, 203–210. [CrossRef]
66. Gumaste, S.G.; Gupta, S.S.; Serajuddin, A.T. Investigation of Polymer-Surfactant and Polymer-Drug-Surfactant Miscibility for Solid Dispersion. *AAPS J.* **2016**, *18*, 1131–1143. [CrossRef]
67. Parikh, T.; Gupta, S.S.; Meena, A.K.; Vitez, I.; Mahajan, N.; Serajuddin, A.T. Application of Film-Casting Technique to Investigate Drug-Polymer Miscibility in Solid Dispersion and Hot-Melt Extrudate. *J. Pharm. Sci.* **2015**, *104*, 2142–2152. [CrossRef]

68. Alhijjaj, M.; Belton, P.; Qi, S. An investigation into the use of polymer blends to improve the printability of and regulate drug release from pharmaceutical solid dispersions prepared via fused deposition modeling (FDM) 3D printing. *Eur. J. Pharm. Biopharm.* **2016**, *108*, 111–125. [CrossRef]
69. Kawakami, K. Miscibility analysis of particulate solid dispersions prepared by electrospray deposition. *Int. J. Pharm.* **2012**, *433*, 71–78. [CrossRef]
70. Sanchez-Rexach, E.; Meaurio, E.; Iturri, J.; Toca-Herrera, J.L.; Nir, S.; Reches, M.; Sarasua, J.R. Miscibility, interactions and antimicrobial activity of poly(ε-caprolactone)/chloramphenicol blends. *Eur. Polym. J.* **2018**, *102*, 30–37. [CrossRef]
71. Al-Obaidi, H.; Lawrence, M.J.; Al-Saden, N.; Ke, P. Investigation of griseofulvin and hydroxypropylmethyl cellulose acetate succinate miscibility in ball milled solid dispersions. *Int. J. Pharm.* **2013**, *443*, 95–102. [CrossRef]
72. Solanki, N.G.; Lam, K.; Tahsin, M.; Gumaste, S.G.; Shah, A.V.; Serajuddin, A.T.M. Effects of Surfactants on Itraconazole-HPMCAS Solid Dispersion Prepared by Hot-Melt Extrusion I: Miscibility and Drug Release. *J. Pharm. Sci.* **2019**, *108*, 1453–1465. [CrossRef]
73. Xi, L.; Song, H.; Wang, Y.; Gao, H.; Fu, Q. Lacidipine Amorphous Solid Dispersion Based on Hot Melt Extrusion: Good Miscibility, Enhanced Dissolution, and Favorable Stability. *AAPS Pharm. Sci. Tech.* **2018**, *19*, 3076–3084. [CrossRef]
74. Hu, X.Y.; Lou, H.; Hageman, M.J. Preparation of lapatinibditosylate solid dispersions using solvent rotary evaporation and hot melt extrusion for solubility and dissolution enhancement. *Int. J. Pharm.* **2018**, *552*, 154–163. [CrossRef]
75. Paudel, A.; Van den Mooter, G. Influence of Solvent Composition on the Miscibility and Physical Stability of Naproxen/PVP K 25 Solid Dispersions Prepared by Cosolvent Spray-Drying. *Pharm. Res.* **2012**, *29*, 251–270. [CrossRef]
76. Hildebrand, J.; Scott, R.L. *The Solubility of Nonelectrolytes*, 3rd ed.; Reinhold: New York, NY, USA, 1950.
77. Hildebrand, J.; Scott, R.L. *Regular Solutions*; Prentice-Hall: Englewood Cliffs, NJ, USA, 1962.
78. Scatchard, G. Equilibria in Non-electrolyte Solutions in Relation to the Vapor Pressures and Densities of the Components. *Chem. Rev.* **1931**, *8*, 321–333. [CrossRef]
79. Van Krevelen, D.W.; Te Nijenhuis, K. *Properties of Polymers: Their Correlation with Chemical Structure; Their Numerical Estimation and Prediction from Additive Group Contributions*; Elsevier: Amsterdam, The Netherlands, 2009.
80. Rey-Mermet, C.; Ruelle, P.; Nam-Trân, H.; Buchmann, M.; Kesselring, U.W. Significance of partial and total cohesion parameters of pharmaceutical solids determined from dissolution calorimetric measurements. *Pharm. Res.* **1991**, *8*, 636–642. [CrossRef]
81. Şen, M.; Güven, O. Determination of solubility parameter of poly(N-vinyl 2-pyrrolidon/ethylene glycol dimethacrylate) gels by swelling measurements. *J. Polym. Sci. Part B* **1998**, *36*, 213–219. [CrossRef]
82. Bozdogan, A.E. A method for determination of thermodynamic and solubility parameters of polymers from temperature and molecular weight dependence of intrinsic viscosity. *Polymer* **2004**, *45*, 6415–6424. [CrossRef]
83. Adamska, K.; Voelkel, A. Inverse gas chromatographic determination of solubility parameters of excipients. *Int. J. Pharm.* **2005**, *304*, 11–17. [CrossRef]
84. Small, P.A. Some factors affecting the solubility of polymers. *J. Appl. Chem.* **1953**, *3*, 71–80. [CrossRef]
85. Hoy, K.L. New values of the solubility parameters from vapor pressure data. *J. Paint Technol.* **1970**, *42*, 76–118.
86. Van Krevelen, D.W.; Hoftyzer, P.J. *Properties of Polymers: Their Estimation and Correlation with Chemical Structure*, 2nd ed.; Elsevier: Amsterdam, The Netherlands, 1976; pp. 129–159.
87. Fedors, R.F. A method for estimating both the solubility parameters and molar volumes of liquids. *Polym. Eng. Sci.* **1974**, *14*, 147–154. [CrossRef]
88. Hansen, C.M. The universality of the solubility parameter. *Ind. Eng. Chem. Prod. Res. Dev.* **1969**, *8*, 2–11. [CrossRef]
89. Hoy, K.L. *The Hoy Tables of Solubility Parameters*; Union Carbide Corporation, Solvents & Coatings Materials, Research & Development Department: South Charleston, WV, USA, 1985.
90. Hoy, K.L. Solubility parameter as a design parameter for water-borne polymers and coatings. *J. Coated Fabrics* **1989**, *19*, 53–67. [CrossRef]
91. Mavrovouniotis, M.L. Estimation of Properties from Conjugate Forms of Molecular Structures: The ABC Approach. *Ind. Eng. Chem. Res.* **1990**, *29*, 1943–1953. [CrossRef]

92. Stefanis, E.; Constantinou, L.; Panayiotou, C. A Group-Contribution Method for Predicting Pure Component Properties of Biochemical and Safety Interest. *Ind. Eng. Chem. Res.* **2004**, *43*, 6253–6261. [CrossRef]
93. Stefanis, E.; Panayiotou, C. Prediction of hansen solubility parameters with a new group-contribution method. *Int. J. Thermophys.* **2008**, *29*, 568–585. [CrossRef]
94. Stefanis, E.; Panayiotou, C. A new expanded solubility parameter approach. *Int. J. Pharm.* **2012**, *426*, 29–43. [CrossRef]
95. Just, S.; Sievert, F.; Thommes, M.; Breitkreutz, J. Improved group contribution parameter set for the application of solubility parameters to melt extrusion. *Eur. J. Pharm. Biopharm.* **2013**, *85*, 1191–1199. [CrossRef]
96. Lydersen, A.L. *Estimation of Critical Properties of Organic Compounds*; Engineering Experiment Station Report 3; College of Engineering, University of Wisconsin: Madison, WI, USA, 1955.
97. Maniruzzaman, M.; Pang, J.; Morgan, D.J.; Douroumis, D. Molecular modeling as a predictive tool for the development of solid dispersions. *Mol. Pharm.* **2015**, *12*, 1040–1049. [CrossRef]
98. Piccinni, P.; Tian, Y.; McNaughton, A.; Fraser, J.; Brown, S.; Jones, D.S.; Li, S.; Andrews, G.P. Solubility parameter-based screening methods for early-stage formulation development of itraconazole amorphous solid dispersions. *J. Pharm. Pharmacol.* **2016**, *68*, 705–720. [CrossRef]
99. Bagley, E.B.; Nelson, T.P.; Scigliano, J.M. Three-dimensional solubility parameters and their relationship to internal pressure measurements in polar and hydrogen bonding solvents. *J. Paint Technol.* **1971**, *43*, 35.
100. Greenhalgh, D.J.; Williams, A.C.; Timmins, P.; York, P. Solubility parameters as predictors of miscibility in solid dispersions. *J. Pharm. Sci.* **1999**, *88*, 1182–1190. [CrossRef]
101. Forster, A.; Hempenstall, J.; Tucker, I.; Rades, T. Selection of excipients for melt extrusion with two poorly water-soluble drugs by solubility parameter calculation and thermal analysis. *Int. J. Pharm.* **2001**, *226*, 147–161. [CrossRef]
102. Chan, S.Y.; Qi, S.; Craig, D.Q. An investigation into the influence of drug-polymer interactions on the miscibility, processability and structure of polyvinylpyrrolidone-based hot melt extrusion formulations. *Int. J. Pharm.* **2015**, *496*, 95–106. [CrossRef]
103. Donnelly, C.; Tian, Y.; Potter, C.; Jones, D.S.; Andrews, G.P. Probing the effects of experimental conditions on the character of drug-polymer phase diagrams constructed using Flory-Huggins theory. *Pharm. Res.* **2015**, *32*, 167–179. [CrossRef]
104. Bansal, K.; Baghel, U.S.; Thakral, S. Construction and Validation of Binary Phase Diagram for Amorphous Solid Dispersion Using Flory-Huggins Theory. *AAPS Pharm. Sci. Tech.* **2016**, *17*, 318–327. [CrossRef]
105. Lu, J.; Cuellar, K.; Hammer, N.I.; Jo, S.; Gryczke, A.; Kolter, K.; Langley, N.; Repka, M.A. Solid-state characterization of Felodipine-Soluplus amorphous solid dispersions. *Drug Dev. Ind. Pharm.* **2016**, *42*, 485–496. [CrossRef]
106. Purohit, H.S.; Taylor, L.S. Miscibility of Itraconazole-Hydroxypropyl Methylcellulose Blends: Insights with High Resolution Analytical Methodologies. *Mol. Pharm.* **2015**, *12*, 4542–4553. [CrossRef]
107. He, Y.; Ho, C. Amorphous Solid Dispersions: Utilization and Challenges in Drug Discovery and Development. *J. Pharm. Sci.* **2015**, *104*, 3237–3258. [CrossRef]
108. Rubinstein, M.; Colby, R.H. *Polymer Physics*; Oxford University Press: New York, NY, USA, 2003.
109. Yang, M.; Wang, P.; Gogos, C. Prediction of acetaminophen's solubility in poly(ethylene oxide) at room temperature using the Flory-Huggins theory. *Drug Dev. Ind. Pharm.* **2013**, *39*, 102–108. [CrossRef]
110. Prudic, A.; Ji, Y.; Sadowski, G. Thermodynamic phase behavior of API/polymer solid dispersions. *Mol. Pharm.* **2014**, *11*, 2294–2304. [CrossRef]
111. Lehmkemper, K.; Kyeremateng, S.O.; Heinzerling, O.; Degenhardt, M.; Sadowski, G. Long-Term Physical Stability of PVP- and PVPVA-Amorphous Solid Dispersions. *Mol. Pharm.* **2017**, *14*, 157–171. [CrossRef]
112. Lehmkemper, K.; Kyeremateng, S.O.; Heinzerling, O.; Degenhardt, M.; Sadowski, G. Impact of Polymer Type and Relative Humidity on the Long-Term Physical Stability of Amorphous Solid Dispersions. *Mol. Pharm.* **2017**, *14*, 4374–4386. [CrossRef]
113. Lehmkemper, K.; Kyeremateng, S.O.; Bartels, M.; Degenhardt, M.; Sadowski, G. Physical stability of API/polymer-blend amorphous solid dispersions. *Eur. J. Pharm. Biopharm.* **2018**, *124*, 147–157. [CrossRef]
114. Luebbert, C.; Sadowski, G. Moisture-induced phase separation and recrystallization in amorphous solid dispersions. *Int. J. Pharm.* **2017**, *532*, 635–646. [CrossRef]
115. Prudic, A.; Kleetz, T.; Korf, M.; Ji, Y.; Sadowski, G. Influence of copolymer composition on the phase behavior of solid dispersions. *Mol. Pharm.* **2014**, *11*, 4189–4198. [CrossRef]

116. Hancock, B.C.; Shamblin, S.L.; Zografi, G. Molecular mobility of amorphous pharmaceutical solids below their glass transition temperatures. *Pharm. Res.* **1995**, *12*, 799–806. [CrossRef]
117. Dünweg, B. Molecular dynamics algorithms and hydrodynamic screening. *J. Chem. Phys.* **1993**, *99*, 6977–6982. [CrossRef]
118. Karplus, M.; Petsko, G.A. Molecular dynamics simulations in biology. *Nature* **1990**, *347*, 631–639. [CrossRef]
119. Cui, Y. Using molecular simulations to probe pharmaceutical materials. *J. Pharm. Sci.* **2011**, *100*, 2000–2019. [CrossRef]
120. Gupta, J.; Nunes, C.; Vyas, S.; Jonnalagadda, S. Prediction of solubility parameters and miscibility of pharmaceutical compounds by molecular dynamics simulations. *J. Phys. Chem. B* **2011**, *115*, 2014–2023. [CrossRef]
121. Anwar, J.; Khan, S.; Lindfors, L. Secondary crystal nucleation: Nuclei breeding factory uncovered. *Angew. Chem. Int. Ed.* **2015**, *54*, 14681–14684. [CrossRef]
122. Gupta, J.; Nunes, C.; Jonnalagadda, S. A molecular dynamics approach for predicting the glass transition temperature and plasticization effect in amorphous pharmaceuticals. *Mol. Pharm.* **2013**, *10*, 4136–4145. [CrossRef]
123. Barmpalexis, P.; Karagianni, A.; Katopodis, K.; Vardaka, E.; Kachrimanis, K. Molecular modelling and simulation of fusion-based amorphous drug dispersions in polymer/plasticizer blends. *Eur. J. Pharm. Sci.* **2019**, *130*, 260–268. [CrossRef]
124. Edueng, K.; Mahlin, D.; Bergström, C.A.S. The Need for Restructuring the Disordered Science of Amorphous Drug Formulations. *Pharm. Res.* **2017**, *34*, 1754–1772. [CrossRef]
125. Maus, M.; Wagner, K.G.; Kornherr, A.; Zifferer, G. Molecular dynamics simulations for drug dosage form development: Thermal and solubility characteristics for hot-melt extrusion. *Mol. Simul.* **2008**, *34*, 1197–1207. [CrossRef]
126. Macháčková, M.; Tokarský, J.; Čapková, P. A simple molecular modeling method for the characterization of polymeric drug carriers. *Eur. J. Pharm. Sci.* **2013**, *48*, 316–322.
127. Barmpalexis, P.; Karagianni, A.; Kachrimanis, K. Molecular simulations for amorphous drug formulation: Polymeric matrix properties relevant to hot-melt extrusion. *Eur. J. Pharm. Sci.* **2018**, *119*, 259–267. [CrossRef]
128. Ouyang, D. Investigating the molecular structures of solid dispersions by the simulated annealing method. *Chem. Phys. Lett.* **2012**, *554*, 177–184. [CrossRef]
129. Xiang, T.X.; Anderson, B.D. Molecular dynamics simulation of amorphous indomethacin-poly (vinylpyrrolidone) glasses: Solubility and hydrogen bonding interactions. *J. Pharm. Sci.* **2013**, *102*, 876–891. [CrossRef]

 © 2019 by the authors. Licensee MDPI, Basel, Switzerland. This article is an open access article distributed under the terms and conditions of the Creative Commons Attribution (CC BY) license (http://creativecommons.org/licenses/by/4.0/).

Review

Overview of the Manufacturing Methods of Solid Dispersion Technology for Improving the Solubility of Poorly Water-Soluble Drugs and Application to Anticancer Drugs

Phuong Tran [1], Yong-Chul Pyo [1], Dong-Hyun Kim [1], Sang-Eun Lee [1], Jin-Ki Kim [2,*] and Jeong-Sook Park [1,*]

1. College of Pharmacy, Chungnam National University, 99 Daehak-ro, Yuseong-gu, Daejeon 34134, Korea; phuongtran24288@gmail.com (P.T.); himchani46@naver.com (Y.-C.P.); dong_bal@naver.com (D.-H.K.); nnininn@hanmail.net (S.-E.L.)
2. College of Pharmacy and Institute of Pharmaceutical Science and Technology, Hanyang University, 55 Hanyangdaehak-ro, Sangnok-gu, Ansan 15588, Korea
* Correspondence: jinkikim@hanyang.ac.kr (J.-K.K.); eicosa@cnu.ac.kr (J.-S.P.); Tel.: +82-42-821-5932 (J.-K.K. & J.-S.P.)

Received: 12 February 2019; Accepted: 15 March 2019; Published: 19 March 2019

Abstract: Approximately 40% of new chemical entities (NCEs), including anticancer drugs, have been reported as poorly water-soluble compounds. Anticancer drugs are classified into biologic drugs (monoclonal antibodies) and small molecule drugs (nonbiologic anticancer drugs) based on effectiveness and safety profile. Biologic drugs are administered by intravenous (IV) injection due to their large molecular weight, while small molecule drugs are preferentially administered by gastrointestinal route. Even though IV injection is the fastest route of administration and ensures complete bioavailability, this route of administration causes patient inconvenience to visit a hospital for anticancer treatments. In addition, IV administration can cause several side effects such as severe hypersensitivity, myelosuppression, neutropenia, and neurotoxicity. Oral administration is the preferred route for drug delivery due to several advantages such as low cost, pain avoidance, and safety. The main problem of NCEs is a limited aqueous solubility, resulting in poor absorption and low bioavailability. Therefore, improving oral bioavailability of poorly water-soluble drugs is a great challenge in the development of pharmaceutical dosage forms. Several methods such as solid dispersion, complexation, lipid-based systems, micronization, nanonization, and co-crystals were developed to improve the solubility of hydrophobic drugs. Recently, solid dispersion is one of the most widely used and successful techniques in formulation development. This review mainly discusses classification, methods for preparation of solid dispersions, and use of solid dispersion for improving solubility of poorly soluble anticancer drugs.

Keywords: solid dispersion; classification; manufacturing methods; bioavailability; anticancer drugs

1. Introduction

Cancer is one of the leading causes of death worldwide, and treatment remains a great challenge. Currently, there are three major cancer treatment strategies of surgery (performed by a surgical oncologist), chemotherapy (use of anticancer drugs), and radiotherapy (delivered by a radiooncologist) [1]. The objective of any treatment is to kill as many cancer cells as possible and minimize death of normal cells. Patients can receive monotherapy or combination therapy. For example, Hwang et al. [2] reported a combination of photodynamic therapy (PDT) and anti-tumor immunity in cancer therapy. Among the three major therapeutic strategies, surgery has been the first line of treatment for many solid tumors. This strategy involves removal of solid tumors by a surgical oncologist under anesthesia. However, patients have to be hospitalized, the entire tumor cannot always be removed, damage can occur to nearby normal tissues, and complications can arise from surgery. Radiotherapy is focused on the tumor and is designed to kill a large proportion of cancer cells within the tumor. As with surgery, this therapy has disadvantages such as damage to surrounding tissues (e.g., lung, heart) and inconvenience for patients (e.g., in some cases, it must be delivered daily, 5 days per week, for 1–2 months). In addition, radiotherapy causes hair loss. As such, use of anticancer drugs (chemotherapy) is currently preferred for treatment of both localized and metastasized cancers. Chemotherapy can kill many cancer cells throughout the body, eradicate microscopic disease at the edges of tumors that may not be seen by a surgeon, and be used in combination with other therapies. Tumor-targeted delivery and controlled release of drugs are two important strategies for improving therapeutic efficacy and reducing side effects.

Cancer-targeting strategies for drug delivery include passive and active targeting strategies [3]. Active targeting focuses anticancer drugs to ligands or receptors in the target region. For example, folic acid (FA) has been used as a targeting ligand to the folate receptor in various tumor sites including lung, ovarian, breast, and colon cancers [4]. Vinothini et al. [5] developed a graphene oxide-methyl acrylate-FA/paclitaxel (GO-MA-FA/PTX) nanocarrier for targeted anticancer drug delivery to breast cancer cells, resulting in reduction of 39% of typical cytotoxic effects. Voeikov et al. [6] prepared dioxadet-loaded nanogels using a block copolymer of polyethylene glycol and polymethacrylic acid (PEG-b-PMAA) as a high-loading capacity (>35% w/w) and high-loading efficiency (>75%) drug delivery system. In another study, trastuzumab, a monoclonal antibody with specific targeting to human epidermal growth factor receptor 2 (HER2) protein, was used in combination with cisplatin for treatment of HER2-overexpressing breast cancer [7]. Passive targeting occurs through interactions with the reticuloendothelial system, allowing for entry into the blood, which is dependent on particle size and surface characteristics. Development of nanoparticle anticancer drugs has improved therapeutic efficacy because the drug can be directly and selectively targeted to cancer cells [8–10]. For example, Kirtane et al. [11] developed a polymer-surfactant nanoparticle composed of a sodium alginate core complexed with doxorubicin and the surfactant aerosol OT for stability. The relative BA of the nanoparticle formulation was higher than that of the pure drug. In another study by Valicherla el al. [12], docetaxel nanoparticles were prepared in a self-emulsifying drug delivery system (SEDDS) to enhance BA and anti-tumor activity. The BA of docetaxel-SEDDS was 3.19-fold higher than that of the pure drug. Furthermore, docetaxel-SEDDS showed 25-fold higher cytotoxic activity than the free drug in vitro.

Intravenous (IV) and oral are the two most popular routes of drug administration. Paclitaxel is an anticancer drug used to treat many types of cancers such as breast, ovarian, lung, and pancreatic cancer. Paclitaxel is administered by IV infusion and sold under the marketed name Taxol 30 mg (5 mL), 100 mg (16.7 mL), and 300 mg (50 mL) [13]. Tamoxifen, an anticancer drug used to treat breast cancer, is sold under the brand name Nolvadex (10 mg and 20 mg tablets) and is formulated for oral administration [14]. IV infusion is the best route of administration for most anticancer drugs because this route leads to 100% BA. However, IV administration is associated with several side effects, short duration of effectiveness, and inconvenience due to hospitalization. Taxol is prepared by solubilizing paclitaxel in ethanol: Cremophor EL (1:1, v/v). This formulation is then diluted 5–20-fold in normal saline, resulting in a final concentration of 0.03–0.60 mg/mL [15]. However, several side effects result from administration of Cremophor EL such as hypersensitivity, nephrotoxicity, and neurotoxicity [16–18]. Despite the continuing interest on anticancer drugs in recent years, their use in clinical anticancer therapy is limited due to nonspecific biodistribution, low therapeutic indices, and poor aqueous solubility. As such, oral administration has received increasing attention, leading to increased numbers of anticancer drugs being developed for oral dosing. Oral drugs can be administered at home, do not induce the same discomfort as IV infusion, and the drug concentration can be maintained for long time periods in cancerous cells. Oral dosage forms rely on drug solubility to achieve the desired concentrations in the systemic circulation. Drugs have to dissolve in the gastrointestinal (GI) fluid and then permeate the membrane of the GI tract into the blood to be effective. However approximately 40% of new chemical entities (NCEs), including anticancer drugs, are poorly water-soluble [19–21]. Due to poor aqueous solubility, the drug cannot completely absorb in the GI tract, resulting in poor bioavailability (BA) and high intra- and inter-individual pharmacokinetic variability.

The Biopharmaceutics Classification System (BCS) divides drugs into four groups as follows: Class I (high solubility, high permeability), Class II (low solubility, high permeability), Class III (high solubility, low permeability), and Class IV (low solubility, low permeability) (Figure 1) [22]. A drug substance is considered highly soluble when the highest single therapeutic dose is soluble in 250 mL or less of aqueous media over the pH range of 1.2–6.8 at 37 ± 1 °C. Permeability is evaluated on the basis of the extent of absorption of a drug from human pharmacokinetic studies. Alternatively, in vitro culture methods can also be used to predict drug absorption in humans. A drug is considered highly permeable when the absolute BA is $\geq$85%. High permeability can also be concluded if $\geq$85% of the administered dose is recovered in urine as unchanged (parent drug) or as the sum of the parent drug, Phase 1 oxidative, and Phase 2 conjugative metabolites. Among four groups, drugs belonging to Class II and IV exhibit poor aqueous solubility, resulting in poor BA. Therefore, enhancing solubility and BA of poorly water-soluble drugs in BCS Classes II and IV is a significant challenge in the pharmaceutical industry.

In the clinic, there are many insoluble drugs with small dose administration such as risperidone (0.25–4 mg), lorazepam (0.5–2 mg), diazepam (2–10 mg), and clonazepam (0.5–2 mg), which do not require increased solubility. However, the solubility of these drugs is usually affected by pH due to physicochemical properties resulting in a decrease in the effective treatment. For example, risperidone is indicated for treatment of schizophrenia at a small dose (0.25 mg, 0.5 mg, 1 mg, 2 mg, 3 mg, or 4 mg) in oral administration. It is a weak base that is practically insoluble in water. Its solubility is pH dependent, with high solubility in acidic pH, and decreasing solubility as pH increases (range from >200 mg/mL at pH 2.1 down to 0.29 mg/mL at pH 7.6 and reaches 0.08 mg/mL at pH 8). After oral administration, risperidone is rapidly absorbed, and approximately 80% of drugs will be absorbed in the GI tract, where the solubility significantly drops. Moreover, risperidone is a metabolized drug, in which approximately 70% and 14% of the dose is excreted in urine and feces, respectively. Therefore, enhancing solubility in simulated intestinal pH to ensure higher drug concentrations at the main absorption site and improve BA is a challenge in drug development.

To improve the solubility and BA of poorly water-soluble drugs, several methods have been developed such as solid dispersion (SD) [23–25], complexation [26], lipid-based systems [27,28], micronization [29,30], nanonization [31–33], and co-crystals [34,35]. Among these, SD is one of the most potent and successful methods. SD is defined as a group of solid products consisting of a hydrophobic drug dispersed in at least one hydrophilic carrier, resulting in enhanced surface area, leading to higher drug solubility and dissolution rate. Improving wettability and dispersibility and reducing aggregation and agglomeration of drug particles result in enhanced drug BA. An SD is typically characterized on the molecular level using Fourier-transform infrared spectroscopy (FTIR), Raman spectroscopy, near-infrared spectroscopy (NIR), and solid-state nuclear magnetic resonance (SSNMR) at the particulate level using powder X-Ray diffraction (PXRD), differential scanning calorimetry (DSC), scanning electron microscopy (SEM), and transmission electron microscopy (TEM) and at the bulk level using density, contact angle, flowability, and Karl Fischer titration [36]. SD can be accomplished by several methods such as solvent evaporation [23], hot-melt extrusion [37], and spray drying [38]. In this review, we mainly discuss classification of drugs, methods for preparation of SD, and use of SD for improving solubility of poorly soluble anticancer drugs.

Figure 1. Biopharmaceutics classification system (BCS).

2. Solid Dispersions

An SD is defined as a group of solid products consisting of a hydrophobic drug dispersed in at least one hydrophilic carrier, resulting in increased surface area and, enhanced drug solubility and dissolution rate. They are classified as follows.

2.1. Carrier-Based Class of Solid Dispersion

Many carriers are used in SD. These carriers determine the final formulation properties and can be categorized into first, second, and third classes (Figure 2).

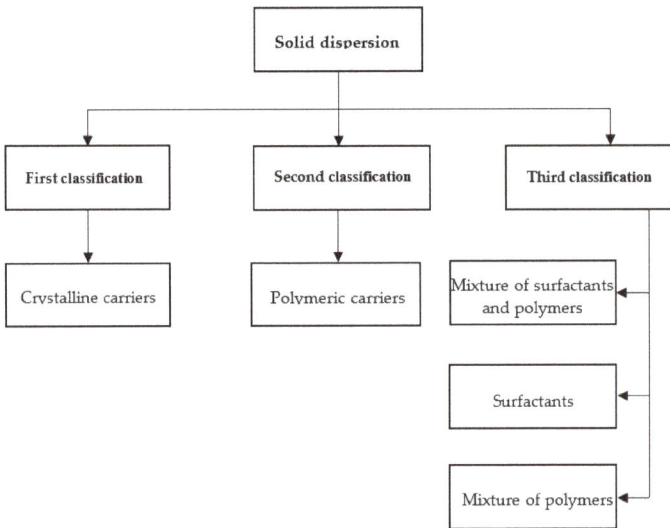

Figure 2. Classification of solid dispersions.

2.1.1. First Class of SD

The first study of SD was conducted by Sekiguchi and Obi in 1961 [39]. They studied absorption of a eutectic mixture of sulfathiazole [39] and chloramphenicol [40] compared with that of the original formulations of the same drugs. The results showed that the use of urea as a hydrophilic carrier increased the absorption of sulfathiazole and chloramphenicol in the eutectic mixture compared to that of the conventional formulations, thereby improving BA. Sugars and their derivatives are carriers with high solubility in water and low toxicity. Levy [41] and Kaning [42] used mannitol as a carrier to develop SD as a solid mixture instead of as a eutectic mixture. The formulation using mannitol as a carrier showed higher dissolution compared to the original formulation of the drug. To enhance the dissolution profile of clotrimazole, Madgulkar et al. [43] prepared an SD by a fusion method using various sugars such as D-mannitol, D-fructose, D-dextrose, and D-maltose as carriers at a different weight ratios to the drug. The results showed that a 100% solution of mannitol showed an 806-fold increase in solubility compared to the conventional drug in water. The dissolution profile of clotrimazole SD was improved at 1:3 drug to mannitol ratio. In conclusion, urea and sugars were first used as crystalline carriers for production of SD. These formulations were thermodynamically unstable, resulting in slow drug release.

2.1.2. Second Class of SD

Because of thermodynamic instability of first class SD [44], second class SDs were introduced using amorphous polymeric carriers [45] instead of urea or sugars. The polymeric carriers can be synthetic or natural polymers. Synthetic polymers include povidone (PVP) [46–50], PEG [51–54], and polymethacrylates [55,56], and natural polymers include hydroxypropylmethylcellulose (HPMC) [57–61], ethyl cellulose [62–64], and starch derivatives such as cyclodextrins (CDs) [65]. A study by Franco et al. [66] showed that using PVP as a carrier in ketoprofen SD increased the dissolution rate of ketoprofen 4.2-fold compared to that of the conventional drug. In a study by Dhandapani and El-gied [67], β-CD was used as a carrier in cefixime SD to improve solubility. The dissolution rate of cefixime SD was 6.77-fold higher than that of the pure drug. In another study, an SD of diclofenac sodium was prepared by solvent evaporation using Eudragit E 100 as the carrier [68]. Solubility of diclofenac sodium from the SD (0.823 mg/mL) was approximately 58.8-fold higher compared with the pure drug (0.014 mg/mL). In dissolution studies, diclofenac sodium released from SD was approximately 60% after 2 h at pH 1.2, while the pure drug release was less than 10% after 2 h. Second class SDs are dispersed in polymeric carriers and achieve a supersaturated state. These formulations have smaller particle sizes and enhanced wettability thereby increasing the aqueous solubility of drugs.

2.1.3. Third Class of SD

The third class of SD was recently developed. In this class, surfactant can be used alone or in the combination with other hydrophilic carriers in the preparation of SD (Figure 2). Surfactants were widely used to improve the solubility and BA of poorly water-soluble drugs and play a crucial role in the pharmaceutical industry. Adsorption of a surfactant on a solid surface can modify the hydrophobicity of the drug, thereby reducing surface tension between two liquids or between a liquid and a solid. In addition, surfactants can also act as wetting agents, detergents, emulsifiers, foaming agents, and dispersants. Several surfactants such as Inulin [69,70], inutec [70], poloxamer 407 [71], Gelucire 44/14 [72], and Compritol 888 ATO [73] are used in preparation of SD. In a study by Panda et al. [74], Gelucire 50/13 and poloxamer 188 were used in the development of a bosentan SD formulation to improve the solubility and dissolution of this drug. The results showed that the solubility of bosentan from the SD formulation increased 8- and 10-fold when using Gelucire 50/13- and poloxamer 188-based SDs, respectively, in comparison with that of the pure drug. Furthermore, over 90% of the drug was released from SD after 1-h in vitro dissolution studies. Karolewicz et al. [75]

prepared an SD of fenofibrate with poloxamer 407 as the carrier at ratios of 10/90, 20/80, 30/70, 40/60, 50/50, 60/40, 70/30, 80/20, and 90/10 using the fusion method. The results showed a 134-fold increase in dissolution rate for SD containing 30/70 w/w fenofibrate/ poloxamer 407. Surfactants are also used in preparation of SD of poorly soluble anticancer drugs such as docetaxel [76], flutamide [77], and lapatinib [78].

2.2. Structure-Based Class of Solid Dispersion

2.2.1. Eutectic Mixtures

A eutectic mixture is a mixture of two components that melt at a single temperature. Components A and B were co-melted at the eutectic point (E) (Figure 3), where the melting point of the mixture was lower than that of component A or B alone. In 1961, Sekiguchi and Obi [39] were the first to prepare a eutectic mixture of sulfathiazole and urea. The results showed that the absorption of sulfathiazole in the eutectic mixture was improved compared to that of the conventional drug.

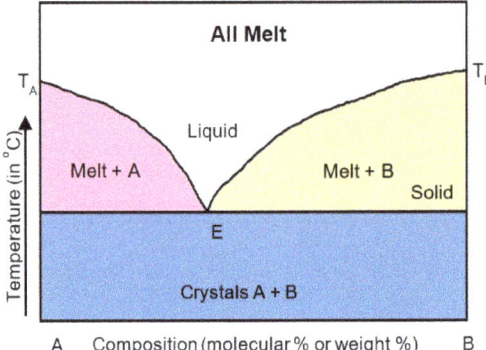

Figure 3. Phase diagram of a eutectic mixture. A, B (drug, carrier), E (eutectic point).

2.2.2. Solid Solution

Herein, SD is a mixture of the drug and a carrier [79]. Solid solution is categorized on the basis of miscibility and molecular size of the components as continuous and discontinuous solid solutions. In continuous solid solutions, the two components can be mixed in all proportions at which the bonding strength between the two components is greater that of the individual components [80]. In discontinuous solid solutions, the solubility of each component is limited in solid solvents [81]. Solid solutions are classified as substitutional (Figure 4A) and interstitial (Figure 4B) based on molecular size. In substitutional solid solutions, solute molecules substitute for solvent molecules in the crystal lattice. In interstitial solid solutions, the dissolved molecules occupy the interstitial spaces between the solvent molecules in the crystal lattice [82].

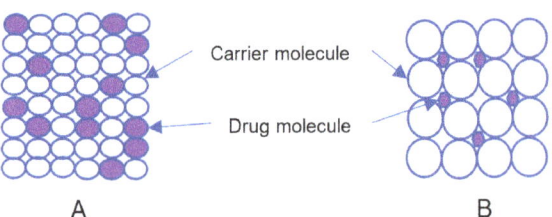

Figure 4. Schematic structure of the solid solution.

2.2.3. Glass Solution/Glass Suspension

A glass solution is a homogeneous system in which the drug molecule is dissolved in a glassy solvent [81,82]. Glass suspension is a homogeneous system in which the drug molecule is suspended in a glassy solvent [83]. The glassy state is characterized by transparency and brittleness below the glass transition temperature for both glass solutions and glass suspensions.

2.3. Advantages of Solid Dispersions

SD was widely used for enhancing dissolvability in water of poorly water-soluble drugs with several advantages as follows:

- One of the most important advantages of SD is drugs interacting with hydrophilic carriers can decrease agglomeration and release in a supersaturation state, resulting in rapid absorption and improved BA [84].
- SD can improve drug wettability and increase the surface area, resulting in enhanced aqueous solubility of drugs.
- SD can be produced as a solid oral dosage form, which is more convenient for patients than other forms like liquid products.
- In addition, SD showed an advantage compared to salt formulation, cocrystallization, and other methods. For example, salt formulations use ionized active pharmaceutical ingredients (APIs) (cationic or anionic form) and are widely used in the pharmaceutical industry due to the broad capacity of design according to desired drug properties. However, not all drugs can ionize with all cations/anions, and phase dissociation or stability issue is inherent in salt formation or cocrystallization. Salt formulation showed several disadvantages such as reduced solubility and dissolution rate, resulting in decreased relative BA (common ion effect for HCl salts); greater regulatory scrutiny for strong acid salts isolated from alkyl alcohols; and increased hygroscopicity, e.g., for Na and, K salts, spray-drying/lyophilization can dissociate strong acid salts. The disadvantages of salt formulation can be resolved when the formulation is produced using an SD.
- Practically, dissolution of drugs is a prerequisite for complete absorption to have the desired therapeutic effect of anticancer drugs after oral administration. Most of the anticancer drugs exhibit poor aqueous solubility causes of dissolution limit resulting low BA and high variability in blood concentration. The limitation of drug dissolution can improve by SD, a technique that induces supersaturated drug dissolution and with that it enhances in vivo absorption.

2.4. Disadvantages of Solid Dispersions

SD is a good technique for improving solubility and BA of hydrophobic drugs. However, some disadvantages are as follows:

- Physical instability.
- SDs show changes in crystallinity and decreased dissolution rate with aging.
- Due to their thermodynamic instability, SD is sensitive to temperature and humidity during storage. These factors can promote phase separation and crystallization of SD by increasing the overall molecular mobility, decreasing the glass transition temperature (T_g) or disrupting interactions between the drug and carrier, resulting in a decreased solubility and dissolution rate of the drug.
- Patients suffering from cancer should continue to use anticancer drugs during treatment. However, the instability of SD during the period of storage can affect drug quality and the effectiveness of treatment.

2.5. Preparation Methods for Solid Dispersions

SD can be prepared by several methods such as solvent evaporation, melting, and supercritical fluid (SCF) technology (Figure 5). The list of drugs investigated for SDs is shown in Table 1, and a list of commercial SDs is shown in Table 2.

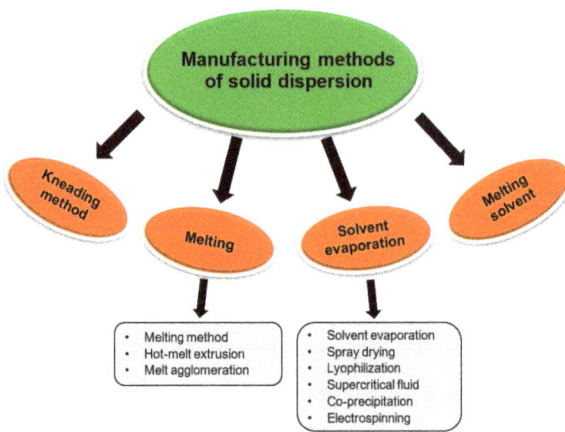

Figure 5. Manufacturing methods of solid dispersion.

2.5.1. Melting Method/Fusion Method

The melting method was first used in 1961 by Sekiguchi and Obi [39]. The basic principle of the melting method is that a physical mixture of a drug and hydrophilic carrier is heated directly until they melt at a temperature slightly above their eutectic point. Then, the melt is cooled and solidified rapidly in an ice bath with stirring. The final solid mass is crushed and sieved. The advantages of this method are simplicity and economy. Several drug SDs have been prepared using this method such as sulfathiazole [39], fenofibrate [75], furosemide [85], albendazole [54], and paclitaxel [86] (Table 1). The melting method has also been used to improve solubility of poorly soluble anticancer drugs. For example, to improve solubility of prednisolone, an SD was prepared by the melting method using PEG 4000 and mannitol as the carriers [87]. The results showed that, at weight ratios of drug: PEG 4000 (1:4) and drug: mannitol (1:7), release of drug from the SD (~85%) increased in comparison with the pure drug (~50%). In a study for improving release of paclitaxel from poly(ε-caprolactone) (PCL)-based-film, an SD of paclitaxel was prepared by the melting method using poloxamer 188 and PEG as the carriers and was then incorporated into PCL films. Drug released from SD was higher than that from the pure drug, with over 90% of drug released from the SD after 1 h at a weight ratio of drug: poloxamer 188 (1:3).

Table 1. List of drugs investigated for solid dispersions.

Methods	Drugs
Melting/fusion method	Sulfathiazole [39], clotrimazole [43], albendazole [54], tacrolimus [61], fenofibrate [75], furosemide [85], paclitaxel [86], manidipine [88], olanzapine [89], diacerein [90]
Solvent evaporation method	Dutasteride [23], tadalafil [50], glimepiride [53], nimodipine [59], diclofenac [68], azithromycin [91], tectorigenin [92], flurbiprofen [93], cilostazol [94], ticagrelor [95], piroxicam [96], indomethacin [97], loratadine [98], abietic acid [99], efavirenz [100], repagnilide [101], prednisolone [102]

Table 1. Cont.

Methods	Drugs
Hot-melt extrusion method	Ritonavir [37], naproxen [46], oleanolic acid [103], efavirenz [104], tamoxifen [105], lafutidine [106], disulfiram [107], bicalutamide [108], itraconazole [109], miconazole [110], glyburide [111]
Lyophilization/Freeze-drying	Nifedipine and sulfamethoxazole [112], celecoxib [113], meloxicam [114], docetaxel [115]
Co-precipitation method	Silymarin [116], celecoxib [117], GDC-0810 [118]
Supercritical fluid method	Ketoprofen [66], irbesartan [119], apigenin [120], carbamazepine [121], glibenclamide [122], carvedilol [123]
Spray-drying method	Nilotinib [124], spironolactone [125], valsartan [126], rebamipide [127], artemether [128], naproxen [129]
Kneading method	Cefixime [67], efavirenz [100], domperidone [130]

Table 2. List of commercial solid dispersions.

Products	Drugs	Polymers	Company
Afeditab®	Nifedipine	Poloxamer or PVP	Elan Corp, Ireland
Cesamet®	Nabilone	PVP	Lilly, USA
Cesamet®	Nabilone	PVP	Valeant Pharmaceuticals, Canada
Certican®	Everolimus	HPMC	Novartis, Switzweland
Gris-PEG®	Griseofulvin	PEG	Novartis, Switzweland
Gris-PEG®	Griseofulvin	PVP	VIP Pharma, Denmark
Fenoglide®	Fenofibrate	PEG	LifeCycle Pharma, Denmark
Nivadil®	Nivaldipine	HPC/HPMC	Fujisawa Pharmaceuticals Co., Ltd
Nimotop®	Nimodipine	PEG	Bayer
Torcetrapib®	Torcetrapib	HPMC AS	Pfizer, USA
Ibuprofen®	Ibuprofen	Various	Soliqs, Germany
Incivek®	Telaprevir	HPMC AS	Vertex
Sporanox®	Itraconazole	HPMC	Janssen Pharmaceutica, Belgium
Onmel®	Itraconazole	HPMC	Stiefel
Prograf®	Tacrolimus	HPMC	Fujisawa Pharmaceuticals Co., Ltd
Cymbalta®	Duloxetine	HPMC AS	Lilly, USA
Noxafil®	Posaconazole	HPMC AS	Merck
LCP-Tacro®	Tacrolimus	HPMC	LifeCycle Pharma, Denmark
Intelence®	Etravirine	HPMC	Tibotec, Yardley, PA
Incivo®	Etravirine	HPMC	Janssen Pharmaceutica, Belgium
Rezulin®	Troglitazone	PVP	Pfizer, USA
Isoptin SRE-240®	Verapamil	Various	Soliqs, Germany
Isoptin SR-E®	Verapamil	HPC/HPMC	Abbott Laboratories, USA
Crestor®	Rosuvastatin	HPMC	AstraZeneca

Table 2. Cont.

Products	Drugs	Polymers	Company
Zelboraf®	Vemurafenib	HPMC AS	Roche
Zortress®	Everolimus	HPMC	Novartis, Switzweland
Kalydeco®	Ivacaflor	HPMC AS	Vertex
Kaletra®	Lopinavir and Ritonavir	PVP/polyvinyl acetate	Abbott Laboratories, USA

PVP: polyvinylpyrrolidone; HPMC: hydroxypropylmethylcellulose; PEG: polyethyleneglycol; HPC: hydroxypropylcellulose; HMPC AS: hydroxypropylmethylcellulose acetylsuccinate.

2.5.2. Solvent Evaporation Method

The solvent evaporation method is one of the most commonly used methods in the pharmaceutical industry for improving solubility of poorly water-soluble drugs. This method was developed mainly for heat unstable components because drug and carrier are mixed by a solvent instead of heat as in melting method. Therefore, this method allows use of carriers with an excessively high melting point. The basic principle of this method is that drug and carrier are dissolved in a volatile solvent for homogeneous mixing. SD is obtained by evaporating the solvent under constant agitation. Then, the solid SD is crushed and sieved. This method was first applied by Tachibana and Nakamura in 1965 [131]. The formulation was prepared by dissolving a drug (β-carotene) and a carrier (PVP) in an organic solvent (chloroform). After that, the solvent was completely evaporated to form a solid mass, which was then sieved and dried. The main advantage of this method is avoidance of decomposition of drug and carrier because the required temperature for evaporation is low. In 1966, Mayersohn and Gibaldi developed an SD of griseofulvin using PVP as the carrier and chloroform as the solvent [132]. Dissolution of griseofulvin from the SD was 11-times greater than that of the pure drug at a ratio of griseofulvin: PVP (1:20). This method has been used to improve solubility of many drugs such as azithromycin [91], tectorigenin [92], flurbiprofen [93], cilostazol [94], ticagrelor [95], piroxicam [96], indomethacin [97], loratadine [98], diclofenac [68], abietic acid [99], efavirenz [100], and repaglinide [101] (Table 1). An SD of tectorigenin, PVP, and PEG 4000 at a weight ratio of 7:54:9 was prepared using solvent evaporation to increase dissolution and BA [92]. In vitro release of the drug from the SD was 4.35-fold greater than that of the pure drug after 2.5 h. In addition, the oral BA of the drug from the SD was higher than that of the conventional drug as determined by AUC (4.8-fold) and C_{max} (13.1-fold). The solvent evaporation method has often been used to improve solubility of poorly water-soluble anticancer drugs such as paclitaxel [133], docetaxel [76], and others (Table 3). For example, the solubility and dissolution of emulsified SD of docetaxel at 2 h were 34.2- and 12.7-fold greater, respectively, compared to those of the conventional drug [76]. In the study by Adeli [91], azithromycin SD was prepared by the solvent evaporation method using various PEG such as PEG 4000, PEG 6000, PEG 8000, PEG 12,000, and PEG 20,000 as the carriers at different ratios. Using PEG as the hydrophilic carrier in SD, the solubility of drug is improved compared to the pure conventional drug. The best result was obtained from SD containing azithromycin: PEG 6000 (1:7). After 1 h, the amount of azithromycin released from SD was more than 49%.

Table 3. Anticancer drugs investigated for solid dispersions.

Anticancer Drugs	Carriers	Methods	Attributes of Modified Anticancer Drugs	Reference	Years
Bicalutamide	PVP K30	Solvent evaporation	Using PVP K30 as carrier, SD showed the highest cumulative released percentage (about 98% during the initial 10 min) and stability after 6 months	[134]	2006
Docetaxel	HPMC, PEG	Solvent evaporation	The solubility and dissolution of emulsified SD of docetaxel at 2 h were 34.2- and 12.7-fold higher, respectively, compared to the pure conventional drug	[76]	2011
Docetaxel	Poloxamer F68/P85	Freeze-drying	A combination of poloxamer F68 and P85 in the preparation of docetaxel SD not only enhanced solubility, but also improved intestinal permeation	[135]	2016
Etoposide	PEG	Fusion method	The solubility and dissolution of etoposide in SD were higher in comparison with etoposide alone	[136]	1993
Everolimus	HPMC	Co-precipitation	At a ratio of drug to HPMC (1:15), drug release from SD was 75% after 30 min, thereby improving oral absorption of everolimus	[137]	2014
Exemestane	Lipoid® E80S/sodium deoxycholate	Freeze-drying	The exemestane SD showed 4-6-fold increase in absorptive transport compared to the pure drug. In addition, AUC_{0-72h} of exemestane SD was 2.3-fold higher in comparison with that of drug alone	[138]	2017
Flutamide	PVP K30, PEG, Pluronic F127	Lyophilization	The dissolution of flutamide was higher (81.64%) than the drug alone (13.45%) using poloxamer 407 as a carrier	[77]	2010
Lapatinib	Soluplus, poloxamer 188	Solvent evaporation, hot-melt extrusion	Solubility and dissolution of lapatinib SD were enhanced compared to the drug alone. After 15 min, the drug in SD was released at 92% compared to the drug alone (48%)	[78]	2018
Letrozole	CO_2-menthol	Supercritical fluid	Solubility of letrozole SD using supercritical fluid is 7.1 times higher compared to that of the conventional drug	[139]	2018
Megestrol acetate	HPMC, Ryoto sugar ester L1695	Supercritical fluid	The SD with drug: HPMC: Ryoto sugar ester L1695 ratio of 1:2:1 showed over 95% rapid dissolution within 30 min. In addition, AUC and C_{max} (0-24h) of drug in SD were 4.0- and 5.5-fold higher, respectively, compared to those in pure drug	[140]	2015

Table 3. Cont.

Anticancer Drugs	Carriers	Methods	Attributes of Modified Anticancer Drugs	Reference	Years
Oridonin	PVP K17	Supercritical fluid	The dissolution of oridonin SD significantly increased compared to the original drug. In addition, the absorption of oridonin in SD showed 26.4-fold improvement in BA	[141]	2011
Paclitaxel	Poloxamer 188, PEG	Fusion method	Paclitaxel SD was successfully prepared, and the drug release from SD was higher than that of the drug alone	[86]	2013
Paclitaxel	HPMC AS	Solvent method	The solubility and permeability of paclitaxel were not increased simultaneously through supersaturation in vivo	[133]	2018
Prednisolone	HP-β-CD, PEG, PVP, PEG 4000, MNT, SMP, Cremophor	Solvent evaporation, melting method, kneading method	The in vitro dissolution of prednisolone SD was improved compared with the pure drug	[87]	2011
Raloxifene	PVP K30	Spray-drying	The absorption of raloxifene from SD showed 2.6-fold enhanced BA in comparison with the conventional drug	[142]	2013
Sorafenib	Soluplus	Spray-drying	The C_{max} and AUC_{0-48h} of sorafenib in SD formulation increased 1.5- and 1.8-fold, resocetuvely, compared with the pure drug	[143]	2015
Tamoxifen	Soluplus	Hot-melt extrusion	The dissolution and BA of tamoxifen in SD were improved compared with the drug alone	[105]	2018
Vemurafenib	HPMC AS	Solvent-controlled precipitation	The BA of vemurafenib in SD was improved 4~5-fold compared to the conventional drug	[144]	2013

HP-β-CD: hydroxypropyl-β-cyclodextrin, MNT: mannitol, SMP: skimmed milk powder.

2.5.3. Melting Solvent Method (Melt Evaporation)

The melting solvent method was first studied by Goldberg el al. [145]. In their study, an SD was prepared to improve dissolution of griseofulvin using succinic acid as the carrier and methanol as the solvent. The melting solvent method combines melting method and solvent evaporation method. The drug is first dissolved in a suitable solvent and incorporated into the melt of the carrier, and the mixture is then evaporated to dryness. Practically, this method is very useful for drugs with a high melting point. Chen el al. [146] showed a novel monolithic osmotic tablet composed of an SD of 10-hydroxycamptothecin (HCPT) prepared by the melting solvent method with PEG 6000 as the carrier and methanol as the solvent. At 12 h, the cumulative release of drug was over 90%, and the optimized formulation was able to deliver HCPT at a constant rate of 1.21 mg/h for 12 h in simulated intestinal fluid (SIF; pH 6.8).

2.5.4. Melt Agglomeration Process

Melt agglomeration is a process in which a binder acts as a carrier. In this method, the drug, binder, and other excipients are heated to above the melting point of the binder. Alternatively, a dispersion of the drug is sprayed onto the heated binder [147–150]. A diazepam SD was prepared by melt agglomeration method in a high shear mixer to improve the dissolution rate. In this preparation, lactose monohydrate was used as the binder and was melt agglomerated with PEG 3000 or Gelucire 50/13. The binder was added by either pump-on or melt-in procedures. Use of melt agglomeration resulted in a high dissolution rate at a lower drug concentration. The dissolution rates were similar between pump-on and melt-in procedures. In addition, the SD of diazepam containing Gelucire 50/13 showed higher dissolution compared with the SD containing PEG 3000.

2.5.5. Hot-Melt Extrusion Method

Hot-melt extrusion is a common method for improving solubility and oral BA of poorly water-soluble drugs, in which the amorphous SD is formed without solvent, thereby avoiding residual solvents in the formulation [151]. This method is conducted by a combination of the melting method and an extruder, in which a homogeneous mixture of drug, polymer, and plasticizer is melted and then extruded through the equipment. The shapes of products at the outlet of extruder can be controlled and do not require grinding in the final step. For example, the melt extrusion method was used to increase dissolution and oral BA of oleanolic acid [103]. Using PVP VA 64 as the carrier, an SD of oleanolic acid was successfully prepared. Dissolution of this SD was better (about 90% of drug from SD released in the 10 min) in comparison with those of a physical mixture (45% after 2 h) and pure drug (37% after 2 h). In addition, the AUC_{0-24h} (1840 ± 381.8 ng·h/mL) and C_{max} (498.7 ng/mL) of the drug from the SD were enhanced 2.4 times and 5.6 times compared with those of pure drug (761.8 ± 272.2 ng·h/mL and 89.1 ± 33.1 ng/mL, respectively). In another study by Sathigari el al., [104], efavirenz SD was prepared via hot-melt extrusion method using Eudragit EPO or Plasdone S-630 as carriers to improve the dissolution rate of efavirenz. In the dissolution test, because of very low aqueous solubility (3–9 µg/mL), sodium lauryl sulfate (SLS) was added in the dissolution medium. The results showed that the solubility of efavirenz increased substantially (197 µg/mL) in comparison with that of the pure drug. After 30 min, about 96% and 82% of drug was released from SD using Eudragit EPO and Plasdone S-630 as carriers, approximately 2-fold and 1.7-fold higher compared with drug alone, respectively. In addition, the SD was stable after 9 months.

2.5.6. Lyophilization Techniques/Freeze-Drying

Lyophilization is an alternative process to the solvent evaporation method in which the drug and carrier are dissolved in a solvent and then the solution is frozen in liquid nitrogen to form a lyophilized molecular dispersion [152]. This method is typically used for thermolabile products that are unstable in aqueous solutions but stable in the dry state for prolonged storage periods. In a previous study, nifedipine and sulfamethoxazole SD were prepared by Soluplus and PEG 6000 as carriers to evaluate physicochemical and in vitro characteristics [112]. SDs of the two drugs were successfully prepared, and drug dissolution rate were increased. In a study of the anticancer drug exemestane [138], an exemestane-loaded phospholipid/sodium deoxycholate SD was prepared to improve the solubility and oral BA of the drug. The solubility and dissolution rate of exemestane from SD were increased compared to those of the pure drug. The absorptive transport of the SD was 4.6-fold greater in comparison with that of the conventional drug. Furthermore, the AUC_{0-72h} of the drug in the SD was 2.3-fold greater than that of the drug alone. In another study on a flutamide SD prepared by lyophilization to enhance the dissolution rate [77], PVP K30, PEG 6000, and poloxamer 407 were used as the carriers. Among these carriers, dissolution of SD when using poloxamer 407 as the carrier (81.6%) was higher compared with that using other carriers (PVP K30 66.5% and PEG 6000 78.2%) after 30 min and higher compared with that of the pure drug (13.5%).

2.5.7. Electrospinning Method

The electrospinning method is a combination of SD technology and nanotechnology. In this method, solid fibers are produced from a polymeric fluid stream or melt delivered through a millimeter-scale nozzle [153]. The advantage of this method is that the process is simple and inexpensive. This method is suitable for preparing nanofibers and controlling release of biomedicine. A nanofiber of polyvinyl alcohol (PVA):ketoprofen (1:1, w/w) was prepared by the electrospinning method [154]. Dissolution of this nanofiber was significantly ($p < 0.05$) greater than that of ketoprofen alone. In another study, an amorphous formulation of indomethacin and griseofulvin was prepared by the electrospinning method using PVP as the carrier. This formulation was stable for 8 months in a desiccator [155].

2.5.8. Co-Precipitation

In this method, the carrier is first dissolved in solvent to prepare a solution, and the drug is incorporated into the solution with stirring to form a homogeneous mixture. Then, water is added dropwise to the homogenous mixture to induce precipitation. Finally, the precipitate is filtered and dried. In a study by Sonali et al. [116], a silymarin SD was prepared with HPMC E15LV as the carrier with various methods such as kneading, spray-drying, and co-precipitation. The silymarin SD prepared by co-precipitation showed significantly ($p < 0.05$) enhanced dissolution compared with the other two methods. Furthermore, the solubility of silymarin from the SD prepared by co-precipitation improved 2.5-fold in comparison with that of the conventional drug.

2.5.9. Supercritical Fluid (SCF) Technology

SCF was introduced in late 1980s and early 1990s. SCF produces a formulation with a narrow particle size range (microparticles or nanoparticles) without solvent and was reported by Hannay and Hogarth in 1897 as a medium for particle production [156]. A substance is in the supercritical state when the temperature and pressure are above its critical point. SCF can act as solvent or antisolvent in SD. The basic principle of SCF is that the drug and carrier are dissolved in a supercritical solvent (e.g., CO_2) and sprayed through a nozzle into an expansion vessel with lower pressure. The rapid expansion induces rapid nucleation of the dissolved drugs and carriers, leading to the formation of SD particles with a desirable size distribution in a very short time. To date, SCF can be performed by several methods such as rapid expansion from supercritical solution (RESS) [157], gas antisolvent (GAS) [141], supercritical antisolvent (SAS) [119], and solution enhanced dispersion by SCF (SEDS) [158]. The RESS process is conducted as follows: Drug and carrier are dissolved in SCF, then sprayed through an atomizer in an expansion vessel maintained at low pressure, resulting in formation of an SD. The advantage of this method is that it can minimize use of organic solvents for preparation of SD. In SCF technology, CO_2 is a suitable solvent for preparation of SD of insoluble drugs, primarily due to its low critical temperature (31.04 °C) and low critical pressure (7.38 MPa), lack of toxicity, lack of inflammability, and environmental safety [159]. To improve dissolution of irbesartan, Adeli [119] prepared an SD by the SAS method using poloxamer 407 as the carrier. The optimal ratio of drug and carrier was 1:1. As a result, dissolution of the irbesartan-SAS sample was 13 times higher than that of the pure drug. In another study, to enhance the BA of apigenin, apigenin nanocrystals were prepared by the SAS method [120]. The results showed that the C_{max} and AUC of the final formulation increased 3.6-fold and 3.4-fold, respectively, in comparison with that of the drug alone, demonstrating improved BA. In drug development, SCF technology is a potential method for enhancing solubility and BA of poorly water-soluble drugs. One limitation of this method is that most drugs are not soluble in CO_2.

2.5.10. Spray-Drying Method

Spray-drying is one of the oldest methods for drying materials, especially thermally-sensitive materials such as foods and pharmaceuticals. In this method, the drug is dissolved in a suitable solvent, and the carrier is dissolved in water to prepare the feed solution. Then, the two solutions are mixed by sonication or other suitable methods until the solution is clear. In the procedure, the feed solutions were firstly sprayed in a drying chamber via a high-pressure nozzle to form fine droplets. The formed droplets are composed of drying fluid (hot gas) and form particles of nano or micro size [160]. Clinically, the spray-drying method has been widely used for preparation of SD for improving solubility and BA of poorly water-soluble drugs such as nilotinib [124], spironolactone [125], valsartan [126], rebamipide [127], and artemether [128] (Table 1). For example, in a study by Herbrink et al. [124], an SD of nilotinib was prepared by spray-drying to enhance solubility. Soluplus was selected as the best carrier based on in vitro dissolution studies. At a drug: Soluplus (1:7) ratio, the solubility of nilotinib was improved 630-fold in comparison with the pure drug. In another study by Pawar et al. [128], an artemether SD was prepared by spray-drying to improve solubility and dissolution. The results showed that the optimal ratio of drug: carrier (artemether: Soluplus) was 1:3. After 1 h, artemether release from SD was 82%, 4.1-fold higher than the conventional drug (20%). Spray-drying is an efficient technology for preparation of SD for improving solubility and BA of hydrophobic drugs.

2.5.11. Kneading Method

In this method, the carrier is dispersed in water and processed into a paste. Then, the drug is added and kneaded thoroughly. The final kneaded formulation is dried and passed through a sieve if necessary. In a previous study by Dhandapani and El-gied [67], cefixime SD was prepared with β-CD as the carrier using the kneading method. The result showed that the dissolution rate of cefixime from the SD was 6.77-fold greater than that of the pure drug, suggesting a possible improvement in BA. In another study [130], the HP-β-CD was used as the carrier in a domperidone SD. Saturation solubility and in vitro dissolution of domperidone from SD were considerably higher (3-fold) in comparison with the pure drug.

2.5.12. Suitable Methods for Production of SDs of Anticancer Drugs

Anticancer drugs are classified into biologic drugs (monoclonal antibodies) and small molecule drugs (nonbiologic anticancer drugs) based on effectiveness and safety profile. Biologic drugs are administered by intravenous (IV) injection due to their large molecular weight, while small molecule drugs are preferentially administered by gastrointestinal route. Oral administration is currently preferred for treatment of cancer in comparison with IV route because it is convenient, painless, safe, and economic. Oral drugs can be administered at home, do not induce the same discomfort as an IV infusion, and the drug concentration can be maintained for long time periods in cancerous cells. In addition, oral dosage forms are easy to store and transport. Therefore, oral administration has received increasing attention, leading to increased numbers of anticancer drugs being developed for oral dosing.

To date, SD technology is widely used to improve the solubility and BA of anticancer drugs due to its simplicity, economy, and high effectiveness. Most methods are suitable for making SD, in which melting method, solvent evaporation method, SCF technology, and freeze-drying are common for production of SD formulation of anticancer drugs in comparison with other methods. The selected method will be based on physicochemical properties of anticancer drugs.

2.5.13. Lab Scale and Industrial Scale Manufacturing Processes

Several manufacturing methods were used to produce SD. However, not all methods are available for commercial processes. Practically, the melting method and solvent evaporation method are two distinct processes that are widely used on lab and industrial scale.

On the lab scale, for the solvent evaporation method, a rotary evaporator was mostly used to produce SD. Recently, SCF and freeze-drying are also employed. Due to its simplicity and economy, the melting method is popularly used. Currently, several types of equipment from many manufacturers such as Brabender Technologies, Coperion GmbH, Thermo Fisher Scientific, and Leistritz Advanced Technologies Corp are available in the laboratory, in which SD amount can be produced from a few grams to a kilogram.

On the industrial scale, production of SD is not as simple as at the lab scale because it involves a large amount of product from a few to several hundred kilograms. In addition, processes need to be robust, reproducible, and follow good manufacturing practices (GMP). These are difficult to ensure for processes such as solvent cast evaporation or water bath melting process. Spray-drying and freeze-drying are the most representative of the solvent evaporation methods used for manufacturing SD. Moreover, the spray-drying process is easy to scale up from lab scale to industrial scale. Melt agglomeration and hot-melt extrusion are two types of melting processes available on the industrial scale. For instance, hot-melt extrusion is one of the most common methods used on an industrial scale to produce SD using twin-screw extruder with a large diameter of the screw (16–50 mm) compared with small diameter of the screw at lab scale (11–16 mm).

In summary, the selected method for manufacturing process plays an important role in the success of a formulation. On the lab scale, the criteria for selecting the melting method are based on the melting point and thermal stability. For selecting the solvent evaporation method, important factors to consider are properties of the drug, carrier, and an organic solvent. On the industrial scale, the production of SD is limited to only a few manufacturing processes. Hot-melt extrusion is the most common among the melting processes to produce SD. For the evaporation method, the selection criteria are based on solvent toxicity and loading capacity.

2.6. Use of SD for Improving Poorly Soluble Anticancer Drugs

Cancer is one of the leading causes of death worldwide and is defined as a group of diseases involving abnormal cell growth with potential to invade or spread to other parts of the body. The World Health Organization predicted that the burden of cancer will increase to 23.6 million new cases each year by 2030 [161]. In 2018, among 1,735,350 new cancer cases, an estimated 609,640 Americans will die from cancer, corresponding to almost 1,700 deaths per day [162]. Therefore, treatment of cancer is one of the most important issues studied during the past several decades. For anticancer drugs to induce a therapeutic effect, they must first be absorbed and enter the circulation. To ensure complete BA, most anticancer drugs are preferably administered by IV infusion because the entire dose of the drug will directly enter into the circulatory system and instantaneously distribute to its sites of action. However, IV administration inconveniences the patients because they have to visit the hospital to receive treatment. In addition, several side effects may occur during the treatment period. For example, the commercial product paclitaxel (Taxol), which is prepared with Cremophor EL and ethanol as solvents (50:50, v/v), is associated with serious side effects due to Cremophor EL such as severe hypersensitivity, myelosuppression, neutropenia, and neurotoxicity [16–18]. To avoid these side effects, Park et al. [163] prepared paclitaxel SD without Cremophor EL using the supercritical antisolvent method. The solubility of paclitaxel in SD is 10 mg/mL, an almost 10 000-fold increase compared to the conventional drug at a weight ratio of 1/20/40 of paclitaxel/HP-β-CD/HCO-40. In addition, the SD is stable over 6 months.

In the past few decades, many oral formulations of anticancer drugs have been developed. Oral administration is currently the desired route for treatment of cancer because it is convenient, painless, safe, and economic. In addition, oral dosage forms are easy to store and transport. The prerequisite for oral administration is complete and predictable absorption. To achieve this, drugs have to dissolve in water to absorb in GI tract to be effectively taken up in the circulatory system. However, nearly all anticancer drugs are poorly water-soluble, which can lead to incomplete absorption and poor BA, resulting in large inter- and intra-individual variability in drug concentrations in vivo. Thus, improving the solubility of anticancer drugs is a great challenge in development of improved cancer therapies in the pharmaceutical industry [164]. Among several methods such as complexation, lipid-based systems, micronization, nanonization, and co-crystals, SD is the most successful for improving solubility and BA of anticancer drugs. Vemurafenib (Zelboraf® Roche), regorafenib (Stivarga®, Bayer), and everolimus (Afinitor®, Votubia®, Certican®, Novartis) are three commercial anticancer drugs that were prepared by SD [165]. Zelboraf® was prepared from vemurafenib and hypromellose acetate succinate carrier at a weight ratio of 30/70 (w/w). Dissolution of the SD formulation was approximately 30 times higher compared to the crude powder. Stivarga, which contains regorafenib and PVP-25 as the carrier, showed a 4.5-fold increase in dissolution rate compared to that of the drug mixture. In addition, the BA of the drug from the SD was approximately 7 times higher than that of the conventional drug. Afinitor is an SD prepared with drug and HPMC at a weight ratio of 1:40 in which the dissolution rate from the SD was improved about 4-fold in comparison with the pure drug. The anticancer drugs investigated for SD are shown in Table 3. To enhance the solubility and dissolution rate of docetaxel, an SD was prepared and the solubility and dissolution rate at 2 h were 34.2- and 12.7-fold higher in comparison with the crude powder, respectively [76]. In another study by Ren et al. [134], dissolution of bicalutamide was improved by preparing an SD using PVP K30 as the carrier at a weight ratio of drug: PVP K30 (1:5). At this ratio, about 98% of bicalutamide was dissolved during the first 10 min. Recently, mixtures of surfactants were used to prepare SD, resulting in increased solubility and improved permeability of BCS Class IV drugs. Song et al. [135] prepared a docetaxel SD using poloxamer F68 alone or a poloxamer SD using a combination of poloxamer F68 and poloxamer P85. Performance of the two SDs was compared, showing that the SD prepared with only poloxamer F68 increase in solubility (1.39-fold increases in BA), while the SD prepared with a combination of poloxamer F68 and poloxamer P85 showed enhanced solubility and permeability (2.97-fold increase in BA). Thus, SD is a promising technique for improving solubility and BA of poorly water-soluble anticancer drugs.

2.7. Future Prospects

SD is currently considered one of the most effective methods for enhancing the solubility and BA of poorly water-soluble drugs. Even though the issues related to preparation, stability, and storage formulation of drugs may limit the numbers of commercial SD products on the market, SD products are still steadily increasing in clinical settings based on improved manufacturing methods and carriers to solve the above problems.

In recent years, carriers used in the preparation of SD have been developed. Some studies used new carriers, while other studies used more than one carrier for production of SD formulation. Using more than one carrier in the formulation of SD, many effective methods were designed, recrystallization was decreased, and the stability of SD was improved. Some carriers used recently are Inulin®, Gelucire®, Pluronic®, and Soluplus®.

In the manufacturing process, Kinetisol Dispersing (KSD) [166–168] is a novel high-energy mixing process for preparation of SD, in which the drug and carrier are processed by utilizing a series of rapidly rotating blades through a combination of kinetic and thermal energy without the aid of external heating sources. This brings new hope for development of more SD products in the future.

3. Conclusions

In this review, we focused on classification of SD, methods for preparation of SD, and current trends in SD for improving the solubility of poorly water-soluble drugs, including anticancer drugs. IV administration is preferred for anticancer treatment. However, patients are inconvenienced by this route because they have to visit a hospital to receive treatment. Therefore, scientists are working to develop oral dosage forms of anticancer drugs. Oral administration is currently the most common route of administration of drugs because it is convenient for patients. A prerequisite for oral administration is dissolution of the drug in water to allow absorption in the GI tract; however, approximately 40% of NCEs including anticancer drugs are insoluble in water which leads to poor absorption, poor BA, and high intra- and inter-individual variability in blood concentrations. Therefore, improving the solubility of poorly water-soluble drugs is a large challenge in the pharmaceutical industry. To overcome this problem, various methods such as complexation, lipid-based systems, SD, micronization, nanonization, and cocrystallization were developed for clinical use. Among these, SD is one of the most successful methods and is widely used for development of drugs. It is considered a promising technique to overcome problems related to poor aqueous solubility and poor BA. By improving wettability of drugs and surface area, drug solubility and dissolution were increased. In the preparation of SD, understanding the properties of the carrier and drug, and selecting a suitable method play crucial roles in the success of the formulation.

Funding: This research was supported by the Basic Science Research Program through the National Research Foundation of Korea (NRF) funded by the Korea government, Ministry of Science and ICT (NRF-2017R1A2B4006458 and NRF-2019R1H1A2039708).

Conflicts of Interest: The authors declare no conflicts of interest.

References

1. Shewach, D.S.; Kuchta, R.D. Introduction to cancer chemotherapeutics. *Chem. Rev.* **2009**, *109*, 2859–2861. [CrossRef] [PubMed]
2. Hwang, H.S.; Shin, H.; Han, J.; Na, K. Combination of photodynamic therapy (PDT) and anti-tumor immunity in cancer therapy. *J. Pharm. Investig.* **2018**, *48*, 143–151. [CrossRef]
3. Arias, J.L. Drug targeting strategies in cancer treatment: An overview. *Mini Rev. Med. Chem.* **2011**, *11*, 1–17. [CrossRef] [PubMed]
4. Lin, H.-M.; Lin, H.-Y.; Chan, M.-H. Preparation, characterization, and in vitro evaluation of folate-modified mesoporous bioactive glass for targeted anticancer drug carriers. *J. Mater. Chem. B* **2013**, *1*, 6147. [CrossRef]
5. Vinothini, K.; Rajendran, N.K.; Ramu, A.; Elumalai, N.; Rajan, M. Folate receptor targeted delivery of paclitaxel to breast cancer cells via folic acid conjugated graphene oxide grafted methyl acrylate nanocarrier. *Biomed. Pharmacother.* **2019**, *110*, 906–917. [CrossRef]
6. Voeikov, R.; Abakumova, T.; Grinenko, N.; Melnikov, P.; Bespalov, V.; Stukov, A.; Chekhonin, V.; Klyachko, N.; Nukolova, N. Dioxadet-loaded nanogels as a potential formulation for glioblastoma treatment. *J. Pharm. Investig.* **2017**, *47*, 75–83. [CrossRef]
7. Huang, R.; Wang, Q.; Zhang, X.; Zhu, J.; Sun, B. Trastuzumab-cisplatin conjugates for targeted delivery of cisplatin to HER2-overexpressing cancer cells. *Biomed. Pharmacother.* **2015**, *72*, 17–23. [CrossRef]
8. Le, Q.-V.; Choi, J.; Oh, Y.-K. Nano delivery systems and cancer immunotherapy. *J. Pharm. Investig.* **2018**, *48*, 527–539. [CrossRef]
9. Park, O.; Yu, G.; Jung, H.; Mok, H. Recent studies on micro-/nano-sized biomaterials for cancer immunotherapy. *J. Pharm. Investig.* **2017**, *47*, 11–18. [CrossRef]
10. Kim, H.S.; Lee, D.Y. Photothermal therapy with gold nanoparticles as an anticancer medication. *J. Pharm. Investig.* **2017**, *47*, 19–26. [CrossRef]
11. Kirtane, A.R.; Narayan, P.; Liu, G.; Panyam, J. Polymer-surfactant nanoparticles for improving oral bioavailability of doxorubicin. *J. Pharm. Investig.* **2017**, *47*, 65–73. [CrossRef]
12. Valicherla, G.R.; Dave, K.M.; Syed, A.A.; Riyazuddin, M.; Gupta, A.P.; Singh, A.; Wahajuddin; Mitra, K.; Datta, D.; Gayen, J.R. Formulation optimization of docetaxel loaded self-emulsifying drug delivery system to enhance bioavailability and anti-tumor activity. *Sci. Rep.* **2016**, *6*, 26895. [CrossRef]
13. Fda, Cder Taxol (paclitaxel) injection. Available online: https://www.accessdata.fda.gov/drugsatfda_docs/label/2011/020262s049lbl.pdf (accessed on 24 January 2019).
14. Nolvadex (Tamoxifen citrate). Available online: https://www.accessdata.fda.gov/drugsatfda_docs/label/1998/17970.pdf (accessed on 4 September 2018).
15. Singla, A.K.; Garg, A.; Aggarwal, D. Paclitaxel and its formulations. *Int. J. Pharm.* **2002**, *235*, 179–192. [CrossRef]
16. Gréen, H.; Khan, M.S.; Jakobsen-Falk, I.; Åvall-Lundqvist, E.; Peterson, C. Impact of CYP3A5*3 and CYP2C8-HapC on paclitaxel/carboplatin-Induced myelosuppression in patients with ovarian cancer. *J. Pharm. Sci.* **2011**, *100*, 4205–4209. [CrossRef]

17. Picard, M. Management of hypersensitivity reactions to taxanes. *Immunol. Allergy Clin. North Am.* **2017**, *37*, 679–693. [CrossRef]
18. Gornstein, E.L.; Schwarz, T.L. Neurotoxic mechanisms of paclitaxel are local to the distal axon and independent of transport defects. *Exp. Neurol.* **2017**, *288*, 153–166. [CrossRef]
19. Kawabata, Y.; Wada, K.; Nakatani, M.; Yamada, S.; Onoue, S. Formulation design for poorly water-soluble drugs based on biopharmaceutics classification system: Basic approaches and practical applications. *Int. J. Pharm.* **2011**, *420*, 1–10. [CrossRef]
20. Takagi, T.; Ramachandran, C.; Bermejo, M.; Yamashita, S.; Yu, L.X.; Amidon, G.L. A provisional biopharmaceutical classification of the top 200 oral drug products in the united states, great britain, spain, and japan. *Mol. Pharm.* **2006**, *3*, 631–643. [CrossRef]
21. Rodriguez-Aller, M.; Guillarme, D.; Veuthey, J.-L.; Gurny, R. Strategies for formulating and delivering poorly water-soluble drugs. *J. Drug Deliv. Sci. Technol.* **2015**, *30*, 342–351. [CrossRef]
22. Guidance for Industry, Waiver of in Vivo Bioavailability and Bioequivalence Studies for Immediate Release Solid Oral Dosage Forms Based on A Biopharmaceutics Classification System. Available online: https://www.fda.gov/downloads/Drugs/Guidances/ucm070246.pdf (accessed on 9 January 2019).
23. Choi, J.-S.; Lee, S.-E.; Jang, W.S.; Byeon, J.C.; Park, J.-S. Solid dispersion of dutasteride using the solvent evaporation method: Approaches to improve dissolution rate and oral bioavailability in rats. *Mater. Sci. Eng. C* **2018**, *90*, 387–396. [CrossRef]
24. Xu, W.; Sun, Y.; Du, L.; Chistyachenko, Y.S.; Dushkin, A.V.; Su, W. Investigations on solid dispersions of valsartan with alkalizing agents: Preparation, characterization and physicochemical properties. *J. Drug Deliv. Sci. Technol.* **2018**, *44*, 399–405. [CrossRef]
25. Choi, J.-S.; Kwon, S.-H.; Lee, S.-E.; Jang, W.S.; Byeon, J.C.; Jeong, H.M.; Park, J.-S. Use of acidifier and solubilizer in tadalafil solid dispersion to enhance the in vitro dissolution and oral bioavailability in rats. *Int. J. Pharm.* **2017**, *526*, 77–87. [CrossRef] [PubMed]
26. Mohammadian, M.; Salami, M.; Momen, S.; Alavi, F.; Emam-Djomeh, Z.; Moosavi-Movahedi, A.A. Enhancing the aqueous solubility of curcumin at acidic condition through the complexation with whey protein nanofibrils. *Food Hydrocoll.* **2019**, *87*, 902–914. [CrossRef]
27. Chen, X.; McClements, D.J.; Zhu, Y.; Chen, Y.; Zou, L.; Liu, W.; Cheng, C.; Fu, D.; Liu, C. Enhancement of the solubility, stability and bioaccessibility of quercetin using protein-based excipient emulsions. *Food Res. Int.* **2018**, *114*, 30–37. [CrossRef] [PubMed]
28. Kim, C.H.; Lee, S.G.; Kang, M.J.; Lee, S.; Choi, Y.W. Surface modification of lipid-based nanocarriers for cancer cell-specific drug targeting. *J. Pharm. Investig.* **2017**, *47*, 203–227. [CrossRef]
29. Karashima, M.; Sano, N.; Yamamoto, S.; Arai, Y.; Yamamoto, K.; Amano, N.; Ikeda, Y. Enhanced pulmonary absorption of poorly soluble itraconazole by micronized cocrystal dry powder formulations. *Eur. J. Pharm. Biopharm.* **2017**, *115*, 65–72. [CrossRef]
30. Seo, B.; Kim, T.; Park, H.J.; Kim, J.-Y.; Lee, K.D.; Lee, J.M.; Lee, Y.-W. Extension of the hansen solubility parameter concept to the micronization of cyclotrimethylenetrinitramine crystals by supercritical anti-solvent process. *J. Supercrit. Fluids* **2016**, *111*, 112–120. [CrossRef]
31. Wong, J.J.L.; Yu, H.; Lim, L.M.; Hadinoto, K. A trade-off between solubility enhancement and physical stability upon simultaneous amorphization and nanonization of curcumin in comparison to amorphization alone. *Eur. J. Pharm. Sci.* **2018**, *114*, 356–363. [CrossRef] [PubMed]
32. Park, J.-B.; Park, C.; Piao, Z.Z.; Amin, H.H.; Meghani, N.M.; Tran, P.H.L.; Tran, T.T.D.; Cui, J.-H.; Cao, Q.-R.; Oh, E.; et al. pH-independent controlled release tablets containing nanonizing valsartan solid dispersions for less variable bioavailability in humans. *J. Drug Deliv. Sci. Technol.* **2018**, *46*, 365–377. [CrossRef]
33. Chen, B.-Q.; Kankala, R.K.; Wang, S.-B.; Chen, A.-Z. Continuous nanonization of lonidamine by modified-rapid expansion of supercritical solution process. *J. Supercrit. Fluids* **2018**, *133*, 486–493. [CrossRef]
34. Reggane, M.; Wiest, J.; Saedtler, M.; Harlacher, C.; Gutmann, M.; Zottnick, S.H.; Piechon, P.; Dix, I.; Müller-Buschbaum, K.; Holzgrabe, U.; et al. Bioinspired co-crystals of imatinib providing enhanced kinetic solubility. *Eur. J. Pharm. Biopharm.* **2018**, *128*, 290–299. [CrossRef] [PubMed]
35. Huang, Y.; Zhang, B.; Gao, Y.; Zhang, J.; Shi, L. Baicalein–Nicotinamide cocrystal with enhanced solubility, dissolution, and oral bioavailability. *J. Pharm. Sci.* **2014**, *103*, 2330–2337. [CrossRef] [PubMed]
36. Liu, X.; Feng, X.; Williams, R.O.; Zhang, F. Characterization of amorphous solid dispersions. *J. Pharm. Investig.* **2018**, *48*, 19–41. [CrossRef]

37. Zhao, Y.; Xie, X.; Zhao, Y.; Gao, Y.; Cai, C.; Zhang, Q.; Ding, Z.; Fan, Z.; Zhang, H.; Liu, M.; et al. Effect of plasticizers on manufacturing ritonavir/copovidone solid dispersions via hot-melt extrusion: Preformulation, physicochemical characterization, and pharmacokinetics in rats. *Eur. J. Pharm. Sci.* **2019**, *127*, 60–70. [CrossRef]
38. Smeets, A.; Koekoekx, R.; Clasen, C.; Van den Mooter, G. Amorphous solid dispersions of darunavir: Comparison between spray drying and electrospraying. *Eur. J. Pharm. Biopharm.* **2018**, *130*, 96–107. [CrossRef]
39. Sekiguchi, K.; Obi, N. Studies on absorption of eutectic mixture. I. A comparison of the behavior of eutectic mixture of sulfathiazole and that of ordinary sulfathiazole in man. *Chem. Pharm. Bull. (Tokyo)* **1961**, *9*, 866–872. [CrossRef]
40. Sekiguchi, K.; Obi, N.; Ueda, Y. Studies on absorption of eutectic mixture. II. Absorption of fused conglomerates of chloramphenicol and ure in rabbits. *Chem. Pharm. Bull. (Tokyo)* **1964**, *12*, 134–144. [CrossRef]
41. Levy, G. Effect of particle size on dissolution and gastrointestinal absorption rates of pharmaceuticals. *Am. J. Pharm. Sci. Support. Public Health* **1963**, *135*, 78–92.
42. Kanig, J.L. Properties of fused mannitol in compressed tablets. *J. Pharm. Sci.* **1964**, *53*, 188–192. [CrossRef]
43. Madgulkar, A.; Bandivadekar, M.; Shid, T.; Rao, S. Sugars as solid dispersion carrier to improve solubility and dissolution of the BCS class II drug: Clotrimazole. *Drug Dev. Ind. Pharm.* **2016**, *42*, 28–38. [CrossRef]
44. Vippagunta, S.R.; Wang, Z.; Hornung, S.; Krill, S.L. Factors affecting the formation of eutectic solid dispersions and their dissolution behavior. *J. Pharm. Sci.* **2007**, *96*, 294–304. [CrossRef] [PubMed]
45. Urbanetz, N.A. Stabilization of solid dispersions of nimodipine and polyethylene glycol 2000. *Eur. J. Pharm. Sci.* **2006**, *28*, 67–76. [CrossRef]
46. Haser, A.; Cao, T.; Lubach, J.; Listro, T.; Acquarulo, L.; Zhang, F. Melt extrusion vs. spray drying: The effect of processing methods on crystalline content of naproxen-povidone formulations. *Eur. J. Pharm. Sci.* **2017**, *102*, 115–125. [CrossRef]
47. Simonelli, A.P.; Mehta, S.C.; Higuchi, W.I. Dissolution rates of high energy polyvinylpyrrolidone (PVP)-Sulfathiazole coprecipitates. *J. Pharm. Sci.* **1969**, *58*, 538–549. [CrossRef] [PubMed]
48. Motallae, S.; Taheri, A.; Homayouni, A. Preparation and characterization of solid dispersions of celecoxib obtained by spray-drying ethanolic suspensions containing PVP-K30 or isomalt. *J. Drug Deliv. Sci. Technol.* **2018**, *46*, 188–196. [CrossRef]
49. Ghanavati, R.; Taheri, A.; Homayouni, A. Anomalous dissolution behavior of celecoxib in PVP/Isomalt solid dispersions prepared using spray drier. *Mater. Sci. Eng. C* **2017**, *72*, 501–511. [CrossRef] [PubMed]
50. Choi, J.-S.; Lee, S.-E.; Jang, W.S.; Byeon, J.C.; Park, J.-S. Tadalafil solid dispersion formulations based on PVP/VA S-630: Improving oral bioavailability in rats. *Eur. J. Pharm. Sci.* **2017**, *106*, 152–158. [CrossRef] [PubMed]
51. Eloy, J.O.; Marchetti, J.M. Solid dispersions containing ursolic acid in poloxamer 407 and PEG 6000: A comparative study of fusion and solvent methods. *Powder Technol.* **2014**, *253*, 98–106. [CrossRef]
52. Otto, D.P.; Otto, A.; de Villiers, M.M. Experimental and mesoscale computational dynamics studies of the relationship between solubility and release of quercetin from PEG solid dispersions. *Int. J. Pharm.* **2013**, *456*, 282–292. [CrossRef] [PubMed]
53. Reginald-Opara, J.N.; Attama, A.; Ofokansi, K.; Umeyor, C.; Kenechukwu, F. Molecular interaction between glimepiride and soluplus ®-PEG 4000 hybrid based solid dispersions: Characterisation and anti-diabetic studies. *Int. J. Pharm.* **2015**, *496*, 741–750. [CrossRef]
54. Jiménez de los Santos, C.J.; Pérez-Martínez, J.I.; Gómez-Pantoja, M.E.; Moyano, J.R. Enhancement of albendazole dissolution properties using solid dispersions with Gelucire 50/13 and PEG 15000. *J. Drug Deliv. Sci. Technol.* **2017**, *42*, 261–272. [CrossRef]
55. Ceballos, A.; Cirri, M.; Maestrelli, F.; Corti, G.; Mura, P. Influence of formulation and process variables on in vitro release of theophylline from directly-compressed Eudragit matrix tablets. *Farm.* **2005**, *60*, 913–918. [CrossRef]
56. Huang, J.; Wigent, R.; Bentzley, C.; Schwartz, J. Nifedipine solid dispersion in microparticles of ammonio methacrylate copolymer and ethylcellulose binary blend for controlled drug delivery. Effect of drug loading on release kinetics. *Int. J. Pharm.* **2006**, *319*, 44–54. [CrossRef]
57. Fan, N.; He, Z.; Ma, P.; Wang, X.; Li, C.; Sun, J.; Sun, Y.; Li, J. Impact of HPMC on inhibiting crystallization and improving permeability of curcumin amorphous solid dispersions. *Carbohydr. Polym.* **2018**, *181*, 543–550. [CrossRef]

58. Fan, N.; Ma, P.; Wang, X.; Li, C.; Zhang, X.; Zhang, K.; Li, J.; He, Z. Storage stability and solubilization ability of HPMC in curcumin amorphous solid dispersions formulated by Eudragit E100. *Carbohydr. Polym.* **2018**, *199*, 492–498. [CrossRef] [PubMed]
59. Sun, Z.; Zhang, H.; He, H.; Zhang, X.; Wang, Q.; Li, K.; He, Z. Cooperative effect of polyvinylpyrrolidone and HPMC E5 on dissolution and bioavailability of nimodipine solid dispersions and tablets. *Asian J. Pharm. Sci.* **2018**. [CrossRef]
60. Wang, S.; Liu, C.; Chen, Y.; Zhang, Z.; Zhu, A.; Qian, F. A high-sensitivity HPLC-ELSD method for HPMC-AS quantification and its application in elucidating the release mechanism of HPMC-AS based amorphous solid dispersions. *Eur. J. Pharm. Sci.* **2018**, *122*, 303–310. [CrossRef] [PubMed]
61. Xu, H.; Liu, L.; Li, X.; Ma, J.; Liu, R.; Wang, S. Extended tacrolimus release via the combination of lipid-based solid dispersion and HPMC hydrogel matrix tablets. *Asian J. Pharm. Sci.* **2018**. [CrossRef]
62. Verreck, G.; Decorte, A.; Heymans, K.; Adriaensen, J.; Liu, D.; Tomasko, D.; Arien, A.; Peeters, J.; Van den Mooter, G.; Brewster, M.E. Hot stage extrusion of p-amino salicylic acid with EC using CO2 as a temporary plasticizer. *Int. J. Pharm.* **2006**, *327*, 45–50. [CrossRef]
63. Ohara, T.; Kitamura, S.; Kitagawa, T.; Terada, K. Dissolution mechanism of poorly water-soluble drug from extended release solid dispersion system with ethylcellulose and hydroxypropylmethylcellulose. *Int. J. Pharm.* **2005**, *302*, 95–102. [CrossRef]
64. Rashid, R.; Kim, D.W.; ud Din, F.; Mustapha, O.; Yousaf, A.M.; Park, J.H.; Kim, J.O.; Yong, C.S.; Choi, H.-G. Effect of hydroxypropylcellulose and tween 80 on physicochemical properties and bioavailability of ezetimibe-loaded solid dispersion. *Carbohydr. Polym.* **2015**, *130*, 26–31. [CrossRef] [PubMed]
65. García-Zubiri, Í.X.; González-Gaitano, G.; Isasi, J.R. Thermal stability of solid dispersions of naphthalene derivatives with β-cyclodextrin and β-cyclodextrin polymers. *Thermochim. Acta* **2006**, *444*, 57–64. [CrossRef]
66. Franco, P.; Reverchon, E.; De Marco, I. PVP/ketoprofen coprecipitation using supercritical antisolvent process. *Powder Technol.* **2018**, *340*, 1–7. [CrossRef]
67. Dhandapani, N.V.; El-gied, A.A. Solid dispersions of cefixime using β-cyclodextrin: Characterization and in vitro evaluation. *Int. J. Pharmacol. Pharm. Sci.* **2016**, *10*, 1523–1527.
68. Jafari, E. Preparation, characterization and dissolution of solid dispersion of diclofenac sodium using Eudragit E-100. *J. Appl. Pharm. Sci.* **2013**, *3*, 167–170.
69. van Drooge, D.J.; Hinrichs, W.L.J.; Visser, M.R.; Frijlink, H.W. Characterization of the molecular distribution of drugs in glassy solid dispersions at the nano-meter scale, using differential scanning calorimetry and gravimetric water vapour sorption techniques. *Int. J. Pharm.* **2006**, *310*, 220–229. [CrossRef]
70. Srinarong, P.; Hämäläinen, S.; Visser, M.R.; Hinrichs, W.L.J.; Ketolainen, J.; Frijlink, H.W. Surface-active derivative of inulin (inutec®SP1) is a superior carrier for solid dispersions with a high drug load. *J. Pharm. Sci.* **2011**, *100*, 2333–2342. [CrossRef]
71. Majerik, V.; Charbit, G.; Badens, E.; Horváth, G.; Szokonya, L.; Bosc, N.; Teillaud, E. Bioavailability enhancement of an active substance by supercritical antisolvent precipitation. *J. Supercrit. Fluids* **2007**, *40*, 101–110. [CrossRef]
72. Damian, F.; Blaton, N.; Naesens, L.; Balzarini, J.; Kinget, R.; Augustijns, P.; Van den Mooter, G. Physicochemical characterization of solid dispersions of the antiviral agent UC-781 with polyethylene glycol 6000 and Gelucire 44/14. *Eur. J. Pharm. Sci.* **2000**, *10*, 311–322. [CrossRef]
73. Li, F.; Hu, J.; Deng, J.; Su, H.; Xu, S.; Liu, J. In vitro controlled release of sodium ferulate from compritol 888 ATO-based matrix tablets. *Int. J. Pharm.* **2006**, *324*, 152–157. [CrossRef]
74. Panda, T.; Das, D.; Panigrahi, L. Formulation development of solid dispersions of bosentan using Gelucire 50/13 and poloxamer 188. *J. Appl. Pharm. Sci.* **2016**, *6*, 027–033. [CrossRef]
75. Karolewicz, B.; Gajda, M.; Pluta, J.; Górniak, A. Dissolution study and thermal analysis of fenofibrate–Pluronic F127 solid dispersions. *J. Therm. Anal. Calorim.* **2016**, *125*, 751–757. [CrossRef]
76. Chen, Y.; Shi, Q.; Chen, Z.; Zheng, J.; Xu, H.; Li, J.; Liu, H. Preparation and characterization of emulsified solid dispersions containing docetaxel. *Arch. Pharm. Res.* **2011**, *34*, 1909–1917. [CrossRef]
77. Elgindy, N.; Elkhodairy, K.; Molokhia, A.; Elzoghby, A. Lyophilization monophase solution technique for preparation of amorphous flutamide dispersions. *Drug Dev. Ind. Pharm.* **2011**, *37*, 754–764. [CrossRef] [PubMed]

78. Hu, X.-Y.; Lou, H.; Hageman, M.J. Preparation of lapatinib ditosylate solid dispersions using solvent rotary evaporation and hot melt extrusion for solubility and dissolution enhancement. *Int. J. Pharm.* **2018**, *552*, 154–163. [CrossRef] [PubMed]
79. Goldberg, A.H.; Gibaldi, M.; Kanig, J.L. Increasing dissolution rates and gastrointestinal absorption of drugs via solid solutions and eutectic mixtures. I. Theoretical considerations and discussion of the literature. *J. Pharm. Sci.* **1965**, *54*, 1145–1148. [CrossRef] [PubMed]
80. Leuner, C. Improving drug solubility for oral delivery using solid dispersions. *Eur. J. Pharm. Biopharm.* **2000**, *50*, 47–60. [CrossRef]
81. Bhatnagar, P.; Dhote, V.; Mahajan, S.C.; Mishra, P.K.; Mishra, D.K. Solid dispersion in pharmaceutical drug development: From basics to clinical applications. *Curr. Drug Deliv.* **2014**, *11*, 155–171. [CrossRef] [PubMed]
82. Serajuddin, A.T. Solid dispersion of poorly water-soluble drugs: Early promises, subsequent problems, and recent breakthroughs. *J. Pharm. Sci.* **1999**, *88*, 1058–1066. [CrossRef]
83. Sarkari, M.; Brown, J.; Chen, X.; Swinnea, S.; Williams, R.O.; Johnston, K.P. Enhanced drug dissolution using evaporative precipitation into aqueous solution. *Int. J. Pharm.* **2002**, *243*, 17–31. [CrossRef]
84. Vo, C.L.-N.; Park, C.; Lee, B.-J. Current trends and future perspectives of solid dispersions containing poorly water-soluble drugs. *Eur. J. Pharm. Biopharm.* **2013**, *85*, 799–813. [CrossRef] [PubMed]
85. Prasad, R.; Radhakrishnan, P.; Singh, S.K.; Verma, P.R.P. Furosemide - soluplus®solid dispersion: Development and characterization. *Recent Pat. Drug Deliv. Formul.* **2018**, *11*, 211–220. [CrossRef] [PubMed]
86. Shen, Y.; Lu, F.; Hou, J.; Shen, Y.; Guo, S. Incorporation of paclitaxel solid dispersions with poloxamer188 or polyethylene glycol to tune drug release from poly(ε-caprolactone) films. *Drug Dev. Ind. Pharm.* **2013**, *39*, 1187–1196. [CrossRef] [PubMed]
87. Palanisamy, M.; Khanam, J. Solid dispersion of prednisolone: Solid state characterization and improvement of dissolution profile. *Drug Dev. Ind. Pharm.* **2011**, *37*, 373–386. [CrossRef] [PubMed]
88. Chamsai, B.; Limmatvapirat, S.; Sungthongjeen, S.; Sriamornsak, P. Enhancement of solubility and oral bioavailability of manidipine by formation of ternary solid dispersion with d-α-tocopherol polyethylene glycol 1000 succinate and copovidone. *Drug Dev. Ind. Pharm.* **2017**, *43*, 2064–2075. [CrossRef] [PubMed]
89. Krishnamoorthy, V.; Suchandrasen; Prasad, V.P.R. Physicochemical characterization and in vitro dissolution behavior of olanzapine-mannitol solid dispersions. *Brazilian J. Pharm. Sci.* **2012**, *48*, 243–255. [CrossRef]
90. Aggarwal, A.K.; Singh, S. Physicochemical characterization and dissolution study of solid dispersions of diacerein with polyethylene glycol 6000. *Drug Dev. Ind. Pharm.* **2011**, *37*, 1181–1191. [CrossRef]
91. Adeli, E. Preparation and evaluation of azithromycin binary solid dispersions using various polyethylene glycols for the improvement of the drug solubility and dissolution rate. *Brazilian J. Pharm. Sci.* **2016**, *52*, 1–13. [CrossRef]
92. Shuai, S.; Yue, S.; Huang, Q.; Wang, W.; Yang, J.; Lan, K.; Ye, L. Preparation, characterization and in vitro/vivo evaluation of tectorigenin solid dispersion with improved dissolution and bioavailability. *Eur. J. Drug Metab. Pharmacokinet.* **2016**, *41*, 413–422. [CrossRef]
93. Daravath, B.; Tadikonda, R.R.; Vemula, S.K. Formulation and pharmacokinetics of Gelucire solid dispersions of flurbiprofen. *Drug Dev. Ind. Pharm.* **2015**, *41*, 1254–1262. [CrossRef]
94. Mustapha, O.; Kim, K.S.; Shafique, S.; Kim, D.S.; Jin, S.G.; Seo, Y.G.; Youn, Y.S.; Oh, K.T.; Yong, C.S.; Kim, J.O.; et al. Comparison of three different types of cilostazol-loaded solid dispersion: Physicochemical characterization and pharmacokinetics in rats. *Colloids Surfaces B Biointerfaces* **2017**, *154*, 89–95. [CrossRef]
95. Kim, S.-J.; Lee, H.-K.; Na, Y.-G.; Bang, K.-H.; Lee, H.-J.; Wang, M.; Huh, H.-W.; Cho, C.-W. A novel composition of ticagrelor by solid dispersion technique for increasing solubility and intestinal permeability. *Int. J. Pharm.* **2019**, *555*, 11–18. [CrossRef] [PubMed]
96. Al-Hamidi, H.; Obeidat, W.M.; Nokhodchi, A. The dissolution enhancement of piroxicam in its physical mixtures and solid dispersion formulations using gluconolactone and glucosamine hydrochloride as potential carriers. *Pharm. Dev. Technol.* **2015**, *20*, 74–83. [CrossRef] [PubMed]
97. Zhang, W.; Zhang, C.; He, Y.; Duan, B.; Yang, G.; Ma, W.; Zhang, Y. Factors affecting the dissolution of indomethacin solid dispersions. *AAPS PharmSciTech* **2017**, *18*, 3258–3273. [CrossRef]
98. Frizon, F.; de Oliveira Eloy, J.; Donaduzzi, C.M.; Mitsui, M.L.; Marchetti, J.M. Dissolution rate enhancement of loratadine in polyvinylpyrrolidone K-30 solid dispersions by solvent methods. *Powder Technol.* **2013**, *235*, 532–539. [CrossRef]

99. Cuzzucoli Crucitti, V.; Migneco, L.M.; Piozzi, A.; Taresco, V.; Garnett, M.; Argent, R.H.; Francolini, I. Intermolecular interaction and solid state characterization of abietic acid/chitosan solid dispersions possessing antimicrobial and antioxidant properties. *Eur. J. Pharm. Biopharm.* **2018**, *125*, 114–123. [CrossRef]
100. Alves, L.D.S.; de La Roca Soares, M.F.; de Albuquerque, C.T.; da Silva, É.R.; Vieira, A.C.C.; Fontes, D.A.F.; Figueirêdo, C.B.M.; Soares Sobrinho, J.L.; Rolim Neto, P.J. Solid dispersion of efavirenz in PVP K-30 by conventional solvent and kneading methods. *Carbohydr. Polym.* **2014**, *104*, 166–174. [CrossRef]
101. Yin, L.-F.; Huang, S.-J.; Zhu, C.-L.; Zhang, S.-H.; Zhang, Q.; Chen, X.-J.; Liu, Q.-W. In vitro and in vivo studies on a novel solid dispersion of repaglinide using polyvinylpyrrolidone as the carrier. *Drug Dev. Ind. Pharm.* **2012**, *38*, 1371–1380. [CrossRef]
102. Nguyen, M.N.-U.; Van Vo, T.; Tran, P.H.-L.; Tran, T.T.-D. Zein-based solid dispersion for potential application in targeted delivery. *J. Pharm. Investig.* **2017**, *47*, 357–364. [CrossRef]
103. Gao, N.; Guo, M.; Fu, Q.; He, Z. Application of hot melt extrusion to enhance the dissolution and oral bioavailability of oleanolic acid. *Asian J. Pharm. Sci.* **2017**, *12*, 66–72. [CrossRef]
104. Sathigari, S.K.; Radhakrishnan, V.K.; Davis, V.A.; Parsons, D.L.; Babu, R.J. Amorphous-State characterization of efavirenz—Polymer hot-melt extrusion systems for dissolution enhancement. *J. Pharm. Sci.* **2012**, *101*, 3456–3464. [CrossRef]
105. Chowdhury, N.; Vhora, I.; Patel, K.; Bagde, A.; Kutlehria, S.; Singh, M. Development of hot melt extruded solid dispersion of tamoxifen citrate and resveratrol for synergistic effects on breast cancer cells. *AAPS PharmSciTech* **2018**, *19*, 3287–3297. [CrossRef]
106. Fule, R.; Amin, P. Development and evaluation of lafutidine solid dispersion via hot melt extrusion: Investigating drug-polymer miscibility with advanced characterisation. *Asian J. Pharm. Sci.* **2014**, *9*, 92–106. [CrossRef]
107. Zhang, C.; Xu, T.; Zhang, D.; He, W.; Wang, S.; Jiang, T. Disulfiram thermosensitive in-situ gel based on solid dispersion for cataract. *Asian J. Pharm. Sci.* **2018**, *13*, 527–535. [CrossRef]
108. Abu-Diak, O.A.; Jones, D.S.; Andrews, G.P. Understanding the performance of melt-extruded poly(ethylene oxide)–Bicalutamide solid dispersions: Characterisation of microstructural properties using thermal, spectroscopic and drug release methods. *J. Pharm. Sci.* **2012**, *101*, 200–213. [CrossRef]
109. Solanki, N.G.; Lam, K.; Tahsin, M.; Gumaste, S.G.; Shah, A.V.; Serajuddin, A.T.M. Effects of surfactants on itraconazole-HPMCAS solid dispersion prepared by hot-melt extrusion I: Miscibility and drug release. *J. Pharm. Sci.* **2018**. [CrossRef]
110. Guns, S.; Dereymaker, A.; Kayaert, P.; Mathot, V.; Martens, J.A.; Van den Mooter, G. Comparison between hot-melt extrusion and spray-drying for manufacturing solid dispersions of the graft copolymer of ethylene glycol and vinylalcohol. *Pharm. Res.* **2011**, *28*, 673–682. [CrossRef]
111. Alshafiee, M.; Aljammal, M.K.; Markl, D.; Ward, A.; Walton, K.; Blunt, L.; Korde, S.; Pagire, S.K.; Kelly, A.L.; Paradkar, A.; et al. Hot-melt extrusion process impact on polymer choice of glyburide solid dispersions: The effect of wettability and dissolution. *Int. J. Pharm.* **2019**, *559*, 245–254. [CrossRef]
112. Altamimi, M.A.; Neau, S.H. Investigation of the in vitro performance difference of drug-soluplus®and drug-PEG 6000 dispersions when prepared using spray drying or lyophilization. *Saudi Pharm. J.* **2017**, *25*, 419–439. [CrossRef]
113. Jacobsen, A.-C.; Elvang, P.A.; Bauer-Brandl, A.; Brandl, M. A dynamic in vitro permeation study on solid mono- and diacyl-phospholipid dispersions of celecoxib. *Eur. J. Pharm. Sci.* **2019**, *127*, 199–207. [CrossRef]
114. Suzuki, H.; Yakushiji, K.; Matsunaga, S.; Yamauchi, Y.; Seto, Y.; Sato, H.; Onoue, S. Amorphous solid dispersion of meloxicam enhanced oral absorption in rats with impaired gastric motility. *J. Pharm. Sci.* **2018**, *107*, 446–452. [CrossRef]
115. Ngo, A.N.; Thomas, D.; Murowchick, J.; Ayon, N.J.; Jaiswal, A.; Youan, B.-B.C. Engineering fast dissolving sodium acetate mediated crystalline solid dispersion of docetaxel. *Int. J. Pharm.* **2018**, *545*, 329–341. [CrossRef]
116. Sonali, D.; Tejal, S.; Vaishali, T.; Tejal, G. Silymarin-solid dispersions: Characterization and influence of preparation methods on dissolution. *Acta Pharm.* **2010**, *60*, 427–443. [CrossRef]
117. Dhumal, R.; Shimpi, S.; Paradkar, A. Development of spray-dried co-precipitate of amorphous celecoxib containing storage and compression stabilizers. *Acta Pharm.* **2007**, *57*, 287–300. [CrossRef]
118. Hou, H.H.; Rajesh, A.; Pandya, K.M.; Lubach, J.W.; Muliadi, A.; Yost, E.; Jia, W.; Nagapudi, K. Impact of method of preparation of amorphous solid dispersions on mechanical properties: Comparison of coprecipitation and spray drying. *J. Pharm. Sci.* **2019**, *108*, 870–879. [CrossRef]

119. Adeli, E. The use of supercritical anti-solvent (SAS) technique for preparation of irbesartan-Pluronic®F-127 nanoparticles to improve the drug dissolution. *Powder Technol.* **2016**, *298*, 65–72. [CrossRef]
120. Zhang, J.; Huang, Y.; Liu, D.; Gao, Y.; Qian, S. Preparation of apigenin nanocrystals using supercritical antisolvent process for dissolution and bioavailability enhancement. *Eur. J. Pharm. Sci.* **2013**, *48*, 740–747. [CrossRef]
121. Moneghini, M.; Kikic, I.; Voinovich, D.; Perissutti, B.; Filipović-Grčić, J. Processing of carbamazepine–PEG 4000 solid dispersions with supercritical carbon dioxide: Preparation, characterisation, and in vitro dissolution. *Int. J. Pharm.* **2001**, *222*, 129–138. [CrossRef]
122. Tabbakhian, M.; Hasanzadeh, F.; Tavakoli, N.; Jamshidian, Z. Dissolution enhancement of glibenclamide by solid dispersion: Solvent evaporation versus a supercritical fluid-based solvent -antisolvent technique. *Res. Pharm. Sci.* **2014**, *9*, 337–350.
123. Djuris, J.; Milovanovic, S.; Medarevic, D.; Dobricic, V.; Dapčević, A.; Ibric, S. Selection of the suitable polymer for supercritical fluid assisted preparation of carvedilol solid dispersions. *Int. J. Pharm.* **2019**, *554*, 190–200. [CrossRef]
124. Herbrink, M.; Schellens, J.H.M.; Beijnen, J.H.; Nuijen, B. Improving the solubility of nilotinib through novel spray-dried solid dispersions. *Int. J. Pharm.* **2017**, *529*, 294–302. [CrossRef] [PubMed]
125. Al-Zoubi, N.; Odah, F.; Obeidat, W.; Al-Jaberi, A.; Partheniadis, I.; Nikolakakis, I. Evaluation of spironolactone solid dispersions prepared by co-spray drying with soluplus ®and polyvinylpyrrolidone and influence of tableting on drug release. *J. Pharm. Sci.* **2018**, *107*, 2385–2398. [CrossRef]
126. Pradhan, R.; Kim, S.Y.; Yong, C.S.; Kim, J.O. Preparation and characterization of spray-dried valsartan-loaded Eudragit®E PO solid dispersion microparticles. *Asian J. Pharm. Sci.* **2016**, *11*, 744–750. [CrossRef]
127. Pradhan, R.; Tran, T.H.; Choi, J.Y.; Choi, I.S.; Choi, H.-G.; Yong, C.S.; Kim, J.O. Development of a rebamipide solid dispersion system with improved dissolution and oral bioavailability. *Arch. Pharm. Res.* **2015**, *38*, 522–533. [CrossRef] [PubMed]
128. Pawar, J.N.; Shete, R.T.; Gangurde, A.B.; Moravkar, K.K.; Javeer, S.D.; Jaiswar, D.R.; Amin, P.D. Development of amorphous dispersions of artemether with hydrophilic polymers via spray drying: Physicochemical and in silico studies. *Asian J. Pharm. Sci.* **2016**, *11*, 385–395. [CrossRef]
129. Paudel, A.; Van den Mooter, G. Influence of solvent composition on the miscibility and physical stability of naproxen/PVP K 25 solid dispersions prepared by cosolvent spray-drying. *Pharm. Res.* **2012**, *29*, 251–270. [CrossRef]
130. Chaturvedi, S.; Alim, M.; Agrawal, V.K. Solubility and dissolution enhancement of domperidone using 2-hydroxypropyl- β - cyclodextrin by kneading method. *Asian J. Pharm.* **2017**, *11*, 168–175.
131. Tachibana, T.; Nakamura, A. A method for preparing an aqueous colloidal dispersion of organic materials by using water-soluble polymers: Dispersion of B-carotene by polyvinylpyrrolidone. *Kolloid-Zeitschrift Zeitschrift für Polym.* **1965**, *203*, 130–133. [CrossRef]
132. Mayersohn, M.; Gibaldi, M. New method of solid-state dispersion for increasing dissolution rates. *J. Pharm. Sci.* **1966**, *55*, 1323–1324. [CrossRef]
133. Miao, L.; Liang, Y.; Pan, W.; Gou, J.; Yin, T.; Zhang, Y.; He, H.; Tang, X. Effect of supersaturation on the oral bioavailability of paclitaxel/polymer amorphous solid dispersion. *Drug Deliv. Transl. Res.* **2018**. [CrossRef]
134. Ren, F.; Jing, Q.; Tang, Y.; Shen, Y.; Chen, J.; Gao, F.; Cui, J. Characteristics of bicalutamide solid dispersions and improvement of the dissolution. *Drug Dev. Ind. Pharm.* **2006**, *32*, 967–972. [CrossRef] [PubMed]
135. Song, C.K.; Yoon, I.-S.; Kim, D.-D. Poloxamer-based solid dispersions for oral delivery of docetaxel: Differential effects of F68 and P85 on oral docetaxel bioavailability. *Int. J. Pharm.* **2016**, *507*, 102–108. [CrossRef] [PubMed]
136. Du, J.; Vasavada, R.C. Solubility and dissolution of etoposide from solid dispersions of PEG 8000. *Drug Dev. Ind. Pharm.* **1993**, *19*, 903–914. [CrossRef]
137. Jang, S.W.; Kang, M.J. Improved oral absorption and chemical stability of everolimus via preparation of solid dispersion using solvent wetting technique. *Int. J. Pharm.* **2014**, *473*, 187–193. [CrossRef]
138. Kaur, S.; Jena, S.K.; Samal, S.K.; Saini, V.; Sangamwar, A.T. Freeze dried solid dispersion of exemestane: A way to negate an aqueous solubility and oral bioavailability problems. *Eur. J. Pharm. Sci.* **2017**, *107*, 54–61. [CrossRef]
139. Sodeifian, G.; Sajadian, S.A. Solubility measurement and preparation of nanoparticles of an anticancer drug (letrozole) using rapid expansion of supercritical solutions with solid cosolvent (RESS-SC). *J. Supercrit. Fluids* **2018**, *133*, 239–252. [CrossRef]

140. Kim, M.-S.; Ha, E.-S.; Kim, J.-S.; Baek, I.; Yoo, J.-W.; Jung, Y.; Moon, H.R. Development of megestrol acetate solid dispersion nanoparticles for enhanced oral delivery by using a supercritical antisolvent process. *Drug Des. Devel. Ther.* **2015**, 4269. [CrossRef]
141. Li, S.; Liu, Y.; Liu, T.; Zhao, L.; Zhao, J.; Feng, N. Development and in-vivo assessment of the bioavailability of oridonin solid dispersions by the gas anti-solvent technique. *Int. J. Pharm.* **2011**, *411*, 172–177. [CrossRef]
142. Tran, T.H.; Poudel, B.K.; Marasini, N.; Woo, J.S.; Choi, H.-G.; Yong, C.S.; Kim, J.O. Development of raloxifene-solid dispersion with improved oral bioavailability via spray-drying technique. *Arch. Pharm. Res.* **2013**, *36*, 86–93. [CrossRef]
143. Truong, D.H.; Tran, T.H.; Ramasamy, T.; Choi, J.Y.; Choi, H.-G.; Yong, C.S.; Kim, J.O. Preparation and characterization of solid dispersion using a novel amphiphilic copolymer to enhance dissolution and oral bioavailability of sorafenib. *Powder Technol.* **2015**, *283*, 260–265. [CrossRef]
144. Shah, N.; Iyer, R.M.; Mair, H.-J.; Choi, D.; Tian, H.; Diodone, R.; Fahnrich, K.; Pabst-Ravot, A.; Tang, K.; Scheubel, E.; et al. Improved human bioavailability of vemurafenib, a practically insoluble drug, using an amorphous polymer-stabilized solid dispersion prepared by a solvent-controlled coprecipitation process. *J. Pharm. Sci.* **2013**, *102*, 967–981. [CrossRef]
145. Goldberg, A.H.; Gibaldi, M.; Kanig, J.L. Increasing dissolution rates and gastrointestinal absorption of drugs via solid solutions and eutectic mixtures III. *J. Pharm. Sci.* **1966**, *55*, 487–492. [CrossRef]
146. Chen, H.; Jiang, G.; Ding, F. Monolithic osmotic tablet containing solid dispersion of 10-hydroxycamptothecin. *Drug Dev. Ind. Pharm.* **2009**, *35*, 131–137. [CrossRef]
147. van Drooge, D.-J.; Hinrichs, W.L.J.; Dickhoff, B.H.J.; Elli, M.N.A.; Visser, M.R.; Zijlstra, G.S.; Frijlink, H.W. Spray freeze drying to produce a stable delta(9)-tetrahydrocannabinol containing inulin-based solid dispersion powder suitable for inhalation. *Eur. J. Pharm. Sci.* **2005**, *26*, 231–240. [CrossRef]
148. Kaur, J.; Aggarwal, G.; Singh, G.; Rana, A.C. Improvement of drug solubility using solid dispersion. *Int. J. Pharm. Pharm. Sci.* **2012**, *4*, 47–53.
149. Vilhelmsen, T.; Eliasen, H.; Schæfer, T. Effect of a melt agglomeration process on agglomerates containing solid dispersions. *Int. J. Pharm.* **2005**, *303*, 132–142. [CrossRef]
150. Seo, A.; Holm, P.; Kristensen, H.G.; Schaefer, T. The preparation of agglomerates containing solid dispersions of diazepam by melt agglomeration in a high shear mixer. *Int. J. Pharm.* **2003**, *259*, 161–171. [CrossRef]
151. Genina, N.; Hadi, B.; Löbmann, K. Hot melt extrusion as solvent-Free technique for a continuous manufacturing of drug-Loaded mesoporous silica. *J. Pharm. Sci.* **2018**, *107*, 149–155. [CrossRef]
152. Betageri, G. Enhancement of dissolution of glyburide by solid dispersion and lyophilization techniques. *Int. J. Pharm.* **1995**, *126*, 155–160. [CrossRef]
153. Yu, D.-G.; Li, J.-J.; Williams, G.R.; Zhao, M. Electrospun amorphous solid dispersions of poorly water-soluble drugs: A review. *J. Control. Release* **2018**, *292*, 91–110. [CrossRef]
154. Pamudji, J.S.; Khairurrijal; Mauludin, R.; Sudiati, T.; Evita, M. PVA-ketoprofen nanofibers manufacturing using electrospinning method for dissolution improvement of ketoprofen. *Nanotechnol. Appl. Energy Environ.* **2013**, *737*, 166–175. [CrossRef]
155. Lopez, F.L.; Shearman, G.C.; Gaisford, S.; Williams, G.R. Amorphous formulations of indomethacin and griseofulvin prepared by electrospinning. *Mol. Pharm.* **2014**, *11*, 4327–4338. [CrossRef]
156. Hannay, J.B.; Hogarth, J. On the solubility of solids in gases. *Proc. R. Soc. London* **1879**, *30*, 178–188. [CrossRef]
157. Riekes, M.K.; Caon, T.; da Silva, J.; Sordi, R.; Kuminek, G.; Bernardi, L.S.; Rambo, C.R.; de Campos, C.E.M.; Fernandes, D.; Stulzer, H.K. Enhanced hypotensive effect of nimodipine solid dispersions produced by supercritical CO2 drying. *Powder Technol.* **2015**, *278*, 204–210. [CrossRef]
158. Jun, S.W.; Kim, M.-S.; Jo, G.H.; Lee, S.; Woo, J.S.; Park, J.-S.; Hwang, S.-J. Cefuroxime axetil solid dispersions prepared using solution enhanced dispersion by supercritical fluids. *J. Pharm. Pharmacol.* **2005**, *57*, 1529–1537. [CrossRef]
159. Abuzar, S.M.; Hyun, S.-M.; Kim, J.-H.; Park, H.J.; Kim, M.-S.; Park, J.-S.; Hwang, S.-J. Enhancing the solubility and bioavailability of poorly water-soluble drugs using supercritical antisolvent (SAS) process. *Int. J. Pharm.* **2018**, *538*, 1–13. [CrossRef]
160. Singh, A.; Van den Mooter, G. Spray drying formulation of amorphous solid dispersions. *Adv. Drug Deliv. Rev.* **2016**, *100*, 27–50. [CrossRef]
161. World cancer factsheet. Available online: https://www.cancerresearchuk.org/sites/default/files/cs_report_world.pdf (accessed on 15 January 2019).

162. Siegel, R.L.; Miller, K.D.; Jemal, A. Cancer statistics, 2018. *CA. Cancer J. Clin.* **2018**, *68*, 7–30. [CrossRef]
163. Park, J.-H.; Yan, Y.-D.; Chi, S.-C.; Hwang, D.H.; Shanmugam, S.; Lyoo, W.S.; Woo, J.S.; Yong, C.S.; Choi, H.-G. Preparation and evaluation of cremophor-free paclitaxel solid dispersion by a supercritical antisolvent process. *J. Pharm. Pharmacol.* **2011**, *63*, 491–499. [CrossRef]
164. Thanki, K.; Gangwal, R.P.; Sangamwar, A.T.; Jain, S. Oral delivery of anticancer drugs: Challenges and opportunities. *J. Control. Release* **2013**, *170*, 15–40. [CrossRef]
165. Sawicki, E.; Schellens, J.H.M.; Beijnen, J.H.; Nuijen, B. Inventory of oral anticancer agents: Pharmaceutical formulation aspects with focus on the solid dispersion technique. *Cancer Treat. Rev.* **2016**, *50*, 247–263. [CrossRef]
166. Bennett, R.C.; Brough, C.; Miller, D.A.; O'Donnell, K.P.; Keen, J.M.; Hughey, J.R.; Williams, R.O.; McGinity, J.W. Preparation of amorphous solid dispersions by rotary evaporation and KinetiSol dispersing: Approaches to enhance solubility of a poorly water-soluble gum extract. *Drug Dev. Ind. Pharm.* **2015**, *41*, 382–397. [CrossRef] [PubMed]
167. Keen, J.M.; LaFountaine, J.S.; Hughey, J.R.; Miller, D.A.; McGinity, J.W. Development of itraconazole tablets containing viscous KinetiSol solid dispersions: In vitro and in vivo analysis in dogs. *AAPS PharmSciTech* **2018**, *19*, 1998–2008. [CrossRef] [PubMed]
168. DiNunzio, J.C.; Brough, C.; Miller, D.A.; Williams, R.O.; McGinity, J.W. Fusion processing of itraconazole solid dispersions by Kinetisol®dispersing: A comparative study to hot melt extrusion. *J. Pharm. Sci.* **2010**, *99*, 1239–1253. [CrossRef] [PubMed]

© 2019 by the authors. Licensee MDPI, Basel, Switzerland. This article is an open access article distributed under the terms and conditions of the Creative Commons Attribution (CC BY) license (http://creativecommons.org/licenses/by/4.0/).

Article

Polyelectrolytes in Hot Melt Extrusion: A Combined Solvent-Based and Interacting Additive Technique for Solid Dispersions

Felix Ditzinger [1,2], Catherine Dejoie [3], Dubravka Sisak Jung [4] and Martin Kuentz [2,*

- [1] Department of Pharmaceutical Sciences, University of Basel, 4056 Basel, Switzerland; felix.ditzinger@fhnw.ch
- [2] Institute of Pharma Technology, University of Applied Sciences and Arts Northwestern Switzerland, Hofackerstr. 30, 4132 Muttenz, Switzerland
- [3] European Synchrotron Radiation Facility, 38000 Grenoble, France; catherine.dejoie@esrf.fr
- [4] Department of Chemistry, University of Zurich, 8057 Zurich, Switzerland; dubravka.sisak@chem.uzh.ch
- * Correspondence: martin.kuentz@fhnw.ch; Tel.: +41-61-228-5642

Received: 25 February 2019; Accepted: 5 April 2019; Published: 10 April 2019

Abstract: Solid dispersions are important supersaturating formulations to orally deliver poorly water-soluble drugs. A most important process technique is hot melt extrusion but process requirements limit the choice of suitable polymers. One way around this limitation is to synthesize new polymers. However, their disadvantage is that they require toxicological qualification and present regulatory hurdles for their market authorization. Therefore, this study follows an alternative approach, where new polymeric matrices are created by combining a known polymer, small molecular additives, and an initial solvent-based process step. The polyelectrolyte, carboxymethylcellulose sodium (NaCMC), was tested in combination with different additives such as amino acids, meglumine, trometamol, and urea. It was possible to obtain a new polyelectrolyte matrix that was viable for manufacturing by hot melt extrusion. The amount of additives had to be carefully tuned to obtain an amorphous polymer matrix. This was achieved by probing the matrix using several analytical techniques, such as Fourier transform infrared spectroscopy, differential scanning calorimetry, hot stage microscopy, and X-ray powder diffraction. Next, the obtained matrices had to be examined to ensure the homogeneous distribution of the components and the possible residual crystallinity. As this analysis requires probing a sample on several points and relies on high quality data, X-ray diffraction and starring techniques at a synchrotron source had to be used. Particularly promising with NaCMC was the addition of lysine as well as meglumine. Further research is needed to harness the novel matrix with drugs in amorphous formulations.

Keywords: polyelectrolytes; amorphous solid dispersions; hot melt extrusion; polyelectrolyte excipient matrix

1. Introduction

The rising number of poorly water-soluble drugs in the development pipelines as well as on the market encouraged the pharmaceutical industry to develop new formulation techniques. One strategy is the formulation of a drug in an amorphous form as a solid dispersion, which normally leads to drug supersaturation upon oral administration to promote absorption [1–5]. Among the different process techniques for the manufacturing of amorphous solid dispersions, hot melt extrusion (HME) and spray drying are the most common methods [5,6]. These two process techniques mostly use a combination of drug and polymeric compound. However, HME formulations currently available on the market utilize only about six of the pharmaceutically accepted polymers or a combination of these [6]. Contemporary research is primarily focused on finding new combinations of well-established polymers with plasticizers and surfactants [7], or even on designing new monomers for novel synthetic polymers that come with the aforementioned multiple development hurdles to reach the pharmaceutical market [8]. Another approach is the fine tuning of the extrusion process by changing screw configuration, temperature profiles or by employing different downstream processing steps [9,10].

Recently, we introduced the approach to molecularly modify a polymeric matrix by interacting excipients [11]. The difference to a classical mixture approach with excipients is that molecular interactions are specifically targeted by design and cannot be facilitated in an extrusion of the physical mixture. In line with this idea, the current study explores the possibility to use selected additives that can interact ionically or via hydrogen bonding to enable HME of a matrix based on the polyelectrolyte carboxymethylcellulose sodium (NaCMC) for the first time.

NaCMC was recently extruded with polydimethylsiloxane as a polymeric mixture to form material for 3D printing [12] or it is occasionally used in spray drying [13]. The polymer shows good water solubility and extensive swelling behavior, which are both interesting properties for a new modified matrix produced by HME.

The concept of formulating ionic substances to produce a semi-solid or even liquid with a lower melting point is a well-known technique of "ionic liquids" and an important pharmaceutical application in the field of lipid-based formulations [14,15]. Recent publications highlighted the positive implications of salt formation on HME [16,17], but primarily for keeping the drug in amorphous form through the formation of ionic interactions [18,19]. Such an approach is of particular interest, since the direct extrusion of neat unprocessed NaCMC is not applicable, because it decomposes at 252 °C instead of having a melting point [20].

Therefore, this paper studies polymeric films of NaCMC in combination with six interacting small molecular additives that were first transformed into a solid excipient dispersion through solvent evaporation. In a second processing step, HME was performed. The solvent evaporation step (involving a medium with a high dielectric constant) enabled targeted ionic interactions between polyelectrolyte NaCMC and the ionizable additive [18]. The main reason, why a solvent evaporation step was conducted prior to extrusion was that the compounds used would not be feasible for extrusion as otherwise neat powders because of their high melting points.

As the first group of coformers to be studied with NaCMC, the basic amino acids, histidine, lysine, and arginine, were chosen, as they have been proven to interact with acidic groups of mostly drugs in various studies and consequently improved formulation properties such as amorphous stability, miscibility and plasticizing effects [21–29]. The second group of substances consisted of water-soluble inactive substances, which were also hypothesized to likely form an interaction with NaCMC after solvent evaporation and extrusion. The chosen coformers were urea, meglumine and trometamol (TRIS).

Powder X-ray diffraction (PXRD) was applied to determine the maximum amount of additive that is still feasible for successful miscibility and an extrusion process to form an amorphous product. Two limiting factors had to be considered during the described processing: on the one hand, the unfavorable extrusion properties of NaCMC, which required a high amount of additive to enable the extrusion and on the other hand, the crystalline structure of the additives, which would lead to a crystalline product in high concentrations because of insufficient miscibility. While the preliminary measurements could be carried out using the laboratory diffractometer, conclusive results could only be obtained by using the data collected at a synchrotron source. Namely, to ensure the amorphous formulation, it was necessary to collect high quality PXRD data that is sensitive to extremely low amounts of crystalline phases in the sample. Secondly, to examine the distribution of the additive in the sample, the sample had to be probed on several points, which again required a specific sample stage at a synchrotron source.

Further assessment included thermal analysis by differential scanning calorimetry, which was complemented by hot stage microscopy and hot stage attenuated total reflectance Fourier transform infrared spectroscopy (ATR-FTIR) to show crystallinity and form changes upon heating [18]. The HSM images were used as a complimentary analysis of the thermal miscibility and melting behavior of the evaporates during the extrusion [10,30].

This paper highlights the capability of different small molecular additives to enable the formulation of a polymeric compound, which would otherwise not be suitable for extrusion. Such a combination resulted in the development of a new modified excipient matrix for HME that formulators will find helpful to cope with challenging pharmaceutical compounds.

2. Materials and Methods

2.1. Materials

Carboxymethylcellulose sodium salt (low viscosity), urea, meglumine, TRIS, L-lysine, L-aspartic acid, and L-histidine were bought from Sigma Aldrich (St. Louis, MO, USA). Purified water, which was used for the solvent evaporation, was taken from a MilliQ Millipore filter system (Millipore Co., Bedford, MA, USA).

2.2. Methods

2.2.1. Preparation of Hot Melt Extrudates

Binary mixtures of NaCMC and the additive (according to the composition given in Sections 3.1.1 and 3.2.1) were mixed in a mortar and dissolved in MilliQ water in a round bottom flask. Afterwards, the water was removed by a rotary evaporator (Rotavapor Büchi, Flawil, Switzerland), which resulted in a transparent film. This film was cut into smaller pieces and extruded on co-rotating screws with a 9-mm diameter and 180 mm in length in a ZE9 ECO twin screw extruder by ThreeTec (Birren, Switzerland). A screw speed of 80 rpm was applied at a temperature of 130 °C through all three heating zones. The final extrudates were cooled to room temperature and stored in falcon tubes.

2.2.2. Laboratory Powder X-ray Diffraction (PXRD)

Mixtures were studied for their potential amorphous form by PXRD on a D2 Phaser diffractometer (Bruker AXS GmbH, Karlsruhe, Germany) with a 1-D Lynxeye detector. The instrument was equipped with a Ge-monochromator (Cu Kα radiation) providing X-ray radiation at a wavelength of 1.541 Å. During the measurements, a voltage of 30 kV and a current of 10 mA were used. The increment and time per step were set to 0.020 ° and 1 s, respectively. The measurements were scanning a range of 5° to 40° (2θ).

2.2.3. Differential Scanning Calorimetry (DSC)

Samples were further assessed by a differential scanning calorimeter on a DSC 3 (Mettler Toledo, Greifensee, Switzerland). The samples were cut in small pieces and 5 to 9 mg was placed in a 40 µL aluminum pan with a pierced lid. A heating rate of 10 °C/min from −10 °C to 140 °C was applied, while the surrounding sample cell was purged with nitrogen 200 mL/min. Moreover, the combination of heating, cooling and heating cycles was used to fully evaluate the samples. For the assessment of the initial form, the first heating was used. The thermograms and glass transition temperatures (T_gs) were analyzed with the STARe Evaluation-Software Version 16 (Mettler Toledo, Greifensee, Switzerland). All thermograms show exothermic events as upward peaks.

2.2.4. Hot Stage Attenuated Total Reflectance Fourier Transform Infrared Spectroscopy (ATR-FTIR)

A Cary 680 Series FTIR spectrometer (Agilent Technologies, Santa Clara, CA, USA) was used, which was equipped with a heatable attenuated total reflectance accessory (Specac Limited, Orprington,

UK) and the control panel 6100+ by WEST (West Control Solutions, Gurnee, IL, USA). The scanning range of 4000–600 cm^{-1} was selected with 1500 scans over a period of 30 min and a resolution of 4 cm^{-1}. The heating rate was set to 5 °C/min going from 30 °C to 130 °C. For the evaluation, a spectrum was extracted and evaluated by the software ACD/Spectrus Processor 2016.1.1 (Advanced Chemistry Development, Canada) every minute (i.e., every 5 °C). Every spectrum shows a 5 °C temperature increase going from the front to the back of the figures. The increase of peaks towards higher temperatures in the area of 2000 cm^{-1} is related to the heat implications on the ATR crystal. For the hot stage FTIR analysis, the solvent evaporated films were used, whereas the FTIR spectra at room temperature were recorded from the physical mixture, solvent evaporates, and extrudates.

2.2.5. Synchrotron Powder X-ray Diffraction

X-ray powder diffraction data were recorded at the ID22 beamline at the European Synchrotron Radiation Facility (ESRF, France) using a two-dimensional detector (PerkinElmer XRD 1611CP3) and an incident X-ray energy of 60 keV (λ = 0.20678 Å, Qmax = 24 Å^{-1}). A beam size of about 0.5 mm × 0.5 mm was used. Reference samples were packed in 0.7-mm diameter borosilicate capillaries. Extrudate samples were mounted directly on capillary supports and measured as is. In order to minimize any possible radiation damage, samples were cooled down to 100 K using an Oxford Cryosystem Cryostream. To improve the overall statistics, 200 two-dimensional images were recorded (2 s per frame) and averaged. The one-dimensional diffraction patterns were retrieved after integration using the PyFAI software [31]. Five diffraction patterns on five different locations were recorded on each extrudate sample in order to check for heterogeneity.

2.2.6. Hot Stage Microscopy (HSM)

The HSM analysis employed a Leica DMRM at magnifications of 100×, which is also displayed as a scale bar in the images. The microscope was equipped with a temperature-controlled microscope stage from Linkram. This analysis was used for the evaluation of the behavior of the formulation upon heating in the extruder and to complement the DSC analysis [9,17,32]. For a close relation to the extrusion process, the temperature ramp was set from room temperature (RT) to 130 °C. During this ramp, the temperature was kept steady and images were taken at RT, 90 °C, and 130 °C. The obtained images were converted into black and white to highlight the melting process.

3. Results and Discussion

3.1. Amino Acids as Additives

3.1.1. Characterization of the Formulations

Formulations containing the additives arginine and lysine were found to be amorphous after evaporation as well as extrusion. In contrary, it was not possible to convert histidine to an amorphous form neither with evaporation nor with extrusion. Table 1 highlights the different aspects, which were essential during processing of the formulation such as a qualitative evaluation of technical feasibility during HME. The extrusion was evaluated compared to a standard extrusion of the polymer, PVPVA 64, which is considered arbitrarily as ideal for extrusion. Such extrusion behavior is influenced by melt viscosity, thermoplasticity, and degradation [10].

Table 1. Properties of amino acid / polyelectrolyte matrices.

Additive	Maximum Amorphous Amount	Molar Fraction (Monomeric) *	T_g After Evaporation	T_g After Extrusion	Extrudability **
Amino Acid + NaCMC					
Lysine	50% (w/w)	0.64	30.27 °C	30.62 °C	++
Arginine	33% (w/w)	0.43	35.36 °C	33.15 °C	+
Histidine	20% (w/w)	0.30	36.59 °C	-	- -

* For the calculation, the molar weight of the NaCMC monomer was used. ** Technical feasibility was qualitatively assessed and details are given in the text.

The optimal amounts of additives necessary to produce an amorphous polymer matrix are presented in Table 1, expressed as loadings in weight/weight as well as the calculated molar fractions of the formulation components. Lysine resulted in the highest amount of additive, which was formulated in an amorphous form in combination with NaCMC, whereas histidine being less feasible for the evaporation and the later extrusion could only be incorporated in the lowest molar ratio used in this study. This is also reflected by very poor extrusion behavior as well as the disappearance of the T_g in the DSC measurements of the corresponding extrudates, which may be explained by recrystallization from amorphous state as crystallinity was found in the extruded histidine formulation (Figure 1). For the above-mentioned table, it has to be mentioned that lower amounts of additive during a previous formulation development were leading to worse extrusion performances, which underlines the insufficient extrusion performance of neat NaCMC.

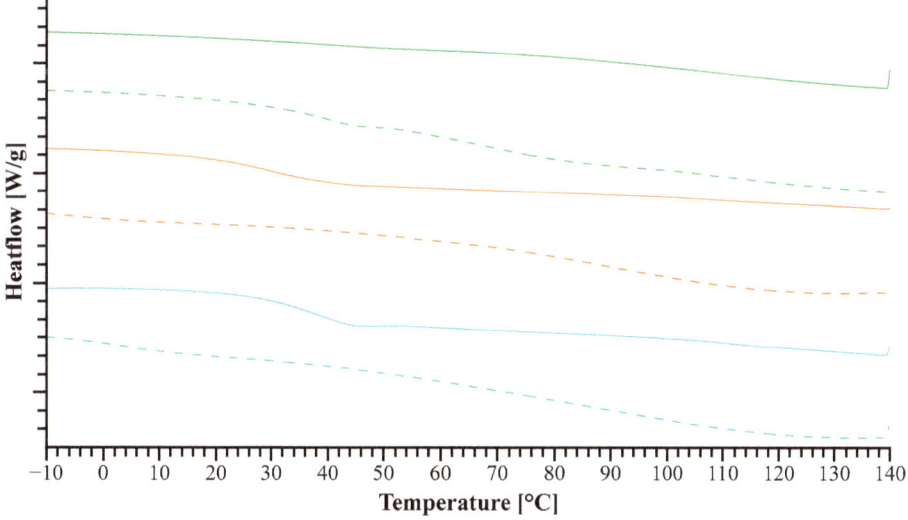

Figure 1. The solid lines represent the differential scanning calorimetry (DSC) thermograms of the extrudates and the dotted lines represent the evaporates which were used for the later extrusion. The amino acids added to sodium carboxymethylcellulose (NaCMC) are arginine (cyan), lysine (orange), and histidine (green).

In detail, the dotted lines in Figure 1, representing the solvent evaporates, show only slight indications of a T_g in all samples. Whereas only the thermograms of extrudates containing lysine and arginine show the presence of a clear T_g in the extrudates (see Table 1). This can be associated with an amorphous form of the additive in the formulation [33] and gives a first indication of formed molecular interactions [22,26]. These two additives also formed more prominent T_gs during the extrusion, which entails a higher amount of amorphous additive in the formulation. Consequently, such a processing was beneficial for the formation of an amorphous modified matrix of NaCMC. However, this still needed further measurements for confirmation.

As mentioned before, the T_g in the histidine extrudates disappeared after extrusion, which suggested that the amorphous form changed during extrusion, leading to a crystalline fraction as indicated by the diffraction peaks in the corresponding PXRD analysis (Figure S3). Although a T_g was detectable for lysine and arginine after the extrusion in the DSC, it has to be kept in mind that the substances used show rather high individual melting points, which would have led to degradation during the thermal measurement.

Therefore, to obtain high quality data that is sensitive to extremely low amounts of crystalline phase in the sample, it was necessary to perform the diffraction and scattering experiments at a synchrotron source.

Thus, synchrotron X-ray diffraction offered a more thorough assessment of the amorphous form to complement the DSC and benchtop PXRD data, which indicated that the raw substances were crystalline except for the polyelectrolyte NaCMC (Figure S1). PXRD data collected at the synchrotron source featured Bragg peaks that could be related directly to the crystallinity of the respective additive. Pronounced crystallinity evidenced in the histidine evaporate was in accordance with the initial X-ray and DSC assessment and was still detectable after extrusion, which is pointed out by the peaks at 1.07 Å^{-1}, 1.71 Å^{-1}, 2.11 Å^{-1}, 2.60 Å^{-1}, 2.81 Å^{-1}, 3.06 Å^{-1}, 3.61 Å^{-1} (Figure 2). Moreover, the measurement at five different locations throughout the extrudate showed the inhomogeneous distribution of the crystalline additive in the extrudate (Figure 2), which can potentially lead to more recrystallization. The diffraction pattern of the arginine extrudate indicated a more homogeneous distribution of the additive compared to histidine, although peaks at 3.05 Å^{-1} still underline some partial crystallinity of the extrudate, which was detectable neither in the initial benchtop PXRD assessment nor by DSC.

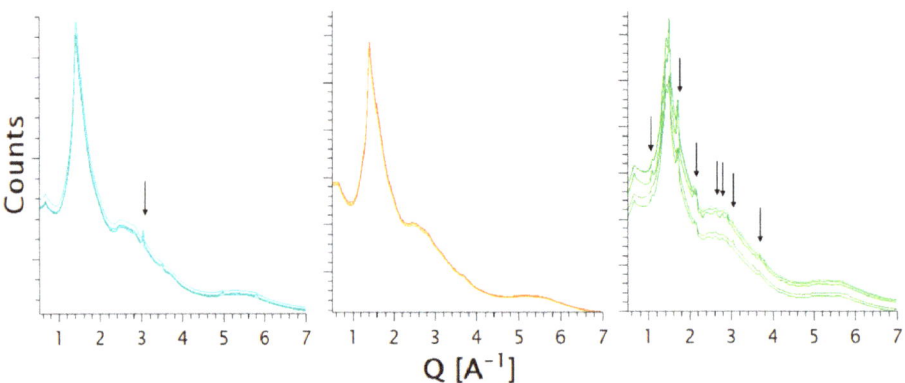

Figure 2. X-ray synchrotron results (i.e., arbitrary counts versus Q vector, Q = 4πsin(θ)/λ) are displayed from the extrudates with amino acid coformers. The amino acids added to NaCMC are from left to right: arginine (cyan), lysine (orange), and histidine (green). Each diffraction pattern corresponds to a measured area in the extrudate.

The FTIR spectra of arginine/NaCMC in Figure 3B exhibit reduced guanidyl vibrations of arginine at 1675 cm^{-1} and 1614 cm^{-1}, which can be associated with the interaction between the ionized arginine side chain and the negatively charged NaCMC [22].

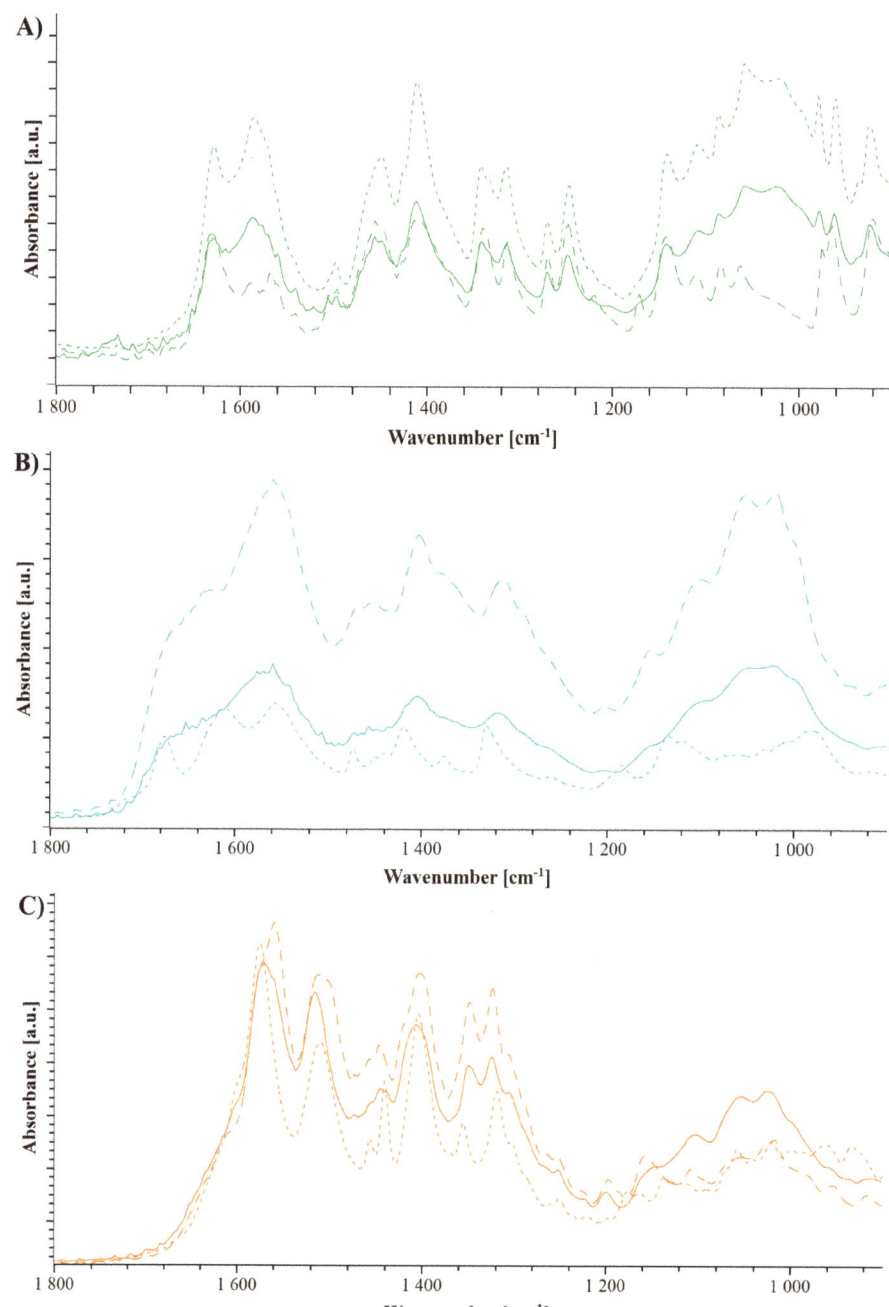

Figure 3. Fourier transform infrared spectroscopy (FTIR) spectrograms of NaCMC and histidine (**A**, green), arginine (**B**, cyan) and lysine (**C**, orange). Dotted lines represent the physical mixture, dashed lines represent the solvent evaporates and the extrudates are shown in solid lines.

For the coformer lysine, only smaller shifts in the FTIR spectrum are present in the evaporate and the extrudate including the shoulder of the COO$^-$ bond at 1607 cm^{-1}, which is less pronounced in the extrudate than in the physical mixture [12,34]. In addition, a slight shift and a pronounced broadening of the peaks at 1570 cm^{-1} and 1540 cm^{-1} [35] both highlight the interaction of the carboxylic group of NaCMC (Figure 3C). The analysis of histidine/NaCMC in Figure 3A shows pronounced similarities of the physical mixture and extrudate. This supported the previous findings of the extrusion leading to a change in the solvent evaporate with recrystallization of histidine [21].

As mentioned in the previous section, the NaCMC was completely amorphous prior to processing. Therefore, the observed peak broadening and shifts are related to the amorphization of the additive.

3.1.2. Heat Assisted Characterization

The hot stage microscopy was applied to better understand the processes occurring during the extrusion of the evaporated films and to complement the results of extrusion performance using the different additives. It should be noted that bright structures in the HSM images are not necessarily related to crystallinity as they can also highlight an increase in capillarity of the samples.

Thus, Figure 4 top shows an increasing number of capillaries building up in the polyelectrolyte film containing arginine, which can be directly associated with the positive extrusion performance. In this case, even though the HSM suggests a successful extrusion, as highlighted in the previous section, the arginine extrudate still contained crystallinity. This could be explained by the insufficient mixing behavior of the two excipients, which is underlined by the minimal changes visible in the heat-resolved FTIR. In Figure 4 bottom, only minor changes in the FTIR are visible during the heating.

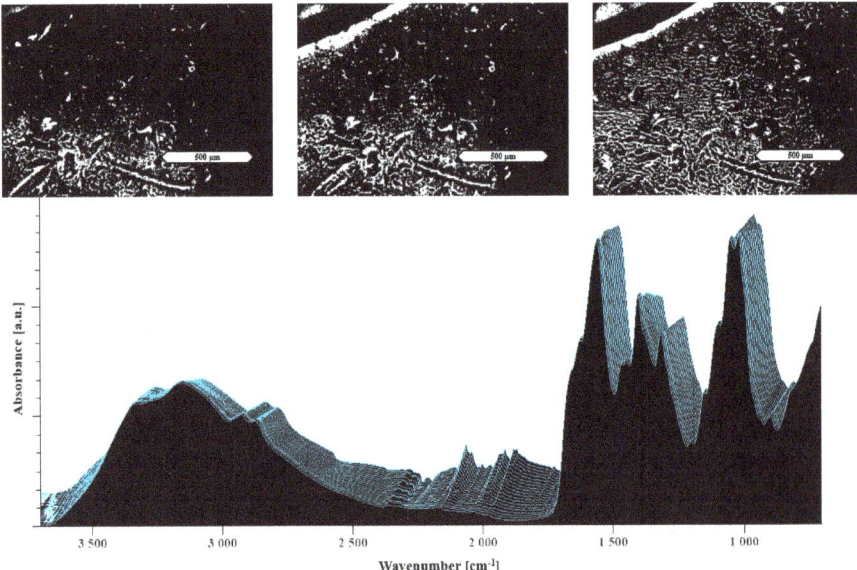

Figure 4. Hot stage microscopy (HSM) images at the top and temperature-resolved FTIR of 33% arginine in NaCMC at the bottom. The images show HSM images taken at RT, 90 °C and 130 °C (from left to right). The displayed scale bar refers to 500 µm. Each spectrum was measured at temperatures from 30 °C (measured first in the front) to 130 °C (in the back) by increasing steps of 5 °C.

The evaporated film containing lysine showed no crystals in the microscopic images and small indications of melting in the images taken at 130 °C in comparison to RT (Figure S4). Even in case of minor melting events, the torque in the extruder facilitates the plasticizing and melting of the evaporate during the extrusion. Therefore, the analysis of films represents a kind of "worst case scenario" regarding shear forces. It is still possible to successfully obtain an extrudable amorphous formulation as in the given case of lysine. The heat-resolved FTIR spectra in Figure S4 at the bottom show an increase in the peak at 1560 cm^{-1} and 1516 cm^{-1}, which are related to the carboxylic groups of NaCMC [35]. Such an observation can be interpreted as an increase of the interaction between NaCMC and lysine.

The HSM images of the histidine evaporate at RT showed pronounced crystallinity, which was in accordance with the PXRD diffraction patterns (Figure S2). Moreover, the images taken at the operating temperature of the extruder (130 °C) did not show any reduction in crystallinity or a phase transition, which could be associated with a glass transition. This is supported by the measurable crystallinity and immiscibility in the extrudate (Figure 5 in green) [18].

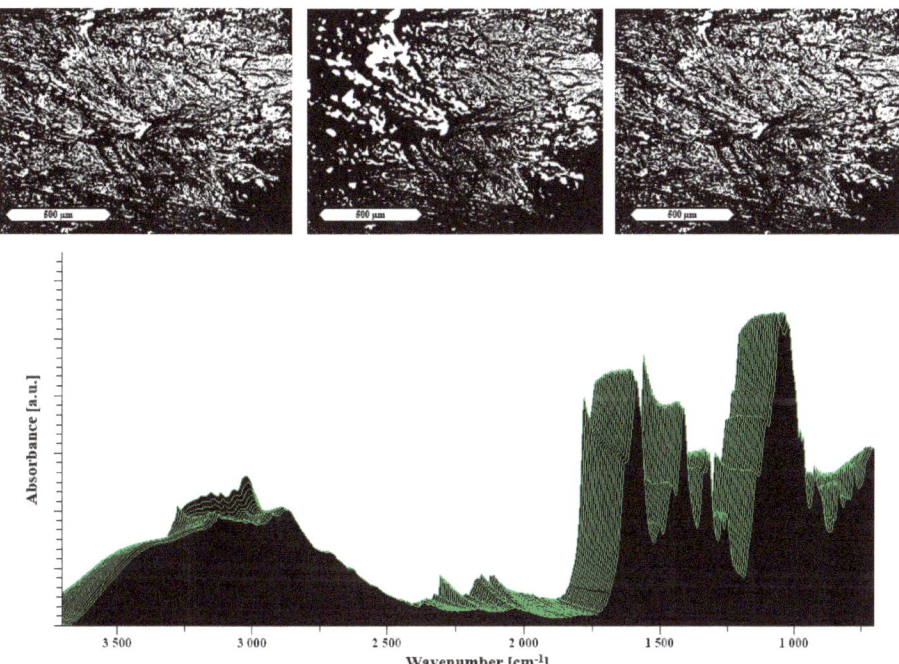

Figure 5. The images at the top show HSM images of histidine/NaCMC evaporated films taken at RT, 90 °C and 130 °C (from left to right). At the bottom, a temperature-resolved FTIR spectrum is shown. The displayed scale bar refers to 500 µm. Each spectrum was measured at temperatures from 30 °C (measured first in the front) to 130 °C (in the back) by increasing steps of 5 °C.

The thermal evaluation of the evaporated films aligned the prior solid-state characterization as well as the actual behavior in the extruder, meaning the formulations containing arginine and lysine, which were successfully incorporated in a concentration of 33% and 50%, respectively, performed well in the extruder and could only be differentiated by a synchrotron X-ray measurement showing slight crystallinity in the arginine formulation. By contrast, the histidine formulations demonstrated poor melting behavior as well as pronounced crystallinity after extrusion. Moreover, the distribution of histidine was insufficient throughout the extrudate, which leads to differences in the diffraction pattern evidenced by the synchrotron X-ray measurement.

3.2. Additives Other than Amino Acids

3.2.1. Characterization of the Formulations

Analogous to previous results, it was necessary to combine solvent evaporation and HME in the mixtures of NaCMC and the further tested coformers. This was suggested by the X-ray diffraction patterns of the formulations following solvent evaporation. The X-ray diffraction pattern of the TRIS/NaCMC solvent evaporates showed Bragg peaks that indicated the presence of TRIS in a crystalline form (Figure S2). However, TRIS was completely transferred into an amorphous form following a subsequent HME step (Figure S3). The urea and meglumine formulations were amorphous after the solvent evaporation and did not recrystallize to a detectable extent based on the benchtop PXRD results (Figure S3).

Table 2 presents a comparison of the maximal amount of additives for which the polymer matrix was kept in an amorphous state. The differences in molar weight have to be taken into account for such an evaluation, leading to a comparable molar fraction of meglumine and urea and a lower loading as well as molar fraction of TRIS (Table 2). This observation was a first indicator of the different technical feasibility of the various additives to obtain suitable modified matrices of NaCMC. A higher loading of TRIS led to crystallinity after evaporation as well as extrusion. Therefore, a lower loading had to be chosen, which resulted in non-ideal extrusion performance as described in the introduction. The additives meglumine and urea could be incorporated at much higher molar ratios and positively influenced the extrusion process.

Table 2. Properties of the other additive polyelectrolyte matrices.

Additive	Maximum Amorphous Amount	Molar Fraction (Monomeric) *	T_g After Evaporation	T_g After Extrusion	Extrudability **
Other Additive + NaCMC					
Meglumine	50% (w/w)	0.57	5.58 °C	9.18 °C	+
Urea	20% (w/w)	0.52	37.99 °C	40.36 °C	0
TRIS	25% (w/w)	0.42	-	39.18 °C	-

* For the calculation, the molar weight of the NaCMC monomer was used. ** Technical feasibility was qualitatively assessed and details are given in the text.

The thermograms in Figure 6 indicate the presence of remaining water after the solvent evaporation, given as a broad peak around 100 °C. The T_g of urea can be hardly detected because of small difference in heat capacity at the glass transition and for the TRIS formulation, no T_g could be detected. On the other hand, the extrudate of meglumine shows a rather pronounced T_g and also a shift towards a lower temperature in comparison to all other extrudates, which can be associated with the good miscibility of meglumine and NaCMC [36,37].

All samples were measured at five different areas. However, differences in the patterns can be seen only for the case where TRIS was used as the additive, particularly for differences in scattered intensity and the emergence of Bragg peaks at Q values of 3.03 Å^{-1}, 3.50 Å^{-1}, 4.95 Å^{-1} and 5.81 Å^{-1}. By contrast, the patterns of the extrudates containing meglumine and urea present no observable differences in their X-ray synchrotron results (Figure 7). The absence of Bragg peaks in the patterns collected on samples containing meglumine and urea prove that the obtained polymer matrices were fully amorphous at the molar fractions of 0.57 and 0.52, respectively. The PXRD patterns, collected on the sample containing TRIS at the synchrotron source, showed indications of crystallinity, which were not detectable in the patterns of the laboratory diffractometer. Such crystallinity could be a sign of recrystallization after the extrusion as well as residual crystallinity. Both sources of crystallinity are related to the instability of an amorphous form [38].

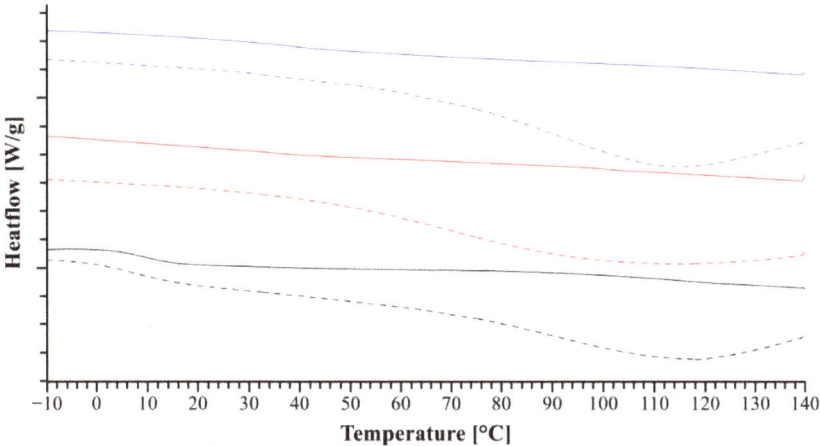

Figure 6. The solid lines represent the thermograms of the extrudate and the dotted lines represent the evaporates, which were used for the later extrusion. The additives used in addition to NaCMC are meglumine (black), TRIS (red) and urea (blue).

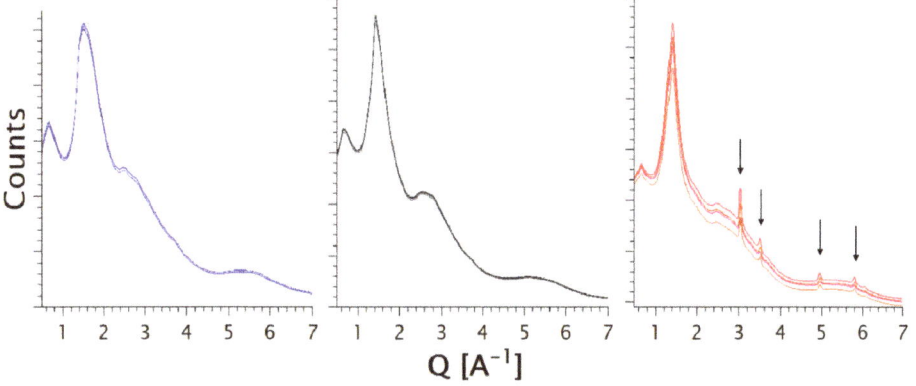

Figure 7. X-ray synchrotron results (i.e., arbitrary counts versus Q vector) of coformers other than amino acids. The additives in addition to NaCMC are: urea (blue), meglumine (black), and TRIS (red). Each diffraction pattern corresponds to a measured area in the extrudate.

In the FTIR spectra, the combination of NaCMC and meglumine shows the previously discussed broadening due to the amorphization [39,40]. It can be seen how following solvent evaporation, distinct peaks are observed, which are broadened in one peak at 1000 cm^{-1} and 1400 cm^{-1}. Moreover, the previous discussed carboxylic peak at 1570 cm^{-1} of NaCMC is more pronounced and broader, which indicates a change in the intermolecular binding of the polyelectrolyte [34,35].

The FTIR spectrum of urea in line with the spectrum of NaCMC/meglumine shows an increase of the peak at 1580 cm^{-1}, which indicates the same interaction with the carboxylic group as meglumine [35].

The spectrum of TRIS showed specific peak broadening as a result of the amorphous formulation (Figure 8). However, this broadening is overlapping a lot with the peaks formed because of a potential interaction. The increase of broader peak between 1400 cm^{-1} and 1600 cm^{-1} can be interpreted as an indication of an interaction [12,16,41]. However, a further, more sensitive analysis is required for a precise statement about the interaction.

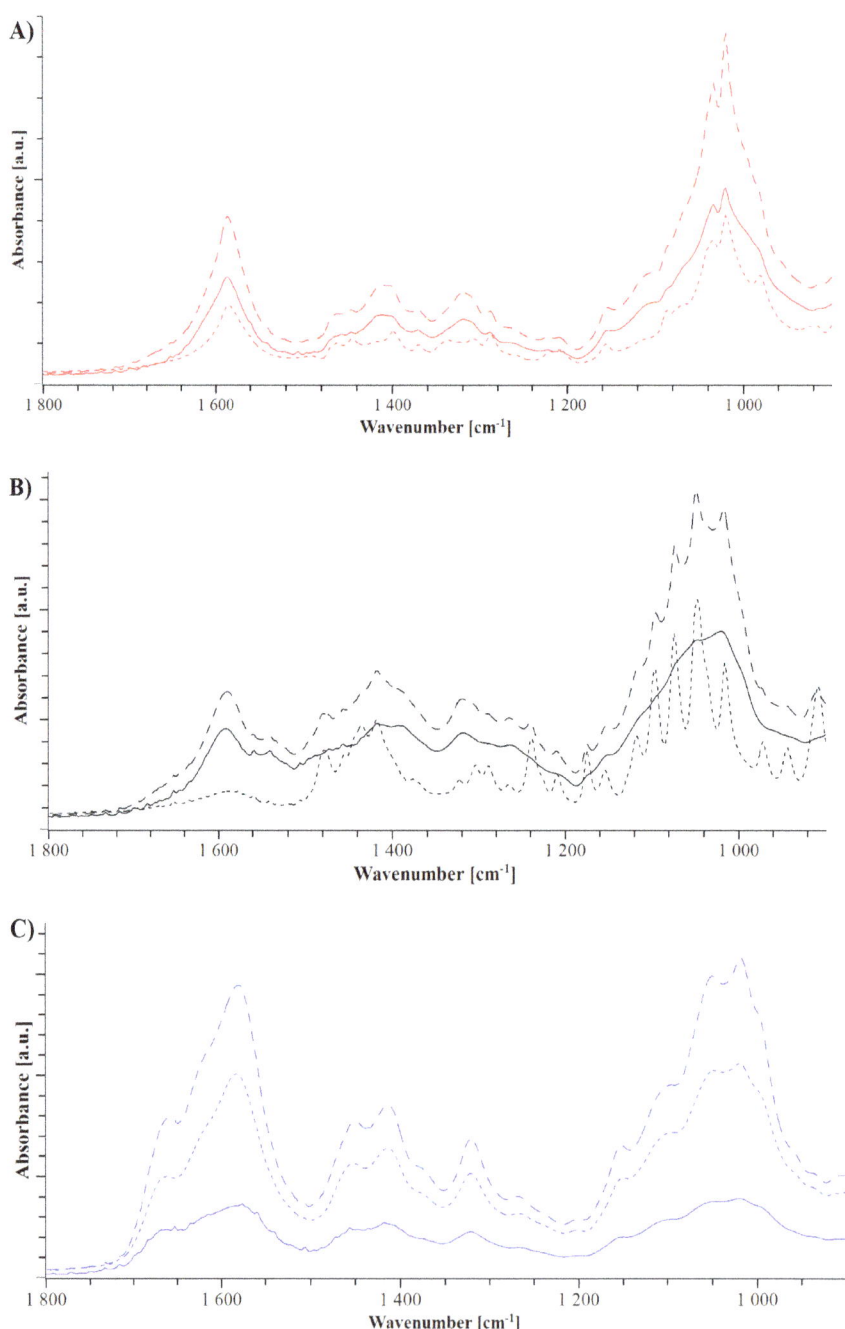

Figure 8. FTIR spectrograms of NaCMC and TRIS (**A**, red), meglumine (**B**, black), and urea (**C**, blue). Dotted lines represent the physical mixture, dashed lines represent the solvent evaporates and the extrudates are shown in solid lines.

Interestingly, the observed changes in the FTIR spectra indicate a likely change in the hydrogen bonding structure because of the processing rather than the formation of a distinct salt.

3.2.2. Heat Assisted Characterization

The microscopic images at the top of Figure 9 present the melting process of the meglumine formulation, which can be connected to the thermoplastic behavior in the extruder. The prominent peak broadening around 90 °C in the area above 3000 cm^{-1} is an indicator of a successful amorphization because of differences in the molecular arrangement as well as the near range order [22,40]. This finding is in line with the start of a melting process in the HSM image at 90 °C. Moreover, these findings are in accordance with the performance observed during extrusion of the meglumine formulation.

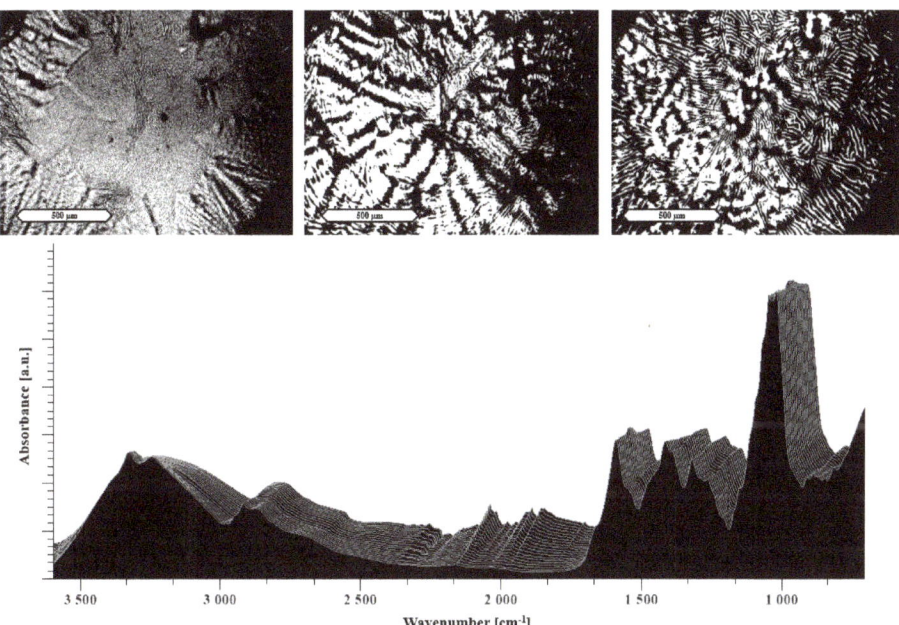

Figure 9. The images at the top show HSM images of 50% meglumine/NaCMC solvent evaporates taken at RT, 90 °C and 130 °C (from left to right). At the bottom, a temperature-resolved FTIR spectrum is shown. The displayed scale bar refers to 500 µm. Each spectrum was measured at temperatures from 30 °C (measured first in the front) to 130 °C (in the back) by increasing steps of 5 °C.

HSM images of the TRIS formulation show only minor changes. Although, an indication of minor "bubble-shaped" features was recorded that may be associated with small melting events taken place in the formulation (Figure 10). The FTIR spectrum supports the observation of a change with higher temperatures. The broadening of the peaks above 3000 cm^{-1} can not only be associated with the successful amorphization in the TRIS sample [40], as discussed before (Table 1), it furthermore shows the decrease of hydrogen bonding throughout the heating process [35,42], leading to a lack of intramolecular interaction in the NaCMC, thereby resulting in more available interaction sites for TRIS.

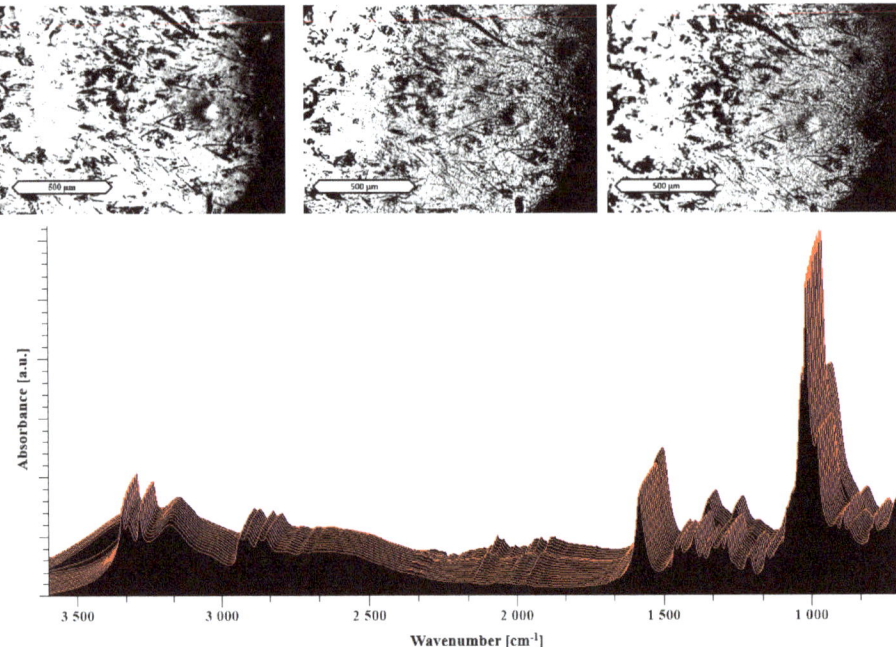

Figure 10. The images at the top show HSM images of 25% TRIS/NaCMC evaporated films taken at RT, 90 °C and 130 °C (from left to right). At the bottom, a temperature-resolved FTIR spectrum is shown. The displayed scale bar refers to 500 µm. Each spectrum was measured at temperatures from 30 °C (measured first in the front) to 130 °C (in the back) by increasing steps of 5 °C.

As described before, the urea formulation formed an amorphous stable formulation after evaporation and extrusion (Figure S5). The HSM images show a melting process over the temperature range recorded. However, in the heat-resolved FTIR, only minor changes in peak intensity can be observed. In line with Figure 9, this suggests that urea has a plasticizing effect without showing a pronounced interaction with NaCMC. A possible reason for that is the lack of ionizable groups in the urea molecule.

4. Conclusions

The application of additives and targeted molecular interactions together with an initial solvent step enabled the extrusion of the polyelectrolyte, NaCMC. Differences in melting behavior and loading highlighted the suitability of the investigated additives to form as a fully amorphous polyelectrolyte matrix after extrusion. The additives, lysine and meglumine, in a concentration of 50% (w/w), have proven to be beneficial for extrusion and formation of a fully amorphous polymer matrix. Moreover, the application of synchrotron X-ray diffraction helped to further differentiate between the formulations by examining the distribution of the additive throughout the matrix and residual crystallinity in the sample. PXRD data collected at the synchrotron source proved the amorphous state of the lysine, meglumine, and urea formulations compared to arginine and TRIS, for which crystallinity was not detectable by means of benchtop PXRD or DSC.

While recent work has shown that the formulation of an amorphous ionic interaction is possible by hot melt extrusion [17], the current concept presented a not extrudable polymer, which was altered by interacting additives in a modified matrix feasible for HME. We refrained from naming the obtained systems as ionic liquids because this would suggest exclusively ionic coformer interactions with the polymer. Moreover, ionic liquids have an arbitrary defined melting characteristic of <100 °C, which is not required for pharmaceutical application as solid dispersions. However, it is expected that the modified matrices share much of the molecular attractiveness of ionic liquids. Further studies may harness the potential benefits of the solvent evaporates for pharmaceutical HME, reaching from new systems for amorphous drug stabilization over the generation of drug supersaturation to the precipitation inhibition of poorly water-soluble compounds.

Supplementary Materials: The following are available online at http://www.mdpi.com/1999-4923/11/4/174/s1, Figure S1: X-ray diffraction pattern of physical mixtures, Figure S2: X-ray diffraction pattern of the solvent evaporates, Figure S3: X-ray diffraction pattern of the extrudates, Figure S4: Heat-resolved FTIR and HSM images of lysine/NaCMC, Figure S5: Heat-resolved FTIR and HSM images of urea/NaCMC.

Author Contributions: Conceptualization, F.D. and M.K.; Methodology, M.K., C.D., F.D. and D.S.J.; software, C.D., F.D. and D.S.J.; writing—review and editing, M.K., C.D., F.D. and D.S.J.; visualization, F.D.; supervision, M.K.; project administration, M.K.; funding acquisition, M.K.

Funding: This project has received funding from the European Union's Horizon 2020 Research and Innovation Program under grant agreement No. 674909.

Acknowledgments: The authors like to thank Alfons Pascual and Markus Grob for the cooperation on the hot stage microscopy.

Conflicts of Interest: The authors declare that they have no conflicts of interest to disclose.

References

1. Hancock, B.C.; Parks, M. What is the True Solubility Advantage for Amorphous Pharmaceuticals? *Pharm. Res.* **2000**, *17*, 397–404. [CrossRef]
2. Serajuddin, A.T.M. Solid dispersion of poorly water-soluble drugs: Early promises, subsequent problems, and recent breakthroughs. *J. Pharm. Sci.* **1999**, *88*, 1058–1066. [CrossRef]
3. Hancock, B.C.; Zografi, G. Characteristics and Significance of the Amorphous State in Pharmaceutical Systems. *J. Pharm. Sci.* **1997**, *86*, 1–12. [CrossRef] [PubMed]
4. Janssens, S.; Van den Mooter, G. Review: Physical chemistry of solid dispersions. *J. Pharm. Pharmacol.* **2009**, *61*, 1571–1586. [CrossRef] [PubMed]
5. Leuner, C.; Dressman, J. Improving drug solubility for oral delivery using solid dispersions. *Eur. J. Pharm. Biopharm.* **2000**, *50*, 47–60. [CrossRef]
6. Wyttenbach, N.; Kuentz, M. Glass-forming ability of compounds in marketed amorphous drug products. *Eur. J. Pharm. Biopharm.* **2017**, *112*, 204–208. [CrossRef] [PubMed]
7. Wilson, M.R.; Jones, D.S.; Andrews, G.P. The development of sustained release drug delivery platforms using melt-extruded cellulose-based polymer blends. *J. Pharm. Pharmacol.* **2017**, *69*, 32–42. [CrossRef] [PubMed]
8. Thiry, J.; Krier, F.; Evrard, B. A review of pharmaceutical extrusion: Critical process parameters and scaling-up. *Int. J. Pharm.* **2015**, *479*, 227–240. [CrossRef] [PubMed]
9. Breitenbach, J. Melt extrusion: From process to drug delivery technology. *Eur. J. Pharm. Biopharm.* **2002**, *54*, 107–117. [CrossRef]
10. Repka, M.A.; Bandari, S.; Kallakunta, V.R.; Vo, A.Q.; McFall, H.; Pimparade, M.B.; Bhagurkar, A.M. Melt extrusion with poorly soluble drugs—An integrated review. *Int. J. Pharm.* **2018**, *535*, 68–85. [CrossRef]
11. Ditzinger, F.; Scherer, U.; Schönenberger, M.; Holm, R.; Kuentz, M. Modified Polymer Matrix in Pharmaceutical Hot Melt Extrusion by Molecular Interactions with a Carboxylic Coformer. *Mol. Pharm.* **2019**, *16*, 141–150. [CrossRef] [PubMed]
12. Calcagnile, P.; Cacciatore, G.; Demitri, C.; Montagna, F.; Esposito Corcione, C. A Feasibility Study of Processing Polydimethylsiloxane–Sodium Carboxymethylcellulose Composites by a Low-Cost Fused Deposition Modeling 3D Printer. *Materials* **2018**, *11*, 1578. [CrossRef]
13. Cho, H.-J.; Jee, J.-P.; Kang, J.-Y.; Shin, D.-Y.; Choi, H.-G.; Maeng, H.-J.; Cho, K. Cefdinir Solid Dispersion Composed of Hydrophilic Polymers with Enhanced Solubility, Dissolution, and Bioavailability in Rats. *Molecules* **2017**, *22*, 280. [CrossRef]
14. Sahbaz, Y.; Williams, H.D.; Nguyen, T.-H.; Saunders, J.; Ford, L.; Charman, S.A.; Scammells, P.J.; Porter, C.J.H. Transformation of Poorly Water-Soluble Drugs into Lipophilic Ionic Liquids Enhances Oral Drug Exposure from Lipid Based Formulations. *Mol. Pharm.* **2015**, *12*, 1980–1991. [CrossRef]

15. Stoimenovski, J.; MacFarlane, D.R.; Bica, K.; Rogers, R.D. Crystalline vs. Ionic Liquid Salt Forms of Active Pharmaceutical Ingredients: A Position Paper. *Pharm. Res.* **2010**, *27*, 521–526. [CrossRef]
16. Bookwala, M.; Thipsay, P.; Ross, S.; Zhang, F.; Bandari, S.; Repka, M.A. Preparation of a crystalline salt of indomethacin and tromethamine by hot melt extrusion technology. *Eur. J. Pharm. Biopharm.* **2018**, *131*, 109–119. [CrossRef]
17. Parikh, T.; Serajuddin, A.T.M. Development of Fast-Dissolving Amorphous Solid Dispersion of Itraconazole by Melt Extrusion of its Mixture with Weak Organic Carboxylic Acid and Polymer. *Pharm. Res.* **2018**, *35*, 127. [CrossRef]
18. Newman, A.; Reutzel-Edens, S.M.; Zografi, G. Coamorphous Active Pharmaceutical Ingredient–Small Molecule Mixtures: Considerations in the Choice of Coformers for Enhancing Dissolution and Oral Bioavailability. *J. Pharm. Sci.* **2018**, *107*, 5–17. [CrossRef]
19. Kasten, G.; Nouri, K.; Grohganz, H.; Rades, T.; Löbmann, K. Performance comparison between crystalline and co-amorphous salts of indomethacin-lysine. *Int. J. Pharm.* **2017**, *533*, 138–144. [CrossRef]
20. Rowe, R.C.; Sheskey, P.J.; Fenton, M.E. *Handbook Of Pharmaceutical Excipients: Pharmaceutical Excipients*; American Pharmacists Association: Washington, DC, USA, 2012; ISBN 0853696187.
21. Kasten, G.; Löbmann, K.; Grohganz, H.; Rades, T. Co-former selection for co-amorphous drug-amino acid formulations. *Int. J. Pharm.* **2018**. [CrossRef] [PubMed]
22. Löbmann, K.; Laitinen, R.; Strachan, C.; Rades, T.; Grohganz, H. Amino acids as co-amorphous stabilizers for poorly water-soluble drugs – Part 2: Molecular interactions. *Eur. J. Pharm. Biopharm.* **2013**, *85*, 882–888. [CrossRef] [PubMed]
23. Laitinen, R.; Priemel, P.A.; Surwase, S.; Graeser, K.; Strachan, C.J.; Grohganz, H.; Rades, T. Theoretical Considerations in Developing Amorphous Solid Dispersions. In *Amorphous Solid Dispersions*; Springer: New York, NY, USA, 2014; pp. 35–90.
24. Laitinen, R.; Löbmann, K.; Grohganz, H.; Priemel, P.; Strachan, C.J.; Rades, T. Supersaturating drug delivery systems: The potential of co-amorphous drug formulations. *Int. J. Pharm.* **2017**, *532*, 1–12. [CrossRef] [PubMed]
25. Jensen, K.T.; Larsen, F.H.; Cornett, C.; Löbmann, K.; Grohganz, H.; Rades, T. Formation Mechanism of Coamorphous Drug–Amino Acid Mixtures. *Mol. Pharm.* **2015**, *12*, 2484–2492. [CrossRef]
26. Ueda, H.; Wu, W.; Löbmann, K.; Grohganz, H.; Müllertz, A.; Rades, T. Application of a Salt Coformer in a Co-Amorphous Drug System Dramatically Enhances the Glass Transition Temperature: A Case Study of the Ternary System Carbamazepine, Citric Acid, and L-Arginine. *Mol. Pharm.* **2018**, *15*, 2036–2044. [CrossRef] [PubMed]
27. Dengale, S.J.; Grohganz, H.; Rades, T.; Löbmann, K. Recent advances in co-amorphous drug formulations. *Adv. Drug Deliv. Rev.* **2016**, *100*, 116–125. [CrossRef] [PubMed]
28. Wu, W.; Löbmann, K.; Rades, T.; Grohganz, H. On the role of salt formation and structural similarity of co-formers in co-amorphous drug delivery systems. *Int. J. Pharm.* **2018**, *535*, 86–94. [CrossRef]
29. Karagianni, A.; Kachrimanis, K.; Nikolakakis, I. Co-Amorphous Solid Dispersions for Solubility and Absorption Improvement of Drugs: Composition, Preparation, Characterization and Formulations for Oral Delivery. *Pharmaceutics* **2018**, *10*, 98. [CrossRef]
30. Li, L.; AbuBaker, O.; Shao, Z.J. Characterization of Poly(Ethylene Oxide) as a Drug Carrier in Hot-Melt Extrusion. *Drug Dev. Ind. Pharm.* **2006**, *32*, 991–1002. [CrossRef] [PubMed]
31. Ashiotis, G.; Deschildre, A.; Nawaz, Z.; Wright, J.P.; Karkoulis, D.; Picca, F.E.; Kieffer, J. The fast azimuthal integration Python library: pyFAI. *J. Appl. Crystallogr.* **2015**, *48*, 510–519. [CrossRef]
32. Forster, A.; Hempenstall, J.; Tucker, I.; Rades, T. Selection of excipients for melt extrusion with two poorly water-soluble drugs by solubility parameter calculation and thermal analysis. *Int. J. Pharm.* **2001**, *226*, 147–161. [CrossRef]
33. Lu, Q.; Zografi, G. Phase behavior of binary and ternary amorphous mixtures containing indomethacin, citric acid, and PVP. *Pharm. Res.* **1998**, *15*, 1202–1206. [CrossRef] [PubMed]
34. Yadollahi, M.; Namazi, H. Synthesis and characterization of carboxymethyl cellulose/layered double hydroxide nanocomposites. *J. Nanopart. Res.* **2013**, *15*, 1563. [CrossRef]
35. Fan, L.; Peng, M.; Zhou, X.; Wu, H.; Hu, J.; Xie, W.; Liu, S. Modification of carboxymethyl cellulose grafted with collagen peptide and its antioxidant activity. *Carbohydr. Polym.* **2014**, *112*, 32–38. [CrossRef]

36. Ellenberger, D.; O'Donnell, K.P.; Williams, R.O. Optimizing the Formulation of Poorly Water-Soluble Drugs. In *Formulating Poorly Water Soluble Drugs*; Springer International Publishing: Cham, Switzerland, 2016; pp. 41–120.
37. Gupta, P.; Bansal, A.K. Molecular interactions in celecoxib-PVP-meglumine amorphous system. *J. Pharm. Pharmacol.* **2005**, *57*, 303–310. [CrossRef]
38. Edueng, K.; Mahlin, D.; Larsson, P.; Bergström, C.A.S. Mechanism-based selection of stabilization strategy for amorphous formulations: Insights into crystallization pathways. *J. Control. Release* **2017**, *256*, 193–202. [CrossRef]
39. Telang, C.; Mujumdar, S.; Mathew, M. Improved physical stability of amorphous state through acid base interactions. *J. Pharm. Sci.* **2009**, *98*, 2149–2159. [CrossRef]
40. Heinz, A.; Strachan, C.J.; Gordon, K.C.; Rades, T. Analysis of solid-state transformations of pharmaceutical compounds using vibrational spectroscopy. *J. Pharm. Pharmacol.* **2009**, *61*, 971–988. [CrossRef]
41. Mankova, A.A.; Borodin, A.V.; Kargovsky, A.V.; Brandt, N.N.; Kuritsyn, I.I.; Luo, Q.; Sakodynskaya, I.K.; Wang, K.J.; Zhao, H.; Chikishev, A.Y.; et al. Terahertz time-domain and FTIR spectroscopy of tris-crown interaction. *Chem. Phys. Lett.* **2012**, *554*, 201–207. [CrossRef]
42. Hebeish, A.; Sharaf, S. Novel nanocomposite hydrogel for wound dressing and other medical applications. *RSC Adv.* **2015**, *5*, 103036–103046. [CrossRef]

© 2019 by the authors. Licensee MDPI, Basel, Switzerland. This article is an open access article distributed under the terms and conditions of the Creative Commons Attribution (CC BY) license (http://creativecommons.org/licenses/by/4.0/).

Article

Characterization of Amorphous Solid Dispersion of Pharmaceutical Compound with pH-Dependent Solubility Prepared by Continuous-Spray Granulator

Ryoma Tanaka [1,2], Yusuke Hattori [1,3,4], Yukun Horie [3], Hitoshi Kamada [5], Takuya Nagato [5] and Makoto Otsuka [1,3,4,*]

1. Graduate School of Pharmaceutical Sciences, Musashino University, 1-1-20 Shin-machi, Nishi-Tokyo, Tokyo 202-8585, Japan; g1878005@stu.musashino-u.ac.jp (R.T.); yhattori@musashino-u.ac.jp (Y.H.)
2. Department of Pharmaceutics, College of Pharmacy, University of Minnesota, Minneapolis, MN 55455, USA
3. Faculty of Pharmacy, Musashino University, 1-1-20 Shin-machi, Nishi-Tokyo, Tokyo 202-8585, Japan; s1343033@stu.musashino-u.ac.jp
4. Research Institute of Pharmaceutical Sciences, Musashino University, 1-1-20 Shin-machi, Nishi-Tokyo, Tokyo 202-8585, Japan
5. Research & Development Department, Technical Division, Powrex Corporation, 5-5-5 Kitagawara, Itami, Hyogo 664-0837, Japan; kamada@powrex.co.jp (H.K.); t-nagato@powrex.co.jp (T.N.)
* Correspondence: motsuka@musashino-u.ac.jp; Tel./Fax: +81-42-468-8658

Received: 21 February 2019; Accepted: 1 April 2019; Published: 3 April 2019

Abstract: A continuous-spray granulator (CTS-SGR) is a one-step granulation technology capable of using solutions or suspensions. The present research objectives were, (1) to reduce the manufacturing operations for solid dosage formulations, (2) to make amorphous solid dispersion (ASD) granules without pre-preparation of amorphous solids of active pharmaceutical ingredients (API), and (3) to characterize the obtained SGR granules by comprehensive pharmaceutical analysis. Rebamipide (RBM), a biopharmaceutical classification system class IV drug, that has low solubility or permeability in the stomach, was selected as a model compound. Five kind of granules with different concentrations of polyvinylpyrrolidone/vinyl acetate copolymer (PVP-VA) were prepared using a one-step SGR process. All of the SGR granules could be produced in amorphous or ASD form and their thermodynamic stability was very high because of high glass transition temperatures (>178 °C). They were unstable in 20 °C/75%RH; however, their stability was improved according to the proportion of polymer. The carboxy group of RBM was ionized in the granules and interactions appeared between RBM and PVP-VA, with the formation of an ASD confirmed and the solubility was enhanced compared with bulk RBM crystals. The SGR methodology has the possibility of contributing to process development in the pharmaceutical industry.

Keywords: amorphous; solid dispersion; molecular complex; rebamipide; polymer; interaction; stability; characterization; continuous processing; granulation; process development

1. Introduction

Among recently developed pharmaceuticals, many synthesized candidate drug compounds have a low bioavailability, due to their low aqueous solubility and/or permeability [1]. These issues influence discovery stage studies and lead to delays in the development of new drugs. According to the biopharmaceutical classification system (BCS), the drug dissolution profile and solubility of the solid-state form are major factors and they influence gastrointestinal permeability, bioavailability, and clinical response [2]. Hence, it is necessary to improve bioavailability by enhancing the solubility of poorly water-soluble drug compounds by making the most of pharmaceutical technology [3,4]. The main methodologies reported to achieve this include alteration of the solid-state by amorphization [5], increasing the particle surface area by size reduction [6], and the formation of pharmaceutical molecular complexes [7,8].

Among these, amorphous solid dispersion (ASD) is a representative of amorphous molecular complexes [9,10]. Generally, ASD is a solid of a polymer-based material involving a homogeneously dispersed active pharmaceutical ingredient (API) molecules in a disordered state. In some cases, a complex between more than two APIs is dispersed in a polymer material [11,12]. The amorphous state of a compound is more unstable without long-range order, compared with the crystal state [13]. The use of amorphization or disruption of the crystal lattice increases the solubility and, consequently, leads to improved bioavailability. As an advantage of the ASD system, the presence of a polymer can stabilize an amorphous API. Some studies on ASDs have investigated API-polymer interactions and relationships [14,15], manufacturing process development [16,17], and other factors [18–20]. The preparation of an ASD can be chosen dependent on the nature of the drug, for example, the melting method [21,22] and the solvent method [23,24]. These preparations are necessary to provide the final product of ASD formulations, but the manufacturing unit operation cutbacks are ideal from the viewpoint of high-level quality control and cost saving in the recent pharmaceutical industry. In addition, an accomplishment for high drug loading granules or tablets is difficult with ASD formulations because of the necessity for high concentrations, such as 70–80%, of polymer carriers to enhance the solubility, stability, and drug-polymer interaction [25,26]. Polymers have been used extensively to fabricate ASDs; however, the formation of stable ASDs requires high polymer concentrations, limiting their use with low-dose APIs. Therefore, considering an alternative methodology for solid dosage formulations, ASD granules with high drug loading, using restricted manufacturing processes, may be suitable.

A continuous-spray granulator (CTS-SGR) provides a one-step granulation method from solution or suspension. The SGR method can be divided into the following three processes (Figure 1): Granule nucleation by spray drying, layering granulation by continuous spray, and product collection using a size classification system. A two-fluid spray nozzle is placed vertically at the bottom. The SGR also has side air nozzles, which act to brush the adhered powder off the face of the wall, to maintain a continuous flowable state in the SGR. The SGR system resembles an ordinary spray drier and fluid bed granulation, which together result in layering granulation [27,28]. The granule size of spray drying and fluid bed granulation depends on the particle size in the suspension sample, the nozzle orifice diameter, and the airflow. However, spray drying granulation may have technical problems, such as the homogeneity of the product [29]. Additionally, layering granulation requires core particles as a seed for the granule in order to obtain the comparatively large size granules in general. Thus, the formation a granule with high drug loading is difficult because of the concentration of the seed in the granule [30]. On the other hand, SGR provides enhanced handling, such as the flowability of bulk powder by layering granulation without core particles, and can prepare uniformly spherical granules until the desired granule size and shape is reached.

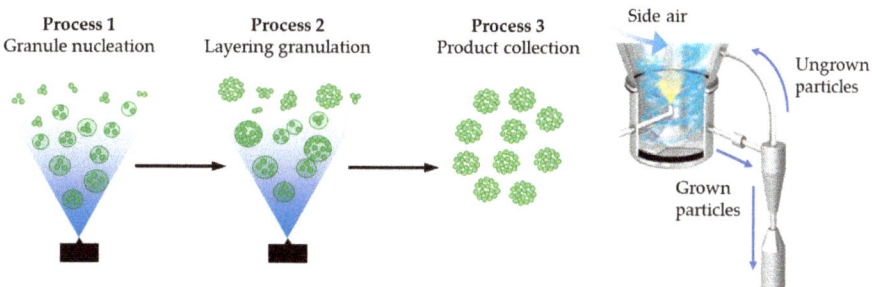

Figure 1. The circulation system of a continuous-spray granulator (CTS-SGR).

Rebamipide (RBM), a gastroprotective agent prescribed for gastric ulcer and gastritis patients, is a weakly acidic BCS class IV drug that is insoluble in acid conditions. However, dissolution in a low pH environment is needed because the precise mechanisms of RBM involve increasing gastric mucosal prostaglandin and gastric mucus production and the site of action of RBM is the stomach [31]. Commercial RBM tablets contain 100 mg of the API and the prescribed dosage is three times per day. The current study aimed to combine a high API loading formulation and enhancement of the solubility of RBM by specific manufacturing processes. For the purpose of achieving this, ASD granules were prepared using SGR as a one-step method for producing high drug loading granules. Five types of granules with different concentrations of polymers (0%, 5%, 10%, 20%, and 30% as weight ratio) were prepared by SGR. The physical properties, stability, thermal behavior, molecular state, and solubility of the obtained SGR samples were investigated. Comprehensive identification was performed to understand the characteristics of the granules and expand the possibility of SGR as a process development.

2. Materials and Methods

2.1. Materials

RBM of pharmaceutical grade was kindly provided by Ohara Pharmaceutical (Shiga, Japan, Figure 2). Sodium hydroxide (NaOH) as an alkylating agent was purchased from Fujifilm Wako Pure Chemical (Osaka, Japan). Polyvinylpyrrolidone/vinyl acetate copolymer (PVP-VA; Kollidon® VA64, Figure 2) as a carrier was a generous gift from BASF (Ludwigshafen, Germany). As the additives for the tablets, magnesium aluminometa silicates (MAS; Neusilin® NS2N) was a gift from Fuji Chemical Industry (Wakayama, Japan) and croscarmellose sodium (CCS; Kiccolate™) and magnesium stearate (Mg-St) were purchased from Asahi Kasei Chemicals (Tokyo, Japan) and Fujifilm Wako Pure Chemical, respectively. All other chemicals were commercially available products of analytical grade.

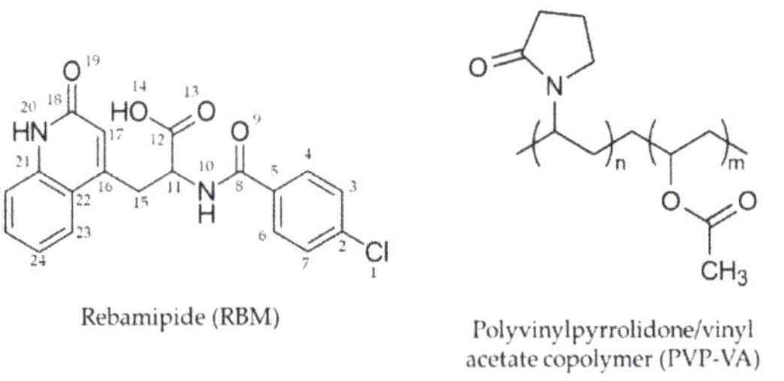

Figure 2. Chemical structures of rebamipide with atom numbering and polyvinylpyrrolidone/vinyl acetate copolymer (Kollidon® VA64).

2.2. Granule Preparation Using a Continuous-Spray Granulator

Table 1 shows the summarized liquid formulation for spraying using SGR. Firstly, RBM, PVP-VA, and NaOH totaling 200 g were dissolved in 1800 g of purified water at 80 °C, with stirring. At this time, the molar ratio of REB and NaOH was fixed at 1:1 and PVP-VA in Runs 1–5 accounted for 0%, 5%, 10%, 20%, and 30% as the weight ratio. Then, the sample solutions were fed into the SGR (CST-SGR-01 without size classification system; Powrex, Hyogo, Japan) at a rate of 10–15 mg/min and sprayed using a two-fluid nozzle with the following conditions: Atomizing air rate was 40–80 NL/min, inlet air temperature was 75 °C, and running time was 120–150 min.

Table 1. Materials used for preparing granules using SGR.

Batch No.	RBM [1] (g)	PVP-VA [2] (g)	NaOH [3] (g)	H$_2$O [4] (g)	Total (g)
1	179.65	0	20.35	1800	2000
2	170.67	10	19.33	1800	2000
3	161.68	20	18.32	1800	2000
4	143.72	40	16.28	1800	2000
5	125.75	60	14.25	1800	2000

[1] Rebamipide, [2] Polyvinylpyrrolidone/vinyl acetate copolymer, [3] Sodium hydroxide, [4] Water.

2.3. Physical Property Measurements

A scanning electron microscope (SEM; JSM-6510LV, JEOL, Tokyo, Japan) was used to characterize the particle state and morphology. Granule samples were sprinkled onto a carbon tape and coated with carbon by a JEC-560 (JEOL). The acceleration voltage, magnification, and working distance were 1.0 kV, ×500, and 8 mm, respectively.

Dried particle size distribution was investigated using a laser light scattering particle analyzer (Mastersizer 3000E with Aero M, Malvern Panalytical, Malvern, UK). Data analysis was done based on algorithms utilizing Mie scattering theory for non-spherical materials. The results were represented as mass median diameter (D50) with a standard deviation of ($n = 5$).

The angle of repose (AR) was evaluated using a modified tilting method [32]. Approximately 30 mg of the granule sample was fed into a sample holder. Then, the holder was slowly tilted until the sample began to slide and the angle of the tilt was measured. The results were described as the mean angle of repose with a standard deviation of ($n = 30$).

The bulk density 1 (ρ_B) was measured by filling a graduated cylinder (50 mL) with a certain amount of each sample, the height of sample was approximately 10 mm. Additionally, the tapped density (ρ_T) was evaluated by tapping down each sample in the cylinder, the tapping was repeated 70 times, and the value of the Hausner ratio (HR) and Carr index (CI) were calculated using the following equations:

$$HR = \frac{\rho_T}{\rho_B} \quad (1)$$

$$CI(\%) = 100\left(1 - \frac{\rho_B}{\rho_T}\right) \quad (2)$$

where B and T are bulk and tapped samples, respectively [33,34]. The results were shown as mean values with a standard deviation of ($n = 3$).

2.4. Stability Testing at Different Humidities by X-Ray Diffractometry

In order to compare the stability of the amorphous state of RBM in each granule type, 4 g of each sample was placed into containers at 30% RH or 75% RH, with an ambient temperature ca. 20 °C, for the specified time periods (6 months at maximum). The samples were promptly measured by X-ray diffraction (XRD) upon removal of the lid at each time point.

The XRD pattern of each sample was collected using RINT-Ultima III (Rigaku, Tokyo, Japan) with Cu Kα radiation (40 kV × 40 mA). The diffraction angle range was from 5° to 45° in 2-theta, with a step of 0.02° and scanned at 15°/min. Relatively large granules were ground using manual grinding in an agate mortar for adequate XRD analysis.

2.5. Thermal Analysis

A differential scanning calorimeter (DSC7000X, Hitachi, Tokyo, Japan) was used for investigating the thermal behavior of granule samples. An approximately 5 mg sample was placed in an aluminum DSC pan. All the measurements were done under a dry nitrogen purge at 30 mL/min and heated from 25 °C to 350 °C at a rate of 5 °C/min. For the purpose of identification of the glass transition temperature (T_g), DSC was operated in the modulated mode with the following conditions: Temperature modulation was ±3 °C, the repetition rate was 0.2 Hz, and the heating rate was 5 °C/min. The value of the glass transition of a binary system was predicted using the following Couchman−Karasz equation:

$$T_{g\ calc}(°C) = \frac{w_1 T_{g1} + K w_2 T_{g2}}{w_1 + K w_2} \quad (3)$$

$$K = \frac{\Delta C_{p2}}{\Delta C_{p1}} \quad (4)$$

where $T_{g\ calc}$ is the theoretical glass transition (°C), w_1 and w_2 are the weight fractions of each component, T_{g1} and T_{g2} are their glass transitions, and ΔC_{p1} and ΔC_{p2} are the change in specific heat capacity at the glass transition [35,36]. Additionally, a positive difference between the measured and calculated glass transition temperatures was obtained as the characteristic parameter of interaction.

2.6. Fourier Transformed Infrared Spectroscopy

Infrared (IR) spectra were accumulated using a Fourier-transform IR spectrometer (FT/IR-4100, Jasco, Tokyo, Japan). The spectral data were collected by powder diffuse reflectance using KBr powder, with 64 scans at 8 cm^{-1} resolution.

2.7. Tablet Preparation and Dissolution Testing

A mixture of resulting granules with specific amounts of PVP-VA, 7.2 mg of MAS (3.6%), and 10.0 mg of CCS (5.0%) was blended with 1.0 mg of Mg-St (0.5%) just before tableting. Each formulation ratio of RBM per tablet was adjusted to the same amount as commercial Mucosta® (114.2 mg as a total API). Additionally, the PVP-VA amount depended on each batch (54.5 mg as total PVP-VA). Specifically, the amounts in batch No. 1−5 granules and additional PVP-VAs in tablets were 127.2 mg (63.6%) and 54.6 mg (27.3%), 136.4 mg (68.2%) and 45.4 mg (22.7%), 145.4 mg (72.7%) and 36.4 mg (18.2%), 163.6 mg (81.8%) and 18.2 mg (9.1%), and 181.8 mg (90.9%) and 0 mg (0.0%), respectively. The mixture was compressed using 8 mm flat-faced punches in a single stroke tablet press (Handtab-100, Ichihashi Seiki, Kyoto, Japan). The tablet weight and hardness were 200 mg and 40 N (60–80 MPa compression pressure), respectively.

Dissolution testing of tablets was carried out in 900 mL water as the test medium (37.5 ± 0.5 °C) using an NTR-3000 apparatus with a paddle speed of 50 rpm (Toyama Sangyo, Osaka, Japan) and a S-2450 spectrophotometer (Shimadzu, Kyoto, Japan). In addition, granule dissolution testing was demonstrated in 900 mL of acidic aqueous solution (pH 1.2 buffer; 37.5 ± 0.5 °C) using a DT-610 apparatus with 100 rpm (Jasco) and a V-530 spectrophotometer (Jasco) because of the low solubility of RBM in an acidic medium. The concentration of API during these tests was determined using a UV/VIS spectrophotometer at 327 nm. The mean values and standard deviations with time were calculated (n = 3).

3. Results and Discussion

3.1. Morphology and Physical Properties

RBM granules, in a dry state, were prepared by SGR. Figure 3 shows SEM images of the obtained samples, which provided insights into the morphology and approximate particle size. The particle size of the 0% PVP-VA sample was the smallest and formed microspheres similar to a spray dried sample, whose size would become the nuclei of the granule. Proportionally to the concentration of PVP-VA, the size of granules increased and the surface became smoother because of the high concentration of polymeric carrier presence as a binder. The 30% PVP-VA sample appeared as heavy and dense granules and the surface rarely had pores, because layering granulation was carried out by continuous spraying of the polymer. In addition, the repeating side airs broke the surface roughness of granules and the spray fluid was extended at the granule surface with drying. Therefore, granule spheroidizing was performed. Table 2 shows a summary of the physical properties of the samples. The bulk density (ρ_B) and tapped density (ρ_T) increased with layering and the flowability was also enhanced. These prepared granules tended to become denser with SGR and a drastic change in density was observed at >10% concentration in the polymer granules. The granule particle size, size distribution, and particle configuration can be controlled by the concentration of the polymeric binder and the processing time of SGR.

Figure 3. SEM images of the 0–30% PVP-VA granules at the same magnification (×500). The scale bar indicates 50 µm.

Table 2. Physical properties of SGR granules.

PVP-VA [1] (%)	ρ_B [2] (g/cm³)	ρ_T [3] (g/cm³)	AR [4] (°)	HR [5]	CI [6] (%)	D50 [7] (µm)
0	0.27 ± 0.02	0.51 ± 0.03	43.9 ± 4.99	1.88 ± 0.02	46.9 ± 0.65	4.18 ± 0.03
5	0.33 ± 0.01	0.55 ± 0.04	40.3 ± 4.59	1.67 ± 0.14	40.0 ± 5.23	11.2 ± 0.12
10	0.52 ± 0.01	0.67 ± 0.00	39.5 ± 3.98	1.29 ± 0.03	22.7 ± 1.76	44.2 ± 0.17
20	0.51 ± 0.02	0.71 ± 0.01	36.3 ± 4.54	1.39 ± 0.05	27.8 ± 2.79	32.5 ± 0.21
30	0.61 ± 0.00	0.76 ± 0.00	27.9 ± 4.16	1.26 ± 0.00	20.4 ± 0.14	72.7 ± 2.99

[1] Weight parentage of polyvinylpyrrolidone/vinyl acetate copolymer in the granule, [2] bulk density, [3] tapped density, [4] angle of repose, [5] Hausner ratio, [6] Carr index, [7] mass median diameter.

3.2. Stability of Amorphous Solid Dispersion

In efforts to gain an insight into the crystallization properties, amorphous state stability testing was performed in conditions of high and low humidity and the thermal analysis investigated the thermodynamic and phase behavior. The SGR granules had been confirmed to have no moisture content, such as free and crystal water, by thermal analysis, then the stability testing was carried out. Figures 4 and 5 show the compiled XRD patterns of storage at the conditions of 20 °C/30% RH and 20 °C/75% RH, respectively. Each XRD pattern on the first day (0 day) was a diffraction halo, indicating RBM in granule was an amorphous state in the polymeric matrix by SGR, despite the various PVP-VA concentrations. The amorphous state was maintained for 6 months at 20 °C/30% RH. Under the high relative humidity conditions (20 °C/75% RH), stability was improved depending on the concentration of PVP-VA. The samples with 0%, 5%, 10%, 20%, and 30% PVP-VA crystallized after 4, 7, 11, and 15 days. Relatively stable ASD granules could be prepared with a high concentration of

PVP-VA. According to Hancock [36], the glass transition temperature of polymer decreases with the increase in water content in the polymer matrix. Hence, the glass transition temperature of amorphous RBM may become lower and the amorphous state is destabilized due to compositing with the polymer. However, the stability of the amorphous state was improved with increasing the polymer content. It suggests that the interaction between RBM and PVP-VA restricts crystallization of RBM. Notably, the diffraction peaks of granules, after crystallization, was different from RBM, as is. There is a possibility that REB formed co-amorphously with sodium ions in the SGR granule and a sodium salt of RBM was formed or crystallized by absorption of moisture. The interaction between RBM and sodium ions is discussed in FT-IR analysis part.

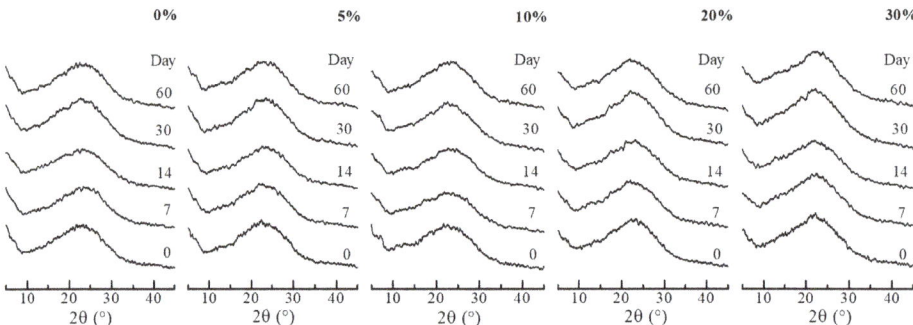

Figure 4. XRD patterns of SGR granules after storage at 20 °C/30% RH. Percentages indicate the weight percentage of PVP-VA in the granules.

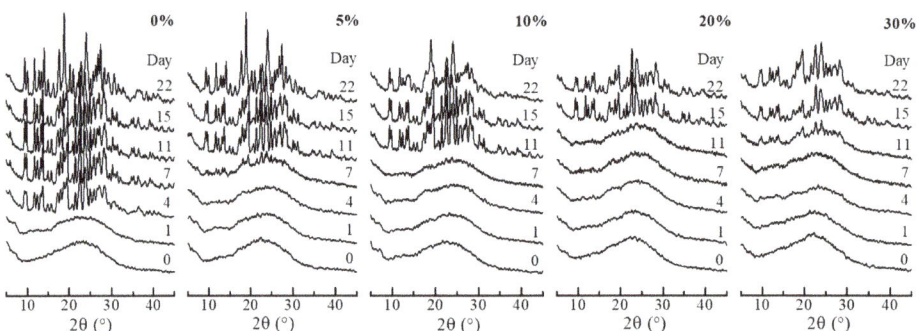

Figure 5. XRD patterns of SGR granules after storage at 20 °C/75% RH. Percentages indicate the weight percentage of PVP-VA in the granules.

Figure 6 shows DSC curves of the SGR granules. Each granule had a baseline shift, according to the glass transition, at approximately 200.0 °C and the crystallization temperatures of 0%, 5%, 10%, 20%, and 30% PVP-VA granules were 263.9 °C, 263.5 °C, 258.6 °C, 255.7 °C, and 246.7 °C, respectively. Hence, it was confirmed that the samples were established as ASD granules. To analyze the glass transition in detail, modulated DSC was employed, and the results are shown in Table 3. Glass transition temperature is well-known as an important value of ASD. The observed glass transition of the 0% PVP-VA granule (amorphous RBM) was 215.4 °C, while unprocessed pure PVP-VA was 107.8 °C, which meant amorphous RBM originally has a high thermodynamic stability. The reason why the glass transition temperature decreased with increasing concentrations of the polymer was due to the addition of PVP-VA to the formulation. Furthermore, the value of ΔT_g represent a specific interaction between the API and the polymeric carrier, because the positive deviation reflects an increase in interactions. The deviation appeared when the number and strength of interactions between homo-materials were lower than between hetero-materials [35–37]. The calculated value showed that SGR granules with PVP-VA have interactions that lead to time scale stabilization, as with the results of XRD. The strength increased with the PVP-VA concentration. On the other hand, the 0% PVP-VA granule interaction was low because of the absence of the polymer. These results of XRD and DSC demonstrated that the thermodynamic stability of the ASD granules by SGR was sufficiently high because the glass transition was over 178.7 °C and also, even pure PVP-VA was high enough (107.8 °C), whereas undesirable crystallization was provoked under the high humidity condition (20 °C/75% RH). The effect of humidity on the crystallization can be prevented by using a high concentration of polymer or granule coating [38,39].

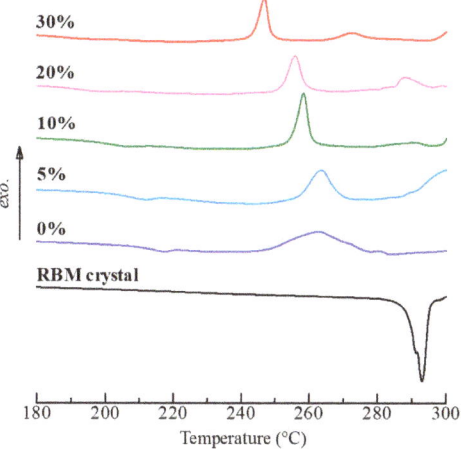

Figure 6. Total heat flow curves of RBM crystals and SGR granules with 0–30% PVP-VA by DSC.

Table 3. Measured and calculated glass transition temperatures of SGR granules.

PVP-VA (%)	$T_{g\ expt}$ [1] (°C)	$T_{g\ calc}$ [2] (°C)	ΔT_g [3] (°C)
0	215.4	n/a	n/a
5	206.9	202.0	4.9
10	203.9	194.3	9.6
20	191.7	175.0	16.7
30	178.7	157.5	21.2

[1] T_g experimental, [2] T_g calculated, [3] T_g experimental − T_g calculated.

3.3. Molecular State

SGR granules have intermolecular interactions between the API and PVP-VA in proportion to the polymer concentration, as described in the preceding section. This section explains in detail the molecular state change of the granules, according to the interaction. Figure 7A shows the IR spectra of ASD granules by SGR. Bulk RBM crystals had an absorption peak at 1735 cm^{-1}, due to the C12=O13 stretching vibration. However, this peak disappeared after the SGR process and different peaks at 1595 cm^{-1} and 1394 cm^{-1}, which correspond to COO$^-$ asymmetric and symmetric stretching vibrations, appeared. These indicated that ionization had occurred due to deprotonation at the carboxy group. Hence, the carboxylate COO$^-$ group is suggested to exist without ionic interaction with sodium ions. The characteristic peak of RBM crystals at 1645 cm^{-1}, which was assigned to the amide I band of C8=O9 and C18=O19, disappeared in the granules. Bulk PVP-VA peaks of C=O stretching vibration in ester and the amide I band were 1724 cm^{-1} and 1655 cm^{-1}, respectively. The other peaks of the amide II and III bands were weak, as shown in Figure 7A.

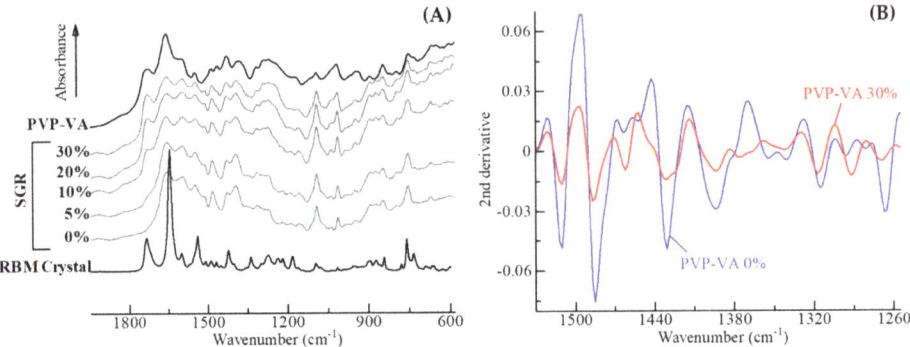

Figure 7. (**A**) Overlays of IR spectra of RBM crystals alone, PVP-VA alone, and SGR granules including 0–30% PVP-VA. (**B**) Representation of the second derivatives of SGR granules.

To detect some weak peaks, second derivative spectra of the SGR granules were calculated (Figure 7B). The second derivative preprocessing allows more specific identification of weak peaks in the original spectra and offers the cancelling of baseline shift [40]. The positive value of the original data was converted into a strong negative peak using the second derivative. The characteristic peaks of RBM at 1480 cm^{-1}, 1510 cm^{-1}, and 1540 cm^{-1} indicated C8-N10 coupled with N10-H and C18-N20 coupled with N20–H vibrations (amide II band). These bands, which were based on amides, appeared strongly in granules, because the interaction between molecules of RBMs may be weakened by the amorphization process. These bands were also shifted or disappeared in proportion to the PVP-VA concentration, the amide of RBM interacts with the matrix though hydrogen bonding. Therefore, the interaction between RBM and PVP-VA was present in SGR granules. The indication corresponded to the result of ΔT_g analysis, shown in Table 3.

3.4. Dissolution Ability

Figure 8A shows the dissolution profiles of the tablets in water. All tablets, including SGR granules, showed enhanced solubility of the RBM crystals and the dissolved concentration reached almost 100% (126.67 μg/mL) at 120 min, whereas RBM crystals dissolved slowly. In Figure 8B, dissolution testing using an acidic medium was performed with dispersible SGR granules, due to the low solubility of weakly acidic RBM. The dissolution profiles in the acidic solution were different compared with in water. All of the SGR granules were more soluble than RBM crystals; however, the profiles depended on the concentration of PVP-VA in the granules. In low PVP-VA formulations, such as 0% and 5%, the dissolved concentration reached approximately 2.22 μg/mL. Granules with 10% PVP-VA reached the highest solubility (11.89 μg/mL) among the samples shown in Figure 8B, which was about 10 times higher than RBM crystals. In granules, including 20% and 30% PVP-VA, gelation was observed and the dissolved amount of RBM, at 120 min, decreased with increasing concentration of the polymer. Generally, the recrystallization and crystal growth were prevented by the interaction with the polymer; thus, the polymer contributes to further improvement of the API solubility [41,42]. However, the wettability and swellability of granules were in proportion to the concentration of polymer, while the dispersibility of granules was decreased. Higher the content rate of the polymer caused aggregation and gelation in the vessel, consequently leading to lower drug release from granules in this study. It should be noted that the dissolution profile may change if the amount of contents was revised. Optimized formulation ratios, depending on respective objectives, are vital for the pharmaceutical industry.

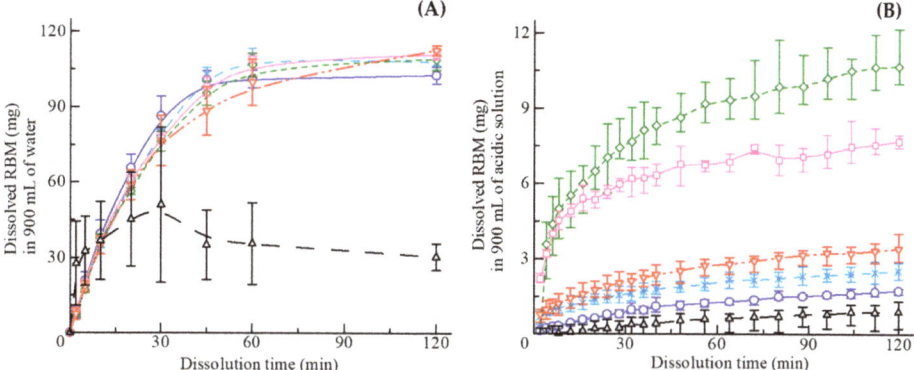

Figure 8. Integrated dissolution profiles in 900 mL of (**A**) tablets in water and (**B**) granules in an acidic solution of RBM crystal (dark triangles), PVP-VA 0% (blue circles), PVP-VA 5% (light blue crosses), PVP-VA 10% (green diamonds), PVP-VA 20% (light red rectangles), and PVP-VA 30% (red inversed triangles). Percentages denote the concentration of PVP-VA in the SGR granules.

The drug release kinetics from ASDs were explained based on mathematical modeling analysis [43–45]. As the result of an R^2 analysis of least-squares fitting, the best-fitted was the Korsmeyer–Peppas model. The model is generally resumed to the following expression:

$$\frac{M_t}{M_\infty} = kt^n \tag{5}$$

where M_t/M_∞ is the amount of drug released on time t per unit area, k is the kinetics constant, and n is the release exponent [43,44]. Each R^2 value of the RBM crystal and the granules including 0–30% PVP-VA in acidic medium was 0.917, 0.989, 0.996, 0.987, 0.930, and 0.997. Additionally, each exponent n value was 1.16, 0.64, 0.34, 0.31, 0.26, and 0.38, respectively. The exponents indicate the diffusional drug release mechanism from a matrix, $n < 0.43$, $0.43 < n < 0.85$, and $0.85 < n$ describes Fickian diffusion, anomalous (non-Fickian) transport, and super case II transport, respectively [43,44]. These results suggested the release mechanisms of the enclosed REB, during the dissolution, were changed into Fickian diffusion due to forming ASD by PVP-VA. The release kinetics from the tablet in water was also explained by the Korsmeyer–Peppas model. Hence, the drug release mechanism after the disintegration of the tablet was the same as the granule.

4. Conclusions

One-step granulation methods such as CTS-SGR could be used to prepare ASD granules with high RBM loading to enhance the solubility of this BCS class IV drug. This method involves no specific preparation and produces the ASD granules using a continuous spraying and layering system. According to the concentration of PVP-VA, the obtained granules were found to be heavy and dense, with a smooth surface. The thermodynamic stability of SGR granules was relatively high and the humidity stability at 20 °C/75% RH depended on the concentration of PVP-VA. Molecular interactions formed between RBM and PVP-VA, including the carboxy group of RBM becoming ionized in SGR granules. Dissolution testing demonstrated the improved water solubility of RBM, even in acidic media due to the formation of an ASD. The SGR method, which directly generates granules from solutions, has the possibility to reduce manufacturing operations. The SGR methodology can contribute to new process development in the pharmaceutical industry.

Author Contributions: The author's contributions are as follows: R.T., Y.H. (Yusuke Hattori), Y.H. (Yukun Horie), and H.K. prepared and analyzed the ASD granules; R.T. wrote the draft of the manuscript; Y.H. (Yusuke Hattori) edited and refined the manuscript; T.N. and M.O. supervised the research project. All the authors designed the protocol of the study and critically reviewed and approved the final version of the manuscript.

Funding: The work was partially supported by a JSPS Overseas Challenge Program for Young Researchers, a Young Researcher Scholarship from Hosokawa Powder Technology Foundation, and a PSJ Nagai Memorial Research Scholarship.

Acknowledgments: We profoundly thank Ohara Pharmaceutical, Fuji Chemical Industry, and BASF for the gifts of rebamipide, Neusilin® NS2N, and Kollidon® VA64, respectively. Spectris kindly supported the particle size distribution measurement using Mastersizer 3000E. We thank Raj Suryanarayanan and Naga Kiran Duggirala fruitful discussions.

Conflicts of Interest: The authors declare that they have no competing interests.

References

1. Ku, M.S.; Dulin, W. A biopharmaceutical classification-based Right-First-Time formulation approach to reduce human pharmacokinetic variability and project cycle time from First-In-Human to clinical Proof-Of-Concept. *Pharm. Dev. Technol.* **2012**, *17*, 285–302. [CrossRef] [PubMed]
2. Amidon, G.L.; Lennernäs, H.; Shah, V.P.; Crison, J.R. A Theoretical Basis for a Biopharmaceutic Drug Classification: The Correlation of in Vitro Drug Product Dissolution and in Vivo Bioavailability. *Pharm. Res.* **1995**, *12*, 413–420. [CrossRef]
3. Kawakami, K. Modification of physicochemical characteristics of active pharmaceutical ingredients and application of supersaturatable dosage forms for improving bioavailability of poorly absorbed drugs. *Adv. Drug Deliv. Rev.* **2012**, *64*, 480–495. [CrossRef]
4. Perrut, M.; Jung, J.; Leboeuf, F. Enhancement of dissolution rate of poorly-soluble active ingredients by supercritical fluid processes: Part I: Micronization of neat particles. *Int. J. Pharm.* **2005**, *288*, 3–10. [CrossRef]
5. Alonzo, D.E.; Zhang, G.G.Z.; Zhou, D.; Gao, Y.; Taylor, L.S. Understanding the Behavior of Amorphous Pharmaceutical Systems during Dissolution. *Pharm. Res.* **2010**, *27*, 608–618. [CrossRef] [PubMed]
6. Liversidge, G.G.; Cundy, K.C. Particle size reduction for improvement of oral bioavailability of hydrophobic drugs: I. Absolute oral bioavailability of nanocrystalline danazol in beagle dogs. *Int. J. Pharm.* **1995**, *125*, 91–97. [CrossRef]
7. Aitipamula, S.; Banerjee, R.; Bansal, A.K.; Biradha, K.; Cheney, M.L.; Choudhury, A.R.; Desiraju, G.R.; Dikundwar, A.G.; Dubey, R.; Duggirala, N.; et al. Polymorphs, Salts, and Cocrystals: What's in a Name? *Cryst. Growth Des.* **2012**, *12*, 2147–2152. [CrossRef]
8. Pitha, J.; Milecki, J.; Fales, H.; Pannell, L.; Uekama, K. Hydroxypropyl-β-cyclodextrin: Preparation and characterization; effects on solubility of drugs. *Int. J. Pharm.* **1986**, *29*, 73–82. [CrossRef]
9. Chiou, W.L.; Riegelman, S. Pharmaceutical Applications of Solid Dispersion Systems. *J. Pharm. Sci.* **1971**, *60*, 1281–1302. [CrossRef]
10. Serajuddin, A.T.M. Solid dispersion of poorly water-soluble drugs: Early promises, subsequent problems, and recent breakthroughs. *J. Pharm. Sci.* **1999**, *88*, 1058–1066. [CrossRef]
11. Karagianni, A.; Kachrimanis, K.; Nikolakakis, I. Co-Amorphous Solid Dispersions for Solubility and Absorption Improvement of Drugs: Composition, Preparation, Characterization and Formulations for Oral Delivery. *Pharmaceutics* **2018**, *10*, 98. [CrossRef] [PubMed]
12. Fung, M.H.; Suryanarayanan, R. Effect of Organic Acids on Molecular Mobility, Physical Stability, and Dissolution of Ternary Ketoconazole Spray-Dried Dispersions. *Mol. Pharm.* **2019**, *16*, 41–48. [CrossRef]
13. Yu, L. Amorphous pharmaceutical solids: Preparation, characterization and stabilization. *Adv. Drug Deliv. Rev.* **2001**, *48*, 27–42. [CrossRef]
14. Fung, M.H.; DeVault, M.; Kuwata, K.T.; Suryanarayanan, R. Drug-Excipient Interactions: Effect on Molecular Mobility and Physical Stability of Ketoconazole–Organic Acid Coamorphous Systems. *Mol. Pharm.* **2018**, *15*, 1052–1061. [CrossRef] [PubMed]
15. Frank, D.S.; Matzger, A.J. Probing the Interplay between Amorphous Solid Dispersion Stability and Polymer Functionality. *Mol. Pharm.* **2018**, *15*, 2714–2720. [CrossRef]

16. Agrawal, A.M.; Dudhedia, M.S.; Zimny, E. Hot Melt Extrusion: Development of an Amorphous Solid Dispersion for an Insoluble Drug from Mini-scale to Clinical Scale. *AAPS PharmSciTech* **2016**, *17*, 133–147. [CrossRef] [PubMed]
17. Kaur, P.; Singh, S.K.; Garg, V.; Gulati, M.; Vaidya, Y. Optimization of spray drying process for formulation of solid dispersion containing polypeptide-k powder through quality by design approach. *Powder Technol.* **2015**, *284*, 1–11. [CrossRef]
18. Chen, J.; Ormes, J.D.; Higgins, J.D.; Taylor, L.S. Impact of Surfactants on the Crystallization of Aqueous Suspensions of Celecoxib Amorphous Solid Dispersion Spray Dried Particles. *Mol. Pharm.* **2015**, *12*, 533–541. [CrossRef]
19. Wilson, V.; Lou, X.; Osterling, D.J.; Stolarik, D.F.; Jenkins, G.; Gao, W.; Zhang, G.G.Z.; Taylor, L.S. Relationship between amorphous solid dispersion in vivo absorption and in vitro dissolution: Phase behavior during dissolution, speciation, and membrane mass transport. *J. Control. Release* **2018**, *292*, 172–182. [CrossRef]
20. Nie, H.; Xu, W.; Taylor, L.S.; Marsac, P.J.; Byrn, S.R. Crystalline solid dispersion-a strategy to slowdown salt disproportionation in solid state formulations during storage and wet granulation. *Int. J. Pharm.* **2017**, *517*, 203–215. [CrossRef]
21. Sekiguchi, K.; Obi, N. Studies on Absorption of Eutectic Mixture. I. A Comparison of the Behavior of Eutectic Mixture of Sulfathiazole and that of Ordinary Sulfathiazole in Man. *Chem. Pharm. Bull.* **1961**, *9*, 866–872. [CrossRef]
22. Patil, H.; Tiwari, R.V.; Repka, M.A. Hot-Melt Extrusion: From Theory to Application in Pharmaceutical Formulation. *AAPS PharmSciTech* **2016**, *17*, 20–42. [CrossRef]
23. Mayersohn, M.; Gibaldi, M. New Method of Solid-State Dispersion for Increasing Dissolution Rates. *J. Pharm. Sci.* **1966**, *55*, 1323–1324. [CrossRef] [PubMed]
24. Takeuchi, H.; Nagira, S.; Yamamoto, H.; Kawashima, Y. Solid dispersion particles of amorphous indomethacin with fine porous silica particles by using spray-drying method. *Int. J. Pharm.* **2005**, *293*, 155–164. [CrossRef]
25. Konno, H.; Taylor, L.S. Influence of Different Polymers on the Crystallization Tendency of Molecularly Dispersed Amorphous Felodipine. *J. Pharm. Sci.* **2006**, *95*, 2692–2705. [CrossRef]
26. Konno, H.; Handa, T.; Alonzo, D.E.; Taylor, L.S. Effect of polymer type on the dissolution profile of amorphous solid dispersions containing felodipine. *Eur. J. Pharm. Biopharm.* **2008**, *70*, 493–499. [CrossRef]
27. Yajima, T.; Nogata, A.; Demachi, M.; Umeki, N.; Itai, S.; Yunoki, N.; Nemoto, M. Particle Design for Taste-Masking Using a Spray-Congealing Technique. *Chem. Pharm. Bull.* **1996**, *44*, 187–191. [CrossRef]
28. Burggraeve, A.; Monteyne, T.; Vervaet, C.; Remon, J.P.; Beer, T. De Process analytical tools for monitoring, understanding, and control of pharmaceutical fluidized bed granulation: A review. *Eur. J. Pharm. Biopharm.* **2013**, *83*, 2–15. [CrossRef]
29. Elversson, J.; Millqvist-Fureby, A.; Alderborn, G.; Elofsson, U. Droplet and Particle Size Relationship and Shell Thickness of Inhalable Lactose Particles During Spray Drying. *J. Pharm. Sci.* **2003**, *92*, 900–910. [CrossRef]
30. Kimura, S.; Uchida, S.; Kanada, K.; Namiki, N. Effect of granule properties on rough mouth feel and palatability of orally disintegrating tablets. *Int. J. Pharm.* **2015**, *484*, 156–162. [CrossRef] [PubMed]
31. Naito, Y.; Yoshikawa, T. Rebamipide: A gastrointestinal protective drug with pleiotropic activities. *Expert Rev. Gastroenterol. Hepatol.* **2010**, *4*, 261–270. [CrossRef] [PubMed]
32. Tanaka, R.; Takahashi, N.; Nakamura, Y.; Hattori, Y.; Ashizawa, K.; Otsuka, M. Performance of an acoustically mixed pharmaceutical dry powder delivered from a novel inhaler. *Int. J. Pharm.* **2018**, *538*, 130–138. [CrossRef]
33. Hausner, H.H. Friction conditions in a mass of metal powder. *Int. J. Powder Metall.* **1967**, *3*, 7.
34. Carr, R.L. Evaluating flow properties of solids. *Chem. Eng.* **1965**, *72*, 163–168.
35. Couchman, P.R.; Karasz, F.E. A Classical Thermodynamic Discussion of the Effect of Composition on Glass-Transition Temperatures. *Macromolecules* **1978**, *11*, 117–119. [CrossRef]
36. Hancock, B.C.; Zografi, G. The Relationship Between the Glass Transition Temperature and the Water Content of Amorphous Pharmaceutical Solids. *Pharm. Res.* **1994**, *11*, 471–477. [CrossRef] [PubMed]
37. Song, Y.; Yang, X.; Chen, X.; Nie, H.; Byrn, S.; Lubach, J.W. Investigation of Drug–Excipient Interactions in Lapatinib Amorphous Solid Dispersions Using Solid-State NMR Spectroscopy. *Mol. Pharm.* **2015**, *12*, 857–866. [CrossRef]

38. Petereit, H.-U.; Weisbrod, W. Formulation and process considerations affecting the stability of solid dosage forms formulated with methacrylate copolymers. *Eur. J. Pharm. Biopharm.* **1999**, *47*, 15–25. [CrossRef]
39. Wu, T.; Sun, Y.; Li, N.; de Villiers, M.M.; Yu, L. Inhibiting Surface Crystallization of Amorphous Indomethacin by Nanocoating. *Langmuir* **2007**, *23*, 5148–5153. [CrossRef] [PubMed]
40. Buckton, G.; Yonemochi, E.; Hammond, J.; Moffat, A. The use of near infra-red spectroscopy to detect changes in the form of amorphous and crystalline lactose. *Int. J. Pharm.* **1998**, *168*, 231–241. [CrossRef]
41. Simonelli, A.P.; Mehta, S.C.; Higuchi, W.I. Dissolution rates of high energy sulfathiazole-povidone coprecipitates II: Characterization of form of drug controlling its dissolution rate via solubility studies. *J. Pharm. Sci.* **1976**, *65*, 355–361. [CrossRef] [PubMed]
42. Gupta, P.; Kakumanu, V.K.; Bansal, A.K. Stability and Solubility of Celecoxib-PVP Amorphous Dispersions: A Molecular Perspective. *Pharm. Res.* **2004**, *21*, 1762–1769. [CrossRef] [PubMed]
43. Ritger, P.L.; Peppas, N.A. A simple equation for description of solute release I. Fickian and non-fickian release from non-swellable devices in the form of slabs, spheres, cylinders or discs. *J. Control. Release* **1987**, *5*, 23–36. [CrossRef]
44. Ritger, P.L.; Peppas, N.A. A simple equation for description of solute release II. Fickian and anomalous release from swellable devices. *J. Control. Release* **1987**, *5*, 37–42. [CrossRef]
45. Mehanna, M.M.; Motawaa, A.M.; Samaha, M.W. In sight into tadalafil – block copolymer binary solid dispersion: Mechanistic investigation of dissolution enhancement. *Int. J. Pharm.* **2010**, *402*, 78–88. [CrossRef]

© 2019 by the authors. Licensee MDPI, Basel, Switzerland. This article is an open access article distributed under the terms and conditions of the Creative Commons Attribution (CC BY) license (http://creativecommons.org/licenses/by/4.0/).

Article

The Self-Assembly Phenomenon of Poloxamers and Its Effect on the Dissolution of a Poorly Soluble Drug from Solid Dispersions Obtained by Solvent Methods

Joanna Szafraniec [1,2,*], Agata Antosik [1], Justyna Knapik-Kowalczuk [3,4], Krzysztof Chmiel [3,4], Mateusz Kurek [1], Karolina Gawlak [2], Joanna Odrobińska [2], Marian Paluch [3,4] and Renata Jachowicz [1]

[1] Department of Pharmaceutical Technology and Biopharmaceutics, Faculty of Pharmacy, Jagiellonian University Medical College, Medyczna 9, 30-688 Krakow, Poland; Agata.Antosik@uj.edu.pl (A.A.); Mateusz.Kurek@uj.edu.pl (M.K.); Mfjachow@cyf-kr.edu.pl (R.J.)

[2] Department of Physical Chemistry and Electrochemistry, Faculty of Chemistry, Jagiellonian University, Gronostajowa 2, 30-387 Krakow, Poland; Gawlak@chemia.uj.edu.pl (K.G.); Odrobinska@chemia.uj.edu.pl (J.O.)

[3] Division of Biophysics and Molecular Physics, Institute of Physics, University of Silesia, Uniwersytecka 4, 40-007 Katowice, Poland; Justyna.Knapik-Kowalczuk@smcebi.edu.pl (J.K.-K.); Krzysztof.Chmiel@smcebi.edu.pl (K.C.); Marian.Paluch@us.edu.pl (M.P.)

[4] Silesian Center for Education and Interdisciplinary Research, 75 Pulku Piechoty 1a, 41-500 Chorzow, Poland

* Correspondence: Joanna.Szafraniec@uj.edu.pl; Tel.: +48-12-62-05-606

Received: 13 February 2019; Accepted: 15 March 2019; Published: 19 March 2019

Abstract: The self-assembly phenomenon of amphiphiles has attracted particular attention in recent years due to its wide range of applications. The formation of nanoassemblies able to solubilize sparingly water-soluble drugs was found to be a strategy to solve the problem of poor solubility of active pharmaceutical ingredients. Binary and ternary solid dispersions containing Biopharmaceutics Classification System (BCS) class II drug bicalutamide and either Poloxamer®188 or Poloxamer®407 as the surface active agents were obtained by either spray drying or solvent evaporation under reduced pressure. Both processes led to morphological changes and a reduction of particle size, as confirmed by scanning electron microscopy and laser diffraction measurements. The increase in powder wettability was confirmed by means of contact angle measurements. The effect of an alteration of the crystal structure was followed by powder X-ray diffractometry while thermal properties were determined using differential scanning calorimetry. Interestingly, bicalutamide exhibited a polymorph transition after spray drying with the poloxamer and polyvinylpyrrolidone (PVP), while the poloxamer underwent partial amorphization. Moreover, due to the surface activity of the carrier, the solid dispersions formed nanoaggregates in water, as confirmed using dynamic light scattering measurements. The aggregates measuring 200–300 nm in diameter were able to solubilize bicalutamide inside the hydrophobic inner parts. The self-assembly of binary systems was found to improve the amount of dissolved bicalutamide by 4- to 8-fold in comparison to untreated drug. The improvement in drug dissolution was correlated with the solubilization of poorly soluble molecules by macromolecules, as assessed using emission spectroscopy.

Keywords: bicaludamide; poloxamer; evaporation; spray drying; dissolution enhancement; nanoaggregates; self-assembly

---

1. Introduction

The issue of poor solubility of active pharmaceutical ingredients (APIs) is one of the biggest limitations for drug development. It is a matter of concern, as the bioavailability depends on the dissolution of drug in the gastrointestinal fluids. The main determinants of the dissolution kinetics *in vivo* are solubility and surface area of the particles. The solubility is a function of the crystal lattice energy and the affinity of solid phase to the solvent. Thus, three groups of strategies that have been implemented to improve the rate of dissolution and solubility rely on: (1) the reduction of the intermolecular forces in solid phase, (2) the enhancement of the solid–solvent interaction, and (3) the increase of the surface area available for solvation (according to the Noyes–Whitney equation) [1].

Due to the fact that almost 50% of currently marketed drugs and over 70% of new chemical entities exhibit low solubility in water, numerous techniques have been developed to overcome this problem [2]. Common strategies include pH adjustment, formation of salts, cosolvency, formation of cocrystals and inclusion complexes, particle size reduction, supercritical fluid technology (SCF), and self-emulsification [3,4]. Recently, nanotechnology has emerged as a technique that leads to the formation of robust delivery systems. Numerous attempts have been applied to obtain several types of delivery systems, i.e., micelles [5], liposomes [6], capsules [7,8], protein nanocontainers [9], and silica-based nanoparticles [10,11]. Poorly water-soluble drugs have been frequently processed with hydrophilic polymers, as the molecular dispersion of drug molecules within the matrix provides better dissolution of the drug. Moreover, when the systems were further formulated into the nanoparticles, the results were more pronounced [12–14].

The main factors affecting the choice of a particular method are the physicochemical characteristics of drugs and carriers. Solid dispersions are commonly formed to enhance the water solubility of APIs; however, the number of marketed products arising from that strategy is rather low. This is a result of the thermal instability of drug and carrier during preparation of systems, a poor *in vitro–in vivo* correlation, and instability during storage [15]. However, the simplicity of preparation, low cost, and great improvements in the dissolution of poorly water-soluble drugs have made the solid dispersions widely investigated. Experimental and theoretical approaches have been involved to determine the thermodynamic properties of APIs dispersed in polymer matrices as well as the mechanisms and factors affecting their stability [16–18].

The concept of solid dispersion—one of the earliest methods of solubility enhancement—was introduced in 1961 by Sekiguchi and Obi, who prepared eutectic mixtures containing microcrystalline drug and a water-soluble carrier [19–22]. Although crystalline forms provide high stability and chemical purity, the lattice energy barrier is the major limitation affecting the dissolution rate. Thus, amorphous carriers such as polyvinylpyrrolidone (PVP) [23,24] and hydroxypropylmethyl cellulose (HPMC) [25,26] have been introduced to prepare amorphous solid dispersions (ASDs). The highly water-soluble amorphous carriers provide stabilization of APIs, increasing the wettability and dispersibility of the drug [27–29]. They limit the precipitation of a drug in water; however, the supersaturation may lead to precipitation and recrystallization of APIs, which negatively affects the bioavailability of the drug. To face this problem, surface active agents or self-emulsifiers such as poloxamers (PLXs) [30,31], Tween 80 [32], or sodium lauryl sulfate (SLS) [33] have been introduced. They improve the dissolution rate as well as physical and chemical stability of the supersaturated system. Surfactants or emulsifiers enhance the miscibility and thus limit the recrystallization rate of the drug. Moreover, they are able to absorb onto the outer layer of drug particles or form micelles encapsulating drug particles, effectively preventing drug precipitation [34]. On the other hand, many surfactants can absorb moisture, which may result in phase separation during storage, an increase in drug mobility, and conversion from the amorphous or metastable form to the more stable crystalline one. They may change the physical properties of the matrix, increase the water content and cause adverse side effects *in vivo*. [35] Thus, their use has to be cautious and their amounts well adjusted.

Among the strategies that allow for obtaining solid dispersions, solvent methods are often used. In these techniques the drug and the carrier are dissolved in a volatile solvent such as ethanol [36] or methylene chloride–ethanol mixture [37] that is further evaporated. It requires sufficient solubility of the drug as well as the carrier in the solvent. Moreover, the type of used solvent, the temperature, and rate of its evaporation are of key importance due to the fact that the concentration of residual solvent needs to be below the detection limit after drying. One of the strategies utilized to fulfill that requirement is the use of low-toxicity solvent mixtures, e.g., water with ethanol, which decreases the amount of each solvent in dry formulation. However, this strategy sometimes fails due to insufficient dissolution of components at a given ratio [35]. Usually, a second drying step is applied to completely removed the solvent as it may lower the glass transition temperature, enhancing the recrystallization tendency.

The common feature of evaporation approaches is the removal of small droplets or thin layers of the solvent from different surfaces. It may lead to the crystal growth of oriented morphology as described for droplet evaporative crystallization or microwave-accelerated evaporative crystallization [38–40]. The crystallization of the celocoxib–PVP mixture was found to generate drug crystals of improved dissolution characteristics [41]. Other approaches such as the evaporative antisolvent method and supercritical carbon dioxide evaporation were applied to the formation of nanoparticles, drug-loaded micelles, and liposomes characterized by improved dissolution of the drug [42,43].

Commonly used solvent methods include vacuum drying using rotary evaporators [44], spray drying [45], or freeze-drying [46], among others. In rotary evaporators, solvents are removed under reduced pressure, limiting thermal decomposition of the components of the mixture as organic solvent evaporation occurs at low temperature. Spray drying combines four processes, i.e., (1) atomization of the liquid containing dissolved or suspended drug, which is transported into the nozzle and then sprayed onto fine droplets, (2) mixing the liquid with the drying gas, (3) evaporation, and finally (4) separation of obtained particles from the gas using cyclone [47]. Generally, the spray drying process can be applied for the generation of amorphous materials as well as a technique for particle engineering, i.e., particle size reduction [48].

In the work reported herein, we study the self-assembly phenomenon of solid dispersions containing either Poloxamer®188 or Poloxamer®407 and its effect on dissolution enhancement of the poorly water soluble drug bicalutamide (BCL). Poloxamers are the nonionic surfactants widely used in pharmaceutical formulations as emulsifiers, wetting agents and solubilizers. They have been introduced into solid dispersions to enhance solubility and dissolution profiles of poorly water-soluble APIs from solid dosage forms [49,50]. Bicalutamide was used as a model drug. It is a non-steroidal antiandrogenic drug assigned to Biopharmaceutics Classification System (BCS) class II because of poor water solubility (below 3.7 mg/L) and high membrane permeability (logP = 2.92) [51–53]. It is known to exhibit polymorphism and undergo mechanical activation upon milling [54–58]. Obtained results indicate that the formation of solid dispersions by means of solvent methods led to the changes of particles in solid state, i.e., morphological features, increased wettability, phase transition (in case of ternary solid dispersions containing PVP) and partial disruption of crystal lattice. Moreover, the formation of nanoaggregates in aqueous media led to the 4- to 8-fold increase in the amount of dissolved bicalutamide. Emission spectroscopy allowed for a correlation of the effect of dissolution changes with the solubilization related to the variations of molecular structure of used poloxamers.

2. Materials and Methods

2.1. Materials

Bicalutamide (BCL, N-[4-cyano-3-(trifluoromethyl)phenyl]-3-[(4-fluorophenyl)sulfonyl]-2-hydroxy-2-methylpropanamide, 99.8%, Hangzhou Hyper Chemicals Limited, Zhejiang, China) was used as a model drug. Poloxamer®188, Poloxamer®407 (BASF, Ludwigshafen am Rhein, Germany), and polyvinylpyrrolidone K29/32 (PVP, Ashland, Covington, KY, USA) were used as excipients. Sodium lauryl sulfate (SLS, BASF, Ludwigshafen am Rhein, Germany) was used to prepare dissolution medium. Ethanol (absolute, 99.8%, pure p.a., Avantor Performance Materials, Gliwice, Poland) and methanol (p.a., Chempur, Piekary Slaskie, Poland) were used as solvents. Cyclohexane (ACS, pure p.a., Avantor Performance Materials, Gliwice, Poland) was used as a dispersant in laser diffraction measurements. Perylene (Pe, p.a., Koch-Light Laboratories Ltd., Colnbrook, UK) and pyrene (98%, Sigma-Aldrich, Darmstadt, Germany) were used in fluorescence emission measurements. All chemicals were used as received. Distilled water was used to prepare all of aqueous solutions.

2.2. Methods of Preparation of Solid Dispersions

2.2.1. Solvent Evaporation (E)

Bicalutamide (2 g) was mixed with either Poloxamer®188 or Poloxamer®407 in a 1:1 and 2:1 wt. ratio, placed in the round-bottomed flask, and dissolved in 200 mL of absolute ethanol. The solution was heated up to 40 °C in the water bath and after complete dissolution of the mixture the solvent was evaporated using a Hei-VAP Value rotavapor (Heidolph, Schwabach, Germany). The rotational speed was equal to 200 rpm and the pressure was reduced stepwise to ca. 40 mbar. The dry solid dispersion was transferred to a container and dried under vacuum prior to further characterization. The systems were further labeled as BCL-PLX188 1:1 (E), BCL-PLX188 2:1 (E), BCL-PLX407 1:1 (E) and BCL-PLX407 2:1 (E), respectively.

2.2.2. Spray Drying (SD)

An ethanolic solution containing bicalutamide mixed with the appropriate carrier or carrier mixture (1:1 and 2:1 wt. ratio, respectively) was spray-dried using a Mini Spray Dryer B-191 (Büchi, Flawil, Switzerland). The process was conducted using following parameters: T_{inlet} = 50–53 °C, T_{outlet} = 39–42 °C, aspirator flow 100%, gas flow rate 600 L/min, liquid flow rate 3.4 mL/min, and a 0.7-mm diameter nozzle. The process was carried out under a constant control and the concentration of ethanol was 10-times lower than the flammability limit. The samples were further dried under vacuum to remove residual solvent. The systems were labeled as BCL-PLX188 1:1 (SD), BCL-PLX188 2:1 (SD), BCL-PLX407 1:1 (SD) and BCL-PLX407 2:1 (SD), BCL-PLX188-PVP 2:1:1 (SD), BCL-PLX188-PVP 4:1:1 (SD), BCL-PLX407-PVP 2:1:1 (SD), and BCL-PLX407-PVP 4:1:1 (SD), respectively.

2.2.3. Scanning Electron Microscopy (SEM)

A Phenom Pro desktop electron microscope (PhenomWorld, Thermo Fisher Scientific, Waltham, MA, USA) equipped with a CeB_6 electron source and backscattered electron detector was used to determine the morphological features of the samples. The acceleration voltage was equal to 10 kV and the magnification was 750× for evaporated samples, 5000× for spray-dried ternary solid dispersions, and 10,000× for zoomed sections. The powder was placed on the conductive adhesive tape previously glued to the specimen mount. The holder for non-conductive samples was used. The excess of sample (loosely bound to the tape) was removed using a stream of argon. The samples were not sputtered prior to the measurement.

2.2.4. Differential Scanning Calorimetry (DSC)

A DSC 1 STARe System (Mettler–Toledo, Greifensee, Switzerland) was used in order to examine the thermal properties of the samples. The measuring device was equipped with a HSS8 ceramic sensor with 120 thermocouples and a liquid nitrogen cooling station. The apparatus was calibrated for temperature and enthalpy using zinc and indium standards. Melting points were determined as the onset of the peak, with the glass transition temperatures as the midpoint of the heat capacity increment. The samples were measured in an aluminum crucible (40 µL). All measurements were carried out with a heating rate equal to 10 K/min.

2.2.5. Powder X-ray Diffraction (PXRD)

The diffraction patterns of the samples were registered using an X-ray diffractometer Mini Flex II (Rigaku, Tokyo, Japan). The angular range 3–70° 2θ was scanned with a scan speed of 5°/min and a step size equal to 0.02. The measurements were carried out using monochromatic Cu Kα radiation

(λ = 1.5418 Å) at ambient temperature. The samples in form of powder were placed in a standard glass sample holder without milling prior the measurement.

2.2.6. Laser Diffraction Measurements

A Mastersizer 3000 equipped with a HydroEV unit (Malvern Instruments, Malvern, UK) was used to determine the particle size distribution. The samples were analyzed by the wet method using cyclohexane (reflective index, RI = 1.426) as a dispersant. The cyclohexane was filtered through the G5 sintered disc filter funnel and placed in the beaker. The rotational speed of the mixer was 1500 rpm. The sample in powder form was added until the obscuration reached the given value (between 5% and 20%) and then the measurement was carried out. A Fraunhofer diffraction theory was applied to find the relationship between particle size and light intensity distribution pattern. Reported data represent the averages from 10 series of measurements for each sample.

2.2.7. Fourier Transform Infrared Spectroscopy (FTIR)

A Nicolet iS10 FT-IR spectrometer (Thermo Fisher Scientific, Waltham, MA, USA) equipped with a Smart iTR™ ATR (Attenuated Total Reflectance) sampling accessory with diamond as an ATR crystal was used to collect the vibrational spectra of powders. Spectra were collected within the range 600–4000 cm^{-1} with 4 cm^{-1} resolution. Presented data represent average from 128 scans for each sample.

2.2.8. Dynamic Light Scattering Measurements (DLS)

The size distribution of aggregates formed by the solid dispersions obtained by either evaporation or spray drying was determined using a Zetasizer Nano ZS instrument (Malvern Instruments, Malvern, UK) working at a 173° detection angle. The distribution analysis was performed at 25 °C using the general purpose mode. The powder was weighted, dissolved in water (c_{PLX} = 2.5 mg/mL) and shaken using a KS 130 Basic orbital shaker (IKA, Staufen im Breisgau, Germany) for 24 h. After that the sample was filtered through the 0.45-µm syringe filter and measured without further dilution. The reported data represent the averages of three series of measurements (10–100 runs each) of hydrodynamic diameter and their standard deviations.

2.2.9. Emission Spectroscopy

A SLM 8100 spectrofluorometer of L-geometry (Aminco, Silver Spring, MD, USA) equipped with a 450W xenon lamp as a light source was used to capture the emission spectra. The microliter quantities of the molecular probes (c ≈ 10^{-4} M), i.e., methanolic perylene solution or ethanolic pyrene solution were slowly injected into a milliliter volume of aqueous PLX188 or PLX407 solutions (c_{PLX} = 5 mg/mL) as well to the solid dispersions solutions previously filtered through a 0.45-µm syringe membrane filter and vigorously stirred. The residues of organic solvents were removed by purging the solution with nitrogen. The samples were equilibrated in the dark for at least 12 h and diluted 10 times before the measurement.

2.2.10. Contact Angle Determination

The wettability of binary systems was assessed by the contact angle measurements performed using a DSA255 drop shape analyzer (Krüss, Hamburg, Germany). The sessile drop technique was used. The droplet of distilled water of volume equal to 2 µL was deposited on the surface of powders compressed using an Atlas™ manual 15Ton hydraulic press (Specac, Kent, UK) with a load pressure of 1.5 tons that was applied for each sample for 15 s.

2.2.11. Dissolution Study

Dissolution of BCL was carried out according to the method recommended by the FDA for BCL tablets (1000 mL of 1% SLS, 37 ± 0.5 °C, 50 rpm) in the pharmacopeial paddle dissolution apparatus Vision Elite 8 (Hanson Research, Chatsworth, CA, USA) equipped with a VisionG2 AutoPlus Autosampler. The sink conditions were maintained. Pure drug and binary systems (solid dispersions, physical mixtures), equivalent of 50 mg of BCL were placed into the beakers. The samples were analyzed spectrophotometrically at 272 nm using a UV-1800 spectrofotometer (Shimatzu, Kioto, Japan) equipped with the flow-through cuvettes. The tests were carried out in triplicate and presented results represents averages with their standard deviations.

3. Results and Discussion

3.1. Solid State Characterization

3.1.1. Size Distribution and Morphology of Particles of Solid Dispersions

The effect of applied processes on the particle size and morphology was studied using both scanning electron microscopy and laser diffraction measurements. The analysis of size distribution of BCL-PLX solid dispersions obtained via evaporation in rotavapor confirmed the heterogeneity of particle size (Figure 1A). Long tails of the distribution curves in the region of small particles were well pronounced. Moreover, the shape of the distribution suggests that several fractions of particles of different sizes were present in the sample, as more than one maximum can be noticed. This is particularly noticeable for the BCL-PLX188 1:1 (E) system, which exhibits a bimodal long-tailed distribution of particle size. This is reflected by great differences in Dx(90) values between the samples, i.e., the point in the size distribution, up to which 90% of the total volume of material in the sample is included. The value was 1190.0 µm for BCL-PLX188 1:1 (E) solid dispersion, while it varied between 630–730 µm for the other evaporated systems. The Dx(10) and Dx(50) values representing the diameter of particles where 10% and a half of the particle population lie below, respectively, did not vary between the corresponding samples; however, the values are greater for BCL-PLX188 1:1 (E) and BCL-PLX407 2:1 (E), which also exhibit long tails in the region of particles that exceeded 1000 µm in length.

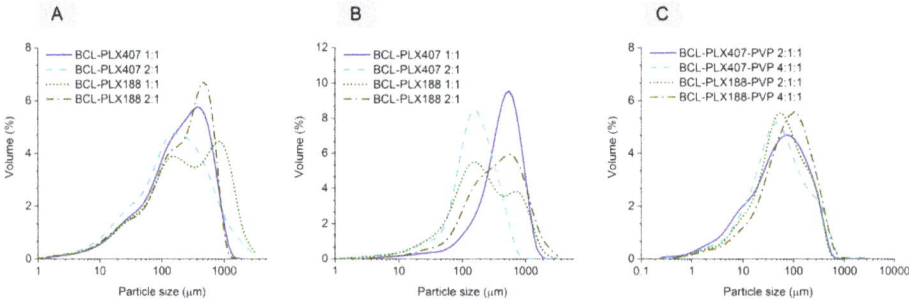

Figure 1. Particle size distribution of solid dispersions obtained by evaporation technique (**A**) and spray drying of binary (**B**) and ternary systems (**C**). BCL: bicalutamide; PLX: poloxamer; PVP: polyvinylpyrrolidone.

In spray drying, the liquid is dispersed in a form of droplets and dried with a hot air. This leads to a formation of particles of consistent size distribution, usually of spherical or ruptured spheres shape of diameter below 10 micrometers. The data presented in Table 1 indicates that spray-dried binary systems exhibited particles of greater size that those obtained via evaporation technique. However, the span values (calculated using Equation (1)) are a bit smaller in case of spray-dried systems, which indicates that the distributions are narrower. The tails of the distribution suggest that the particles aggregated during the process, probably due to the fact of low melting temperature of poloxamer. This may also result from the fact that some amount of drying samples adhered to the inner wall of spray dryer, which additionally leads to the decrease in the process yield. The Dx(90) value of particles of PLX 188-based (SD) solid dispersions are bigger than systems containing PLX407. Interestingly, BCL-PLX407 2:1 (SD) system exhibited the smallest particles among all investigated systems, as seen in Figure 2B and the SEM image (Figure 3). The size distributions of PLX407-based solid dispersions were narrower, with well resolved maxima as compared to those obtained for systems containing PLX188. Moreover, the maximum of particle size distribution of the system containing twice as much bicalutamide as PLX407 in binary solid dispersions was shifted towards bigger particles (Dx(10) = 126.0 μm and Dx(90) = 380.0 μm). All of examined systems exhibited a tailed distribution towards lower values of particle size.

$$Span = \frac{D_x(90) - D_x(10)}{D_x(50)} \quad (1)$$

Table 1. Particle size of solid dispersions obtained using the laser diffraction method. PLX188: Poloxamer®188, PLX407: Poloxamer®407.

Method	Carrier	BCL:polymers wt. Ratio	Dx(50) ± SD (μm)	Span
Evaporation	PLX188	1:1	247.0 ± 55.9	4.698
		2:1	227.0 ± 30.7	2.719
	PLX407	1:1	203.0 ± 23.1	2.971
		2:1	159.0 ± 53.1	4.438
Spray-drying	PLX188	1:1	196.0 ± 27.2	4.405
		2:1	355.0 ± 64.7	2.876
	PLX407	1:1	445.0 ± 28.7	1.771
		2:1	154.0 ± 13.7	2.167
	PLX188-PVP	2:1:1	78.0 ± 0.8	3.339
		4:1:1	55.6 ± 1.5	3.907
	PLX407-PVP	2:1:1	48.6 ± 2.2	5.153
		4:1:1	54.2 ± 2.1	4.098

Interestingly, the addition of PVP to BCL-PLX systems led to the formation of fine powders with the particle size distribution maxima located between 50 and 120 μm. However, the span reaches greater values than for binary solid dispersions, which may be a consequence of the distributions tailed towards smaller particles. Obtained ternary systems were also characterized with better flowability than platelet-like particles of binary solid dispersions. Particle size distributions of all the systems were more unified; moreover, the formation of a fraction of particles of size below 1 μm was also noticeable (Figure 1C).

An SEM analysis indicates that the crystals of neat bicalutamide adopt hexagonal shape and particles of smooth surface exhibiting ca. 160 μm in length [57]. The evaporation process led to the noticeable changes in the surface and morphology of obtained binary solid dispersions. During the rotation of the flask, the surface area of solvent increases. This leads to an enhancement of evaporation rate and fast recrystallization of dissolved bicalutamide. Thus, the formation of sharp-edged aggregates not exceeding 100 μm in case of BCL-PLX407 1:1 (E) and 200 μm for the other systems was observed (Figure 2). The systems comprised particles of wide size distribution, as seen in the SEM images as well as plots obtained using laser diffraction technique (Figure 1A).

A spray drying process usually leads to the formation of spherical particles with a consistent size distribution. The SEM micrographs of ternary solid dispersions show that obtained particles formed spheres of diameter not exceeding several microns, however they tended to agglomerate (Figure 3). In combination with the recrystallization that also occurred it led to the particle size distribution determined by means of laser diffraction measurements being much greater as the Dx(90) values varied between 227 μm and 273 μm.

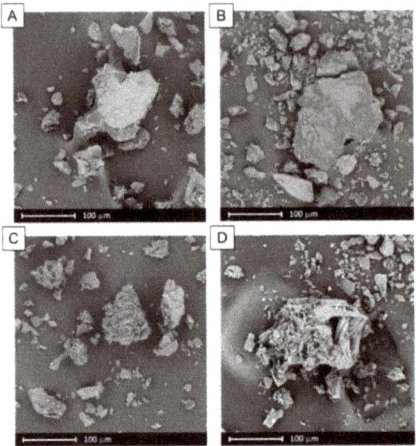

Figure 2. SEM images of binary systems containing bicalutamide and either PLX188 in 1:1 (**A**) and 2:1 (**B**) wt. ratio or PLX407 in 1:1 (**C**) and 2:1 (**D**) wt. ratio obtained using evaporation method.

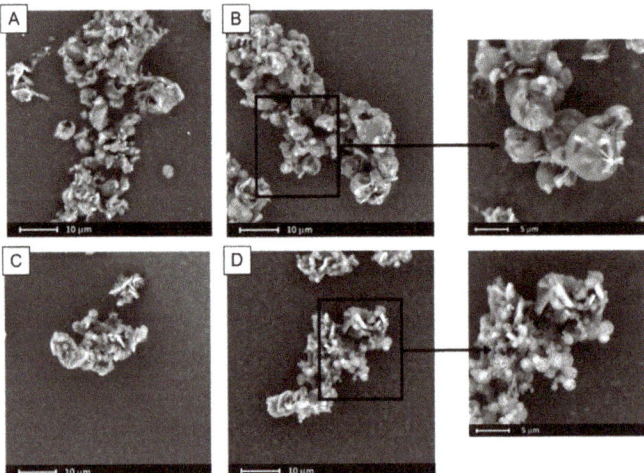

Figure 3. SEM images of ternary systems containing bicalutamide, PVP and either PLX188 in 4:1:1 (**A**) and 2:1:1 (**B**) wt. ratio or PLX407 in 4:1:1 (**C**) and 2:1:1 (**D**) wt. ratio obtained using spray drying.

3.1.2. X-Ray Powder Diffractometry (XRPD)

The XRPD studies were performed to characterize the molecular structure of binary systems. Obtained results indicate that both of applied processes led to the changes in molecular structure of the systems (Figure 4). The diffraction pattern of raw bicalutamide indicated by numerous distinctive Braggs peaks (2θ = 12.18°, 16.88°, 18.92°, 23.82°, 24.66°, and 24.94°) confirms that the drug exhibited highly-ordered arrangement on molecular level. The data confirm that bicalutamide existed as a form I polymorph (according to the 2014 Cambridge Crystallographic Data Centre (CCDC)). The decrease in crystallinity of the drug after co-processing with poloxamers is manifested by the reduction of the relative intensities of peaks. This suggests that the crystal lattice was partially destructed during processing. Moreover, the crystalline diffraction peaks are superimposed on the slightly noticeable amorphous halos. This indicates that the sample is amorphous to a very small extent. No transition to metastable polymorph was observed as no shifts in diffraction peaks appeared. This confirms that used poloxamers did not stabilize the disordered system at low concentration, in agreement with a previously published paper [59].

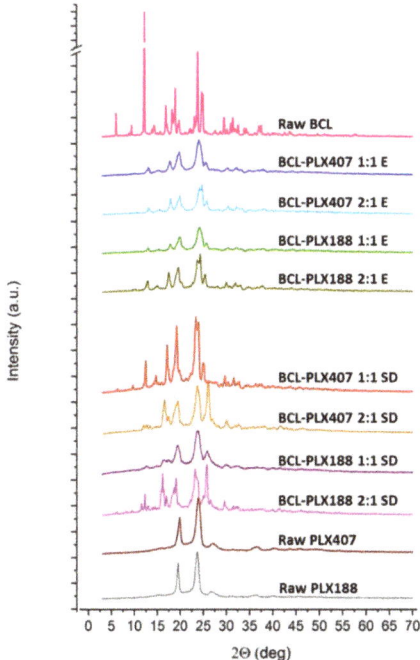

Figure 4. X-ray diffraction patterns of binary systems containing bicalutamide and either PLX 188 or PLX 407 (1:1 and 2:1 wt. ratio) obtained using evaporation technique (E) and spray drying (SD).

Interestingly, the diffraction patterns were more structured in spray-dried systems than evaporated ones. Moreover, the obtained ternary solid dispersions exhibited one more important feature, the transition of BCL from form I into form II polymorph [60]. This is clearly marked in the diffractograms presented in Figure 5 and manifested by the additional intense peak between 25.08° and 25.86° 2θ, which does not appear in the diffractogram of raw BCL. No such solid–solid transition of bicalutamide–poloxamer solid dispersions has been described so far. The diffraction patterns also suggest that ternary systems contain a fraction of amorphous phase as the diffractograms are superimposed on the amorphous halo further assigned to partial amorphization of poloxamers (see Section 3.1.4).

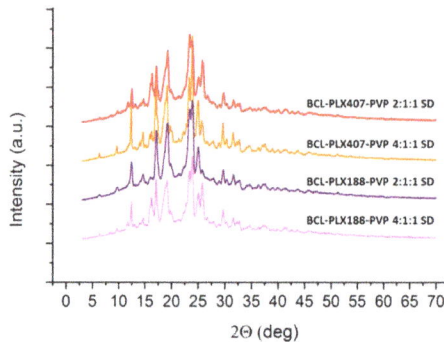

Figure 5. X-ray diffraction patterns of ternary systems containing bicalutamide, PVP, and either PLX 188 or PLX 407 obtained by spray drying.

3.1.3. Vibrational Spectroscopy

FTIR spectroscopy has been applied to determine the molecular structure and possible interactions between BCL and the carriers in solid dispersions. The intensity, shape and position of peaks (the presence of shifts) were evaluated with an emphasis placed on the vibrations within the carbonyl and amine functional groups (Figure 6). Well-resolved bands at 3335 cm^{-1} correspond to N–H stretching vibrations, and the broad band with a maximum at 1687 cm^{-1} originates in C=O stretching vibrations. The spectra of binary solid dispersions do not differ significantly from those of pure drug, suggesting that BCL does not interact with any of used poloxamers or that the strength of the interactions is negligibly small. The new band that appears in the range of 2860–3000 cm^{-1} corresponds to the stretching vibrations of aliphatic C–H group in poloxamers.

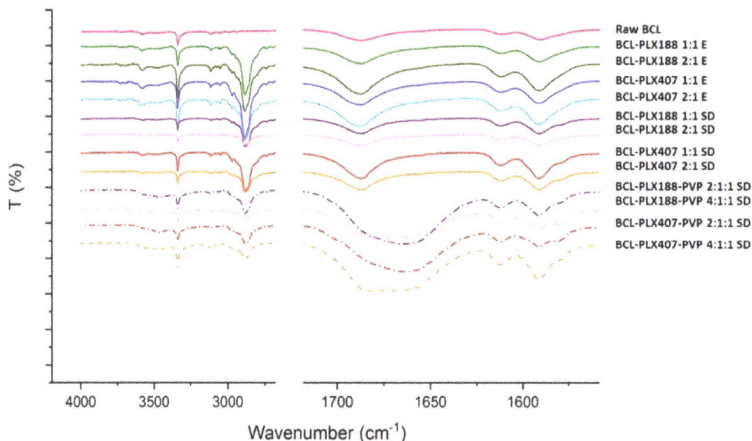

Figure 6. FTIR spectra of raw bicalutamide and binary and ternary solid dispersions.

Noticeable differences were observed for ternary solid dispersions as the band corresponding to carbonyl group vibration is broadened and the maximum red-shifted. This indicates the existence of strong intermolecular interactions between BCL and PVP as we previously showed [57]. Moreover, the O–H band is fuzzy, which confirms partial amorphization of the system.

3.1.4. Thermal Properties of Solid Dispersions

The thermal properties of raw systems (i.e., BCL, PLX188, PLX407, and PVP), the binary formulations containing BCL and either PLX407 or PLX188 polymer (that were obtained by two different methods—evaporation (E) and spray drying (SD)), and the ternary, spray-dried formulations of BCL, PLX, and PVP have been examined by means of the differential scanning calorimetry (DSC) technique. The DSC curves obtained during heating with a rate equal to 10 °C/min are presented in Figure 7.

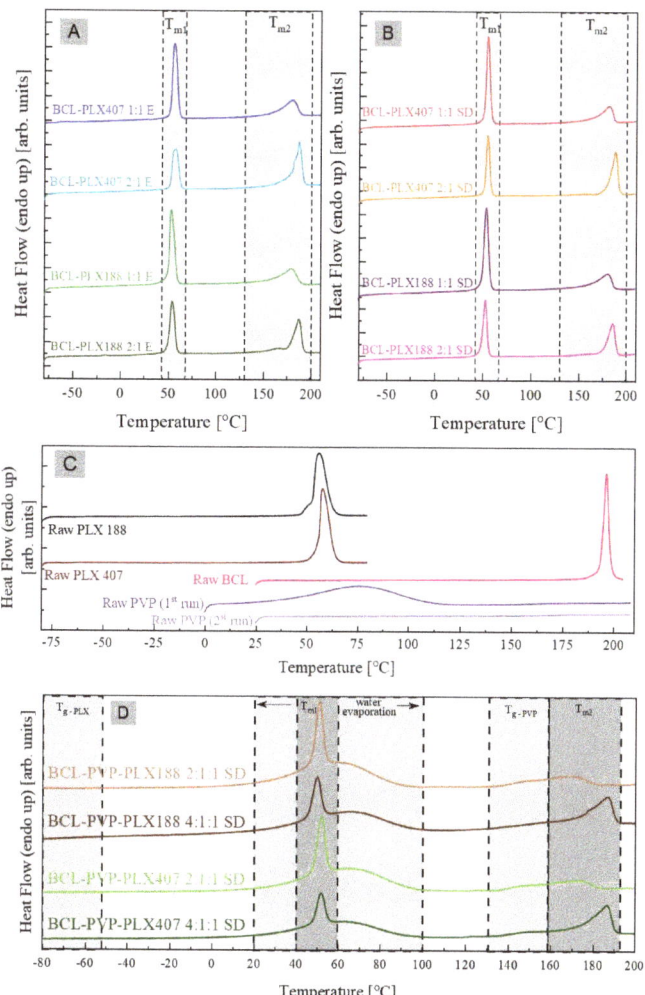

Figure 7. The DSC thermograms of: binary systems containing bicalutamide and either PLX188 or PLX407 obtained using the evaporation method (**A**) and spray drying (**B**), raw bicalutamide and polymers (**C**), and ternary spray-dried systems containing BCL, poloxamer, and PVP (**D**).

As can be seen in the panel (C) of Figure 7, the DSC trace of raw BCL reveals a single sharp peak with an onset at 194 °C. This endothermal process corresponds to the melting of the investigated antiandrogen and is in a perfect agreement with the literature data [61]. Both DSC curves of PLX188 as well as PLX 407 exhibit two thermal events. The first (barely visible on the DSC thermograms presented in Figure 7C) is step-like transition occurring in the vicinity of −60 °C associated with the glass transition of the amorphous part of PLXs (poly(propylene oxide), PEO blocks). The second, located at around 50 °C, is a sharp endothermal peak originating from the melting of the crystalline part of the polymers (poly(propylene oxide), PPO block). Two thermal events have been also observed in the DSC trace of the neat PVP polymer, when measured as received. The first, very broad, thermal event that is located in the range of 20–100 °C is associated with water evaporation (note the absence of this process, when the sample is re-heated). The second step-like transition (barely visible in Figure 7C) occurring in the vicinity of 172 °C is associated with the polymer glass transition.

In the panels (A) and (B) of Figure 7, the DSC traces of binary drug-polymer compositions prepared by evaporation (panel A), and spray drying (panel B) are shown. As can be seen all investigated formulations reveals three thermal events—T_g, T_{m1}, and T_{m2}—in the temperature range from −80 °C to 210 °C. Because the glass transition event (T_g) is almost invisible in the scale of Figure 7, the data from the temperature region: −75 °C to −40 °C are presented in a separate figure (see Figure 8). In Table 2 the values of all investigated thermal events of all examined systems have been collected.

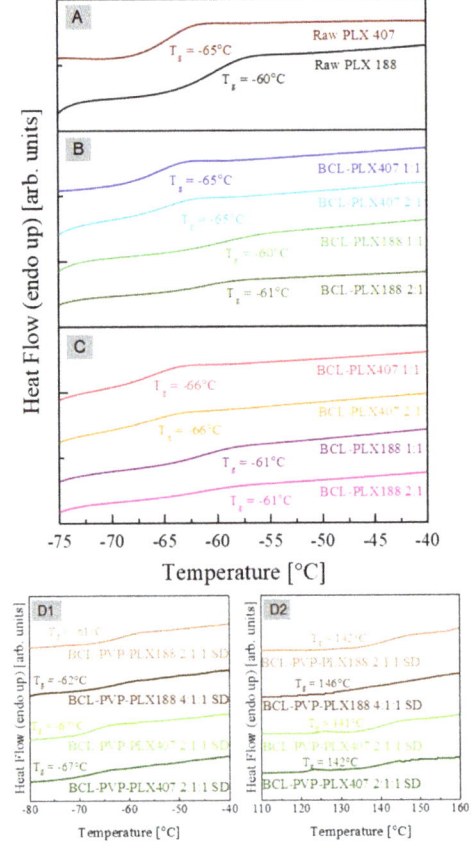

Figure 8. The zoomed fragment of DSC thermograms presented in the Figure 7 of raw bicalutamide and poloxamers (**A**), binary systems containing bicalutamide and either PLX188 or PLX407 obtained using evaporation method (**B**) and spray drying (**C**), and ternary, spray-dried, systems containing BCL, poloxamer, and PVP (**D1,D2**).

Table 2. Comparison of the T_g, T_{m1}, and T_{m2} values of raw BCL, poloxamers, PVP, binary systems containing bicalutamide and either PLX188 or PLX407 (obtained using evaporation method (E) and spray drying (SD)), and ternary systems containing bicalutamide, PLX, and PVP.

System	$T_{g\text{-PLX}}$ (°C) (midpoint)	$T_{g\text{-PVP}}$ (°C) (midpoint)	T_{m1} (°C) (onset)	ΔH_{m1} (J/g)	T_{m2} (°C) (onset)	ΔH_{m2} (J/g)
Raw BCL	-	-	-	-	194	110.8
Raw PVP	-	172	-	-	-	-
Raw PLX 188	−60	-	53	134.9	-	-
Raw PLX 407	−65	-	56	117.3	-	-
BCL-PLX 188 1:1 E	−60	-	50	65.7	160	39.6
BCL-PLX 188 2:1 E	−61	-	49	45.7	176	42.4
BCL-PLX 188 1:1 SD	−61	-	49	66.6	160	39.2
BCL-PLX 188 2:1 SD	−61	-	48	42.3	176	42.6
BCL-PLX 407 1:1 E	−65	-	52	59.6	160	39.5
BCL-PLX 407 2:1 E	−65	-	52	39.2	182	52.4
BCL-PLX 407 1:1 SD	−66	-	52	58.9	165	39.6
BCL-PLX 407 2:1 SD	−66	-	52	38.9	181	49.2
BCL-PLX-PVP 188 2:1:1 SD	−61	142	47	188	135	27
BCL-PLX-PVP 188 4:1:1 SD	−62	146	45	120	134	50
BCL-PLX-PVP 407 2:1:1 SD	−67	141	47	140	135	26
BCL-PLX-PVP 407 4:1:1 SD	−67	142	47	95	137	48

Since (1) the glass transitions of BCL-PLX 188 systems are located at the same temperature as the T_g of raw PLX 188, and (2) the glass transitions of the BCL-PLX 407 systems are located at the same temperature as the T_g of raw PLX 407, one can conclude that the glass transition event registered in binary formulations originates from the amorphous fraction of the PLXs (poly(ethylene oxide) PEO blocks). Comparing the values of the onsets of the thermal events which have been marketed in Figure 7A,B as T_{m1}, one can identify them as the melting of the crystalline part of the polymer which exists in the solid dispersions. The third thermal event that has been registered during the DSC measurements of the BCL-PLX systems is located in the temperature range from 130 °C to 200 °C. This endothermal peak corresponds to the melting of the BCL contained in the solid dispersions. Therefore, one can observed that its enthalpy (ΔH_{m2}) decreases with decreasing amounts of the BCL in the system. As can be seen, the onset of T_{m2} shifts towards lower temperatures with increasing amounts of PLX in the formulation. This might be connected with the dissolution of the drug in a liquid polymer.

In the panel (D) of Figure 7, the DSC traces of ternary drug–polymer–polymer compositions (prepared by spray drying) are shown. As can be seen, the investigated formulations reveal five thermal events which are marked in Figure 7D as T_g-PLX, T_g-PVP, T_{m1}, T_{m2}, and water evaporation. Since both T_g-PLX and T_g-PVP are almost invisible in the scale of Figure 7, the data from the temperature regions −80 °C to −40 °C and 110°C to 160°C are presented in the separate figures (see Figure 7 D1 and D2). From the comparison of the DSC traces of ternary systems to either raw and binary systems one can conclude that: (1) T_g-PLX originates from the amorphous fraction of the PLXs (PEO blocks); (2) T_g-PVP is associated with the glass transition temperature of PVP polymer; (3) T_{m1} reflects the melting of the crystalline part of the PLX polymer which exists in the system; and (4) T_{m2} corresponds to the melting of the BCL. Note that with increasing amount of API in the system, ΔH_{m1} and ΔH_{m2} are changing.

3.1.5. Wettability of Solid Dispersions

Powder wettability is an important issue in pharmaceutical sciences as the solid–liquid interfacial interactions can affect drug dissolution, solubilization, and disintegration [62]. Given the heterogeneity of the surface properties resulting from a specific surface chemistry, variations between polymorphic and amorphous forms have been reported thus far [63]. They affect the level of supersaturation of molecularly-disordered systems and physical stability; thus, the assessment of wetting properties plays a significant role in the systems containing fine particles.

The wetting behavior of raw compounds as well as binary and ternary solid dispersions were assessed by contact angle measurements using the sessile drop technique. The difference between the two used poloxamers is clearly visible (Figure 9). The values of measured contact angle were equal to 56.8 ± 1.8° and 64.7 ± 0.02° for PLX188 and PLX407, respectively. The difference may result from the differences in molecular composition of both polymers, i.e., higher amount of hydrophobic poly(propylene oxide) units and greater molar mass of PLX407 [49]. Interestingly, no significant effects of either the type of applied poloxamer or the process on the wettability of binary solid dispersions were observed. All the systems exhibited improved wettability expressed by the decreased contact angle in comparison with raw BCL (θ = 74.1 ± 0.3°) with slightly higher values determined for systems containing PLX407. Interestingly, the addition of PVP to ternary solid dispersions obtained by spray drying led to well pronounced increase in wetting behavior of the systems. While the values of contact angle for binary systems ranged between 60°–65°, for the systems comprising polyvinylpirrolidone they reached ca. 42°–45°. Moreover, the effect of molecular structure of poloxamers was less significant as lower values of the contact angle were obtained for PLX407-based systems. This is of particular importance as the improved wettability and surface activity of poloxamers can strongly affect the improvement in bicalutamide dissolution.

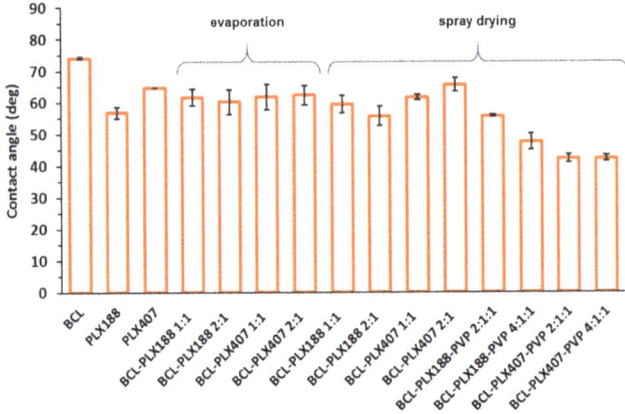

Figure 9. The values of contact angles of raw bicalutamide and poloxamers, binary and ternary systems containing bicalutamide, poloxamers, and PVP obtained using either evaporation method or spray drying.

3.2. Characterization of Solid Dispersions in Solution

3.2.1. Self-Assembly of Poloxamers in Solid Dispersions

The assembly phenomenon of amphiphilic polymers has been intensively studied in recent years [64–66]. Their aggregation leads to the formation of hydrophobic domains that are able to solubilize sparingly water-soluble molecules. This reduces the agglomeration of drug molecules and increases the dissolution of API.

Poloxamers are low-meltable triblock copolymers consisting of hydrophobic chain of poly(propylene oxide) bound with two hydrophilic chains of poly(ethylene oxide) able to solubilize hydrophobic molecules [67]. While PLX188 contains ca. 15% of PPO, PLX407 is composed of ca. 35% of PPO, which may affect the assembly of copolymer in polar media [68].

The size of molecular assemblies of both used poloxamers did not exceed 6 nm, however the diameter of PLX407-based particles is ca. 30% greater than those formed by PLX188, which may result from the higher content of PPO units. Physical mixtures also assembled in particles of diameter below

6 nm; however, the mixtures containing equal amounts of BCL and PLX formed particles of ca. 10–12% greater diameter than those containing the excess of the drug (Figure 10).

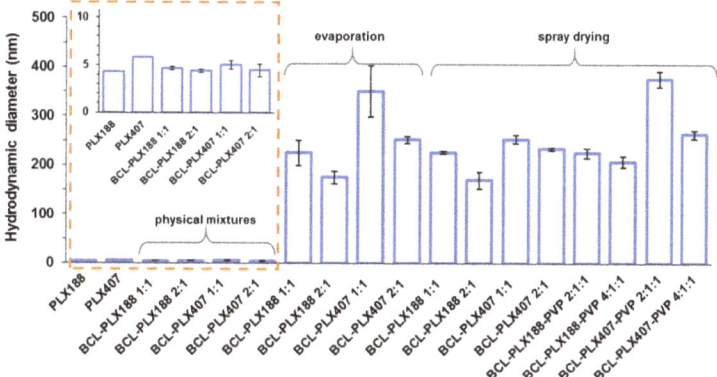

Figure 10. Hydrodynamic diameters of aggregates formed in aqueous solutions of poloxamers, physical mixtures, and binary systems obtained using either the evaporation method or spray drying. Insert: zoomed data corresponding to raw PLXs and physical mixtures.

The mean hydrodynamic diameters of all solid dispersions based on PLX188 are smaller than those containing PLX407, regardless of the method of preparation and the number of system constituents. Moreover, solid dispersions containing 50% of the carrier exhibited aggregates of greater diameter than those with the excess of bicalutamide, similarly to physical mixtures. However, the differences reached up to 39% for the BCL-PLX407 1:1 (E) system. No significant variations occurred between the PLX188-based binary systems of corresponding compositions obtained by the two methods. The systems containing the excess of the drug exhibited particles of ca. 170 nm in diameter, while the size of aggregates formed by 1:1 systems was equal to 225 nm. The great variation in particle size was noticed in BCL-PLX407 systems, especially the evaporated one containing an equal amount of the drug and the carrier. The diameter of these particles reached 350 nm in diameter, while the aggregates of spray-dried solid dispersion did not exceed 250 nm in diameter (Figure 10). Similar behavior was observed for BCL-PLX407-PVP 2:1:1 (SD) which exhibited particles much greater that the other ternary systems. The addition of PVP to solid dispersions did not affect the self-assembly behavior. The differences in hydrodynamic diameters values follows the same trend as for binary systems, with slightly greater values for solid dispersions containing equal amount of BCL and the carriers (PLX and PVP).

The DLS measurements confirmed monomodal and rather narrow distribution of particle size (Figure 11) with maxima slightly shifted towards greater values for PLX407-based systems, regardless of the method of solid dispersion preparation. The long tail of distribution of the BCL-PLX407 1:1 (E) binary system assigned to the formation of several aggregated structures that disrupted the measurement explains the variation in particle size in comparison with the other systems. The size of the formed aggregates was determined to be almost 40% greater than for BCL-PLX407 2:1 (E) system, similarly to BCL-PLX407-PVP 2:1:1 (SD) compared with BCL-PLX407-PVP 4:1:1 (SD) solid dispersions. This confirms that the effect of the applied method of solid dispersion preparation is negligible when considering that the solution and the composition are the most important factors.

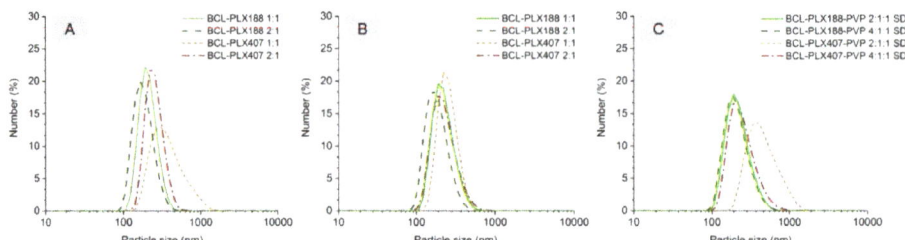

Figure 11. Number-weighted particle size distribution of aqueous solutions of evaporated (**A**), spray-dried binary (**B**), and ternary (**C**) solid dispersions.

3.2.2. Solubilization of Molecular Probes

Fluorescence spectroscopy has been applied to the determination of micropolarity, microviscosity and solubilization ability. Two molecular probes were used due to the act that their fluorescent properties vary depending on physical parameters of nanoassemblies.

Perylene exhibits unique properties, i.e., low solubility in water (c = 1.6×10^{-9} M) and lack of fluorescence in polar environment [69]. The presence of characteristic emission bands indicates that the probe experienced non-polar environment and confirms that the probe is solubilized within hydrophobic packets formed by self-assembled poloxamer molecules (Figure 12).

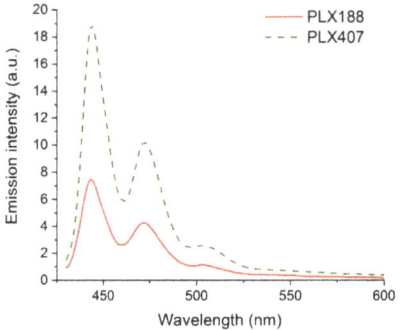

Figure 12. Emission spectra of perylene solubilized by either Poloxamer®188 or Poloxamer®407 solution (λ_{ex} = 405 nm).

The emission spectrum of pyrene yields in the information about the polarity sensed by the probe in the solubilization site. The intensity of the vibronic fine structure of the monomeric form of pyrene depends on the polarity of the microenvironment [70]. In polar media there is an increase in the intensity of the 0-0 band (peak I), whereas band III is affected only slightly [71]. The ratio of the emission intensities I_{III} (at 386 nm) and I_I (at 374 nm) was used to study environmental changes experienced by the probe. The values presented in Table 3 indicate that the probe experienced less polar environment while solubilized within any of examined system in comparison to water. However, the increase in the I_{III} to I_I ratio is rather low in solid dispersions (especially the spray-dried ones), which suggests that the systems formed loosely packed nanoasseblies easily penetrable by water molecules.

Table 3. The I_{III} to I_I ratio calculated based on the fluorescence emission spectra of pyrene solubilized in aqueous solutions of either pure compounds or solid dispersions (λ_{ex} = 330 nm).

System	I_{III}/I_I	System	I_{III}/I_I
Water	0.351	BCL-PLX 188 1:1 (SD)	0.442
Raw PLX188	0.516	BCL-PLX 188 2:1 (SD)	0.430
Raw PLX407	0.503	BCL-PLX 407 1:1 (SD)	0.435
		BCL-PLX 407 2:1 (SD)	0.442
BCL-PLX 188 1:1 (E)	0.508	BCL-PLX 188-PVP 2:1:1 (SD)	0.556
BCL-PLX 188 2:1 (E)	0.523	BCL-PLX 188-PVP 4:1:1 (SD)	0.628
BCL-PLX 407 1:1 (E)	0.441	BCL-PLX 407-PVP 2:1:1 (SD)	0.661
BCL-PLX 407 2:1 (E)	0.512	BCL-PLX 407-PVP 4:1:1 (SD)	0.670

3.2.3. Dissolution Study

The methods aimed at the enhancement of the dissolution of bicalutamide have been already considered in several papers. Solvent evaporation under reduced pressure was applied to obtain solid dispersions containing bicalutamide and PVP in 1:3, 1:4, and 1:5 drug-to-polymer ratios, respectively [59,72]. The formation of binary systems led to the amorphization of BCL; however, a great excess of PVP was required. The authors concluded that such a high proportion of the carrier may lead to the increase in bulkiness and tablet weight during the development of a formulation. Solid dispersions with poloxamer were obtained by melting [73] and supercritical carbon dioxide method [74]. The samples containing 83.8% of the carrier were found to be amorphous, however gelling properties of PLX retarded the dissolution of the drug from the systems containing high concentration of the polymer. The increase in the amount of poloxamer in solid dispersions was concluded not to offer any advantage for the dissolution improvement.

Due to the aforementioned problems caused by the excess of poloxamer, we prepared binary and ternary systems containing either equal amounts of bicalutamide and polymer or twice as much BCL as the carriers. The dissolution profiles presented in Figure 13 showed a significant improvement (from 4- to 8-fold) of the drug dissolution in comparison with raw BCL and BCL-PLX physical mixtures. Only 8.2% of crystalline bicalutamide dissolved after 1 h of the dissolution test. Moreover, the formation of the systems in which BCL was physically mixed with the readily soluble carrier affected the dissolution of the drug only slightly as less than 12.6% of bicalutamide dissolved from physical mixture containing PLX407 and ca. 8% from PLX188-based systems (Figure 13C).

The dissolution profiles of solid dispersions were found to be independent on the applied processes. The amount of bicalutamide dissolved from binary systems processed in 2:1 wt. ratio varied between 36.0% for BCL-PLX407 (SD) and 37.2% for BCL-PLX188 (E) to 44.6% for BCL-PLX188 (SD) and 46.3% for BCL-PLX188 (E). Solid dispersions containing equal amounts of the drug and the carrier exhibited better dissolution than those containing the excess of the drug, as 51.3% of bicalutamide dissolved from BCL-PLX188 (E), 53.3%% from BCL-PLX407 (E), and 54.8% from BCL-PLX407 (SD). The variation was observed only for BCL-PLX188 1:1 (SD) solid dispersion as 69.6% of the drug dissolved after 1 h. Interestingly, dissolution curves obtained for spray-dried systems with both binary and ternary solid dispersions (Figure 13B,D) showed an opposite tendency in comparison to evaporated systems (Figure 13A), as in evaporated systems more bicalutamide dissolved from PLX407-based solid dispersions, while after spray drying, systems containing PLX188 exhibited better dissolution. Importantly, the addition of PVP seems to positively affect BCL dissolution, as 77% of BCL dissolved from both systems containing PLX188, while 75.6% and 57.3% dissolved from the 2:1:1 and 4:1:1 PLX407-based systems, respectively.

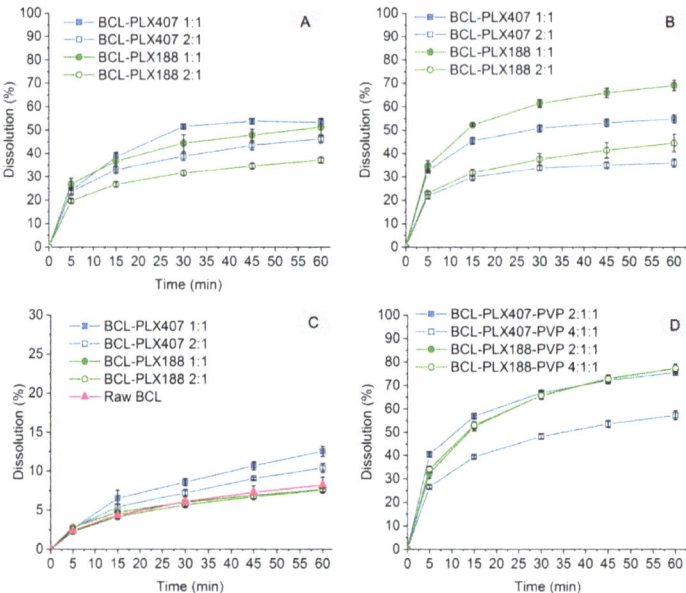

Figure 13. Dissolution of binary and ternary systems containing bicalutamide, poloxamers and PVP (in case of ternary systems) obtained using the evaporation technique (**A**), spray drying (**B,D**), and physical mixing (**C**).

4. Conclusions

The obtained results indicate that co-processing of BCL with PLXs leads to an improvement of bicalutamide dissolution from 4- to 8-times in comparison with the pure drug. That effect was assigned to the formation of nanoaggregates. Surface activity of poloxamers leads to the formation of hydrophobic packets in which bicalutamide was solubilized. Importantly, physical mixtures did not form aggregates with bicalutamide and thus no significant enhancement in drug dissolution was observed. While no variations in dissolution between systems obtained by either spray drying or evaporation processes were noted, some differences in physicochemical characteristics appeared. The most important observation is that the drug partially lost its highly-ordered molecular structure after preparation of solid dispersions. The changes in diffractograms were more pronounced in evaporated systems. The decrease in crystallinity was expressed by the decrease in relative intensity and lack of several peaks. Moreover, the addition of PVP and formation of ternary solid dispersions by spray drying led to the transition of polymorph I into polymorphic form II of bicalutamide. This confirms that the interplay between the process parameters and properties of both drug and carrier is important to obtain solid dispersion of desired characteristics without a great excess of the auxiliary compounds.

The type of polymer was found to affect the size of nanoaggregates formed by solid dispersions in an aqueous medium. The self-assembly of systems containing PLX188 led to the formation of smaller particles, regardless the applied technique of solid dispersion preparation. This may be a result of the composition of the macromolecule, as it contains ca. 15% PPO hydrophobic mers, while PLX407 contains ca. 35%.

Thermal analysis confirmed that poloxamers were partially amorphous in solid dispersions, which indicates that the drug antplasticizes the T_g of the polymer. This would be connected with the dissolution of the drug in a liquid polymer.

Author Contributions: Conceptualization, J.S.; investigation, J.S., A.A, J.K.-K., K.G., M.K., J.O., and K.C.; writing—original draft preparation, J.S. and J.K.-K.; writing—review and editing, R.J. and M.P.; visualization, J.S. and J.K.-K.; supervision, R.J. and M.P.

Funding: This research was funded by the Polish National Science Centre (grant Symfonia 3 no 2015/16/W/NZ7/00404).

Conflicts of Interest: The authors declare no conflict of interest.

References

1. Williams, H.D.; Trevaskis, N.L.; Charman, S.A.; Shanker, R.M.; Charman, W.N.; Pouton, C.W.; Porter, C.J.H. Strategies to Address Low Drug Solubility in Discovery and Development. *Pharmacol. Rev.* **2013**, *65*, 315–499. [CrossRef] [PubMed]
2. Di, L.; Kerns, E.H.; Carter, G.T. Drug-Like Property Concepts in Pharmaceutical Design. *Curr. Pharm. Des.* **2009**, *15*, 2184–2194. [CrossRef] [PubMed]
3. Kawabata, Y.; Wada, K.; Nakatani, M.; Yamada, S.; Onoue, S. Formulation design for poorly water-soluble drugs based on biopharmaceutics classification system: Basic approaches and practical applications. *Int. J. Pharm.* **2011**, *420*, 1–10. [CrossRef] [PubMed]
4. Krupa, A.; Descamps, M.; Willart, J.F.; Jachowicz, R.; Danède, F. High energy ball milling and supercritical carbon dioxide impregnation as co-processing methods to improve dissolution of tadalafil. *Eur. J. Pharm. Sci.* **2016**, *95*, 130–137. [CrossRef] [PubMed]
5. Rios-Doria, J.; Carie, A.; Costich, T.; Burke, B.; Skaff, H.; Panicucci, R.; Sill, K. A Versatile Polymer Micelle Drug Delivery System for Encapsulation and In Vivo Stabilization of Hydrophobic Anticancer Drugs. *J. Drug Deliv.* **2012**, 951741. [CrossRef]
6. Bader, R.A.; Silvers, A.L.; Zhang, N. Polysialic Acid-Based Micelles for Encapsulation of Hydrophobic Drugs. *Biomacromolecules* **2011**, *12*, 314–320. [CrossRef]
7. Luo, R.; Venkatraman, S.S.; Neu, B. Layer-by-Layer Polyelectrolyte–Polyester Hybrid Microcapsules for Encapsulation and Delivery of Hydrophobic Drugs. *Biomacromolecules* **2013**, *14*, 2262–2271. [CrossRef]
8. Szafraniec, J.; Błażejczyk, A.; Kuś, E.; Janik, M.; Zając, G.; Wietrzyk, J.; Chłopicki, S.; Zapotoczny, S. Robust oil-core nanocapsules with hyaluronate-based shells as promising nanovehicles for lipophilic compounds. *Nanoscale* **2017**, *47*. [CrossRef]
9. Han, Y.; Shchukin, D.; Yang, J.; Simon, C.R.; Fuchs, H.; Möhwald, H. Biocompatible Protein Nanocontainers for Controlled Drugs Release. *ACS Nano* **2010**, *4*, 2838–2844. [CrossRef]
10. Juère, E.; Florek, J.; Bouchoucha, M.; Jambhrunkar, S.; Wong, K.Y.; Popat, A.; Kleitz, F. In Vitro Dissolution, Cellular Membrane Permeability, and Anti-Inflammatory Response of Resveratrol-Encapsulated Mesoporous Silica Nanoparticles. *Mol. Pharm.* **2017**, *14*, 4431–4441. [CrossRef]
11. Meka, A.K.; Jenkins, L.J.; Dàvalos-Salas, M.; Pujara, N.; Wong, K.Y.; Kumeria, T.; Mariadason, J.M.; Popat, A. Enhanced Solubility, Permeability and Anticancer Activity of Vorinostat Using Tailored Mesoporous Silica Nanoparticles. *Pharmaceutics* **2018**, *10*, 283. [CrossRef] [PubMed]
12. Mazumder, S.; Dewangan, A.K.; Pavurala, N. Enhanced dissolution of poorly soluble antiviral drugs from nanoparticles of cellulose acetate based solid dispersion matrices. *Asian J. Pharm. Sci.* **2017**, *12*, 532–541. [CrossRef]
13. Kim, M.-S.; Baek, I. Fabrication and evaluation of valsartan-polymer-surfactant composite nanoparticles by using the supercritical antisolvent process. *Int. J. Nanomed.* **2014**, *9*, 5167–5176. [CrossRef] [PubMed]
14. Tran, T.H.; Poudel, B.K.; Marasini, N.; Chi, S.-C.; Choi, H.-G.; Yong, C.S.; Kim, J.O. Preparation and evaluation of raloxifene-loaded solid dispersion nanoparticle by spray-drying technique without an organic solvent. *Int. J. Pharm.* **2013**, *443*, 50–57. [CrossRef] [PubMed]
15. Van den Moter, G. The use of amorphous solid dispersions: A formulation strategy to overcome poor solubility and dissolution rate. *Drug Discov. Today* **2012**, *9*, e79–e85. [CrossRef] [PubMed]
16. Le-Ngoc Vo, C.; Park, C.; Lee, B.-J. Current trends and future perspectives of solid dispersions containing poorly water-soluble drugs. *Eur. J. Pharm. Biopharm.* **2013**, *85*, 799–813. [CrossRef]
17. Maniruzzaman, M.; Pang, J.; Morgan, D.J.; Douroumis, D. Molecular modeling as a predictive tool for the development of solid dispersions. *Mol. Pharm.* **2015**, *12*, 1040–1049. [CrossRef] [PubMed]

18. Vasconcelos, T.; Marques, S.; das Neves, J.; Sarmento, B. Amorphous solid dispersions: Rational selection of a manufacturing process. *Adv. Drug Deliv. Rev.* **2016**, *100*, 85–101. [CrossRef] [PubMed]
19. Sekiguchi, K.; Obi, N. Studies on absorption of eutectic mixture. I. A comparison of the behavior of eutectic mixture of sulfathiazole and that of ordinary sulfathiazole in man. *Chem. Pharm. Bull.* **1961**, *9*, 866–872. [CrossRef]
20. Sekiguchi, K.; Obi, N.; Ueda, Y. Studies on absorption of eutectic mixture. II. Absorption of fused conglomerates of chloramphenicol and urea in rabbits. *Chem. Pharm. Bull.* **1961**, *12*, 134–144. [CrossRef]
21. Goldberg, A.H.; Gibaldi, M.; Kanig, J.L. Increasing dissolution rates and gastrointestinal absorption of drugs via solid solutions and eutectic mixtures. I. Theoretical considerations and discussion of the literature. *J. Pharm. Sci.* **1965**, *54*, 1145–1148. [CrossRef] [PubMed]
22. Chiou, W.L.; Riegelman, S. Pharmaceutical applications of solid dispersion systems. *J. Pharm. Sci.* **1971**, *60*, 1281–1302. [CrossRef] [PubMed]
23. Sharma, A.; Jain, C.P. Preparation and characterization of solid dispersions of carvedilol with PVP K30. *Res. Pharm. Sci.* **2010**, *5*, 49–56. [CrossRef] [PubMed]
24. Gupta, P.; Kakumanu, V.K.; Bansal, A.K. Stability and solubility of celecoxib-PVP amorphous dispersions: A molecular perspective. *Pharm. Res.* **2004**, *21*, 1762–1769. [CrossRef] [PubMed]
25. Lim, H.-T.; Balakrishnan, P.; Oh, D.H.; Joe, K.H.; Kim, Y.R.; Hwang, D.H.; Lee, Y.-B.; Yong, C.S.; Choi, H.-G. Development of novel sibutramine base-loaded solid dispersion with gelatin and HPMC: Physicochemical characterization and pharmacokinetics in beagle dogs. *Int. J. Pharm.* **2010**, *397*, 225–230. [CrossRef]
26. Alonzo, D.E.; Gao, Y.; Zhou, D.; Mo, H.; Zhang, G.G.Z.; Taylor, L.S. Dissolution and precipitation behavior of amorphous solid dispersions. *J. Pharm. Sci.* **2011**, *100*, 3316–3331. [CrossRef] [PubMed]
27. Frank, K.J.; Rosenblatt, K.M.; Westedt, U.; Holig, P.; Rosenberg, J.; Magerlein, M.; Fricker, G.; Brandl, M. Amorphous solid dispersion enhances permeation of poorly soluble ABT-102: True supersaturation vs. apparent solubility enhancement. *Int. J. Pharm.* **2012**, *437*, 288–293. [CrossRef] [PubMed]
28. Sun, D.D.; Lee, P.I. Evolution of supersaturation of amorphous pharmaceuticals: The effect of rate of supersaturation generation. *Mol. Pharm.* **2013**, *10*, 4330–4346. [CrossRef]
29. Chauhan, H.; Hui-Gu, C.; Atef, E. Correlating the behavior of polymers in solution as precipitation inhibitor to its amorphous stabilization ability in solid dispersions. *J. Pharm. Sci.* **2013**, *102*, 1924–1935. [CrossRef]
30. Passerini, N.; Albertini, B.; González-Rodríguez, M.L.; Cavallari, C.; Rodriguez, L. Preparation and characterisation of ibuprofen–poloxamer 188 granules obtained by melt granulation. *Eur. J. Pharm. Sci.* **2002**, *15*, 71–78. [CrossRef]
31. Eloy, J.O.; Marchetti, J.M. Solid dispersions containing ursolic acid in Poloxamer 407 and PEG 6000: A comparative study of fusion and solvent methods. *Powder Technol.* **2014**, *253*, 98–106. [CrossRef]
32. Joshi, H.N.; Tejwani, R.W.; Davidovich, M.; Sahasrabudhe, V.P.; Jemal, M.; Bathala, M.S.; Varia, S.A.; Serajuddin, A. Bioavailability enhancement of a poorly water-soluble drug by solid dispersion in polyethylene glycol–polysorbate 80 mixture. *Int. J. Pharm.* **2004**, *269*, 251–258. [CrossRef] [PubMed]
33. Moes, J.; Koolen, S.; Huitema, A.; Schellens, J.; Beijnen, J.; Nuijen, B. Pharmaceutical development and preliminary clinical testing of an oral solid dispersion formulation of docetaxel (ModraDoc001). *Int. J. Pharm.* **2011**, *420*, 244–250. [CrossRef] [PubMed]
34. Ghebremeskel, A.N.; Vemavarapu, C.; Lodaya, M. Use of surfactants as plasticizers in preparing solid dispersions of poorly soluble API: Stability testing of selected solid dispersions. *Pharm. Res.* **2006**, *23*, 1928–1936. [CrossRef] [PubMed]
35. Chaves, L.L.; Vieira, A.C.C.; Reis, S.; Sarmento, B.; Ferreira, D.C. Quality by Design: Discussing and Assessing the Solid Dispersions Risk. *Curr. Drug Deliv.* **2014**, *11*, 253–269. [CrossRef] [PubMed]
36. Kumar, N.; Jain, A.K.; Singh, C.; Kumar, R. Development, characterization and solubility study of solid dispersion of terbinafine hydrochloride by solvent evaporation method. *Asian J. Pharm.* **2008**, *2*, 154–158. [CrossRef]
37. Tabbakhian, M.; Hasanzadeh, F.; Tavakoli, N.; Jamshidian, Z. Dissolution enhancement of glibenclamide by solid dispersion: Solvent evaporation versus a supercritical fluid-based solvent-antisolvent technique. *Res. Pharm. Sci.* **2014**, *9*, 337–350. [PubMed]
38. Przybyłek, M.; Ziółkowska, D.; Mroczyńska, K.; Cysewski, P. Propensity of salicylamide and ethenzamide cocrystallization with aromatic carboxylic acids. *Eur. J. Pharm. Sci.* **2016**, *31*, 132–140. [CrossRef]

39. Przybyłek, M.; Cysewski, P.; Pawelec, M.; Ziółkowska, D.; Kobierski, M. On the origin of surface imposed anisotropic growth of salicylic and acetylsalicylic acids crystals during droplet evaporation. *J. Mol. Model.* **2015**, *21*, 49. [CrossRef]
40. Al-Ali, M.; Periasamy, S.; Parthasarathy, R. Novel drying of formulated naproxen sodium using microwave radiation: Characterization and energy comparison. *Powder Technol.* **2018**, *334*, 143–150. [CrossRef]
41. Lee, H.; Lee, J. Dissolution enhancement of celecoxib via polymer-induced crystallization. *J. Cryst. Growth* **2013**, *374*, 37–42. [CrossRef]
42. Bajaj, A.; Rao, M.R.P.; Pardeshi, A.; Sali, D. Nanocrystallization by Evaporative Antisolvent Technique for Solubility and Bioavailability Enhancement of Telmisartan. *AAPS Pharm. Sci. Technol.* **2012**, *13*, 1331–1340. [CrossRef] [PubMed]
43. Jiao, Z.; Zha, X.; Wang, Z.; Wang, X.; Fan, W. Response Surface Modeling of Drug-Loaded Micelles Prepared Through Supercritical Carbon Dioxide Evaporation Method Using Box-Behnken Experimental Design. *J. Nanosci. Nanotechnol.* **2019**, *19*, 3616–3620. [CrossRef] [PubMed]
44. Sinha, S.; Ali, M.; Baboota, S.; Ahuja, A.; Kumar, A.; Ali, J. Solid Dispersion as an Approach for Bioavailability Enhancement of Poorly Water-Soluble Drug Ritonavir. *AAPS Pharm. Sci. Technol.* **2010**, *11*, 518–527. [CrossRef] [PubMed]
45. Vehring, R. Pharmaceutical particle engineering via spray drying. *Pharm. Res.* **2008**, *25*, 999–1022. [CrossRef]
46. Ansari, M.T.; Karim, S.; Ranjha, N.M.; Shah, N.H.; Muhammad, S. Physicochemical characterization of artemether solid dispersions with hydrophilic carriers by freeze dried and melt methods. *Arch. Pharm. Res.* **2010**, *33*, 901–910. [CrossRef] [PubMed]
47. Singh, A.; Van den Mooter, G. Spray drying formulation of amorphous solid dispersions. *Adv. Drug Del. Rev.* **2016**, *100*, 27–50. [CrossRef] [PubMed]
48. Paudel, A.; Worku, A.; Meeus, J.; Guns, S.; Mooter, G. Manufacturing of solid dispersions of poorly water soluble drugs by spray drying: Formulation and process considerations. *Int. J. Pharm.* **2013**, *453*, 253–284. [CrossRef]
49. Kolašinac, N.; Kachrimanis, K.; Homšek, I.; Grujić, B.; Đurić, Z.; Ibrić, S. Solubility enhancement of desloratadine by solid dispersion in poloxamers. *Int. J. Pharm.* **2012**, *436*, 161–170. [CrossRef]
50. Shah, T.J.; Amin, A.F.; Parikh, J.R.; Parikh, R.H. Process Optimization and Characterization of Poloxamer Solid Dispersions of a Poorly Water-soluble Drug. *AAPS Pharm. Sci. Technol.* **2007**, *8*, E18–E24. [CrossRef]
51. Masiello, D.; Cheng, S.; Bubley, G.J.; Lu, M.L.; Balk, S.P. Bicalutamide Functions as an Androgen Receptor Antagonist by Assembly of a Transcriptionally Inactive Receptor. *J. Biol. Chem.* **2002**, *227*, 26321–26326. [CrossRef] [PubMed]
52. Kumbhar, D.D.; Pokharkar, V.B. Engineering of nanostructured lipid carrier for the poorly water-soluble drug, bicalutamide: Physicochemical investigations. *Colloids Surf. A* **2013**, *416*, 32–42. [CrossRef]
53. Le, Y.; Chen, J.-F.; Shen, Z.; Yun, J.; Pu, M. Nanosized bicalutamide and its molecular structure in solvents. *Int. J. Pharm.* **2009**, *370*, 175–180. [CrossRef] [PubMed]
54. Vega, D.R.; Polla, G.; Martinez, A.; Mendioroz, E.; Reinoso, M. Conformational polymorphism in bicalutamide. *Int. J. Pharm.* **2007**, *328*, 112–118. [CrossRef] [PubMed]
55. Hu, X.R.; Gu, J.M. N-[4-Cyano-3-trifluoromethyl)phenyl]-3-(4-fluorophenylsulfonyl)-2-hydroxy-2-methylpropionamide. *Acta Cryst. E61* **2005**, 3897–3898. [CrossRef]
56. Perlovich, G.L.; Blokhina, S.V.; Manin, N.G.; Volkova, T.V.; Tkachev, V.V. Polymorphism and solvatomorphism of bicalutamide. Thermophysical study and solubility. *J. Therm. Anal. Calorim.* **2013**, *11*, 655–662. [CrossRef]
57. Szafraniec, J.; Antosik, A.; Knapik-Kowalczuk, J.; Kurek, M.; Syrek, K.; Chmiel, K.; Paluch, M.; Jachowicz, R. Planetary ball milling and supercritical fluid technology as a way to enhance dissolution of bicalutamide. *Int. J. Pharm.* **2017**, *533*, 470–479. [CrossRef] [PubMed]
58. Szafraniec, J.; Antosik, A.; Knapik-Kowalczuk, J.; Chmiel, K.; Kurek, M.; Gawlak, K.; Paluch, M.; Jachowicz, R. Enhanced dissolution of solid dispersions containing bicalutamide subjected to mechanical stress. *Int. J. Pharm.* **2018**, *542*, 18–26. [CrossRef] [PubMed]
59. Srikanth, M.V.; Murali Mohan Babu, G.V.; Sunil, S.A.; Sreenivasa Rao, N.; Ramana Murthy, K.V. In-vitro dissolution rate enhancement of poorly water soluble non-steroidal antiandrogen agent, bicalutamide, with hydrophilic carrier. *J. Sci. Ind. Res.* **2010**, *69*, 629–634.

60. Szafraniec, J.; Antosik, A.; Knapik-Kowalczuk, J.; Gawlak, K.; Kurek, M.; Szlęk, J.; Jamróz, W.; Paluch, M.; Jachowicz, R. Molecular Disorder of Bicalutamide—Amorphous Solid Dispersions Obtained by Solvent Methods. *Pharmaceutics* **2018**, *10*, 194. [CrossRef]
61. Szczurek, J.; Rams-Baron, M.; Knapik-Kowalczuk, J.; Antosik, A.; Szafraniec, J.; Jamróz, W.; Dulski, M.; Jachowicz, R.; Paluch, M. Molecular Dynamics, Recrystallization Behavior, and Water Solubility of the Amorphous Anticancer Agent Bicalutamide and Its Polyvinylpyrrolidone Mixtures. *Mol. Pharm.* **2017**, *14*, 1071–1081. [CrossRef] [PubMed]
62. Puri, K.; Dantuluri, A.K.; Kumar, M.; Karar, N.; Bansal, A.K. Wettability and surface chemistry of crystalline and amorphous forms of a poorly water soluble drug. *Eur. J. Pharm. Sci.* **2010**, *40*, 84–93. [CrossRef] [PubMed]
63. Bérard, V.; Lesniewska, E.; Andrès, C.; Pertuy, D.; Laroche, C.; Pourcelot, Y. Affinity scale between a carrier and a drug in DPI studied by atomic force microscopy. *Int. J. Pharm.* **2002**, *247*, 127–137. [CrossRef]
64. Yusa, S.; Sakakibara, A.; Yamamoto, T.; Morishima, M. Reversible pH-Induced Formation and Disruption of Unimolecular Micelles of an Amphiphilic Polyelectrolyte. *Macromolecules* **2002**, *35*, 5243–5249. [CrossRef]
65. Rymarczyk-Machal, M.; Szafraniec, J.; Zapotoczny, S.; Nowakowska, M. Photoactive graft amphiphilic polyelectrolyte: Facile synthesis, intramolecular aggregation and photosensitizing activity. *Eur. Polym. J.* **2014**, *55*, 76–85. [CrossRef]
66. McKenzie, B.E.; de Visser, J.F.; Friedrich, H.; Wirix, M.J.M.; Bomans, P.H.H.; de With, G.; Holder, S.J.; Sommerdijk, N.A.J.M. Bicontinuous Nanospheres from Simple Amorphous Amphiphilic Diblock Copolymers. *Macromolecules* **2013**, *46*, 9845–9848. [CrossRef]
67. Dumortier, G.; Grossiord, J.L.; Agnely, F.; Chaumeil, J.C. A review of poloxamer 407 pharmaceutical and pharmacological characteristics. *Pharm. Res.* **2006**, *23*, 2709–2728. [CrossRef]
68. Rowe, R.C.; Sheskey, P.J.; Quinn, M.E. (Eds.) *Handbook of Pharmaceutical Excipients*, 6th ed.; Pharmaceuicall Press: London, UK, 2009; pp. 506–509.
69. Miller, M.M.; Wasik, S.P.; Huang, G.L.; Siu, W.Y.; Mackay, D. Relationships between octanol-water partition coefficient and aqueous solubility. *Environ. Sci. Technol.* **1985**, *19*, 522–529. [CrossRef]
70. Capek, I. Fate of excited probes in micellar systems. *Adv. Colloid Interface Sci.* **2002**, *97*, 91–149. [CrossRef]
71. Barry, N.P.E.; Therrien, B. Organic Nanoreactors. Pyrene: The Guest of Honor. In *From Molecular to Supramolecular Organic Compound*; Sadjadi, S., Ed.; Elsevier: Amsterdam, The Netherlands, 2016; pp. 421–461.
72. Ren, F.; Jing, Q.; Tang, Y.; Shen, Y.; Chen, J.; Gao, F.; Cui, J. Characteristics of bicalutamide solid dispersions and improvement of the dissolution. *Drug Dev. Ind. Pharm.* **2006**, *32*, 967–972. [CrossRef]
73. Sancheti, P.P.; Vyas, V.M.; Shah, M.; Karekar, P.; Pore, Y.V. Development and characterization of bicalutamide-poloxamer F68 solid dispersion systems. *Pharmazie* **2008**, *63*, 571–575. [CrossRef] [PubMed]
74. Antosik, A.; Witkowski, S.; Woyna-Orlewicz, K.; Talik, P.; Szafraniec, J.; Wawrzuta, B.; Jachowicz, R. Application of supercritical carbon dioxide to enhance dissolution rate of bicalutamide. *Acta Pol. Pharm.* **2017**, *74*, 1231–1238.

© 2019 by the authors. Licensee MDPI, Basel, Switzerland. This article is an open access article distributed under the terms and conditions of the Creative Commons Attribution (CC BY) license (http://creativecommons.org/licenses/by/4.0/).

Article

Fluoroquinolone Amorphous Polymeric Salts and Dispersions for Veterinary Uses

Hanah Mesallati [1,2], Anita Umerska [1,3] and Lidia Tajber [1,2,*]

1. School of Pharmacy and Pharmaceutical Sciences, Trinity College Dublin, College Green, 2 Dublin, Ireland; mesallah@tcd.ie (H.M.); umerskam@tcd.ie (A.U.)
2. SSPC, Synthesis and Solid State Pharmaceutical Centre, Limerick, Ireland
3. MINT, UNIV Angers, INSERM 1066, CNRS 6021, Universite Bretagne Loire, 4 rue Larrey, CEDEX, 49933 Angers, France
* Correspondence: ltajber@tcd.ie; Tel.: +353-1-896-2787

Received: 18 May 2019; Accepted: 6 June 2019; Published: 9 June 2019

Abstract: Enrofloxacin (ENRO) is a poorly soluble drug used in veterinary medicine. It differs from the more widely used fluoroquinolone ciprofloxacin (CIP) by the presence of an ethyl substituent on its piperazine amino group. While a number of recent studies have examined amorphous composite formulations of CIP, little research has been conducted with ENRO in this area. Therefore, the main purpose of this work was to produce amorphous solid dispersions (ASDs) of ENRO. The solid-state properties of these samples were investigated and compared to those of the equivalent CIP ASDs, and their water uptake behavior, solubility, dissolution, and antibacterial activity were assessed. Like CIP, X-ray amorphous solid dispersions were obtained when ENRO was ball milled with acidic polymers, whereas the use of neutral polymers resulted in semi-crystalline products. Proton transfer from the carboxylic acids of the polymers to the tertiary amine of ENRO's piperazine group appears to occur in the ASDs, resulting in an ionic bond between the two components. Therefore, these ASDs can be referred to as amorphous polymeric salts (APSs). The glass transition temperatures of the APSs were significantly higher than that of ENRO, and they were also resistant to crystallization when exposed to high humidity levels. Greater concentrations were achieved with the APSs than the pure drug during solubility and dissolution studies, and this enhancement was sustained for the duration of the experiments. In addition, the antimicrobial activity of ENRO was not affected by APS formation, while the minimum inhibitory concentrations and minimum bactericidal concentrations obtained with the APS containing hydroxypropyl methylcellulose acetate succinate grade MG (HPMCAS-MG) were significantly lower than those of the pure drug. Therefore, APS formation is one method of improving the pharmaceutical properties of this drug.

Keywords: enrofloxacin; ciprofloxacin; amorphous solid dispersion; amorphous polymeric salt; polymer; ball milling; solubility; dissolution

1. Introduction

Enrofloxacin (ENRO), or 1-cyclopropyl-7-(4-ethylpiperazin-1-yl)-6-fluoro-4-oxo-1,4-dihydroquinoline-3-carboxylic acid, is a fluoroquinolone antibiotic that is licensed for veterinary use. ENRO differs from the more widely known fluoroquinolone ciprofloxacin (CIP) by the presence of an ethyl substituent in the N3 position (Figure 1). Anhydrous CIP generally exists in the zwitterionic state, with a protonated amino group and negatively charged carboxylate group. These oppositely charged groups form head-to-tail ionic bonds with neighboring and adjacent molecules, resulting in a tetramer-like structure [1]. ENRO, on the other hand, is unionized in the solid state and can therefore only form a number of weak C–H•••O and C–H•••N hydrogen bonds [2].

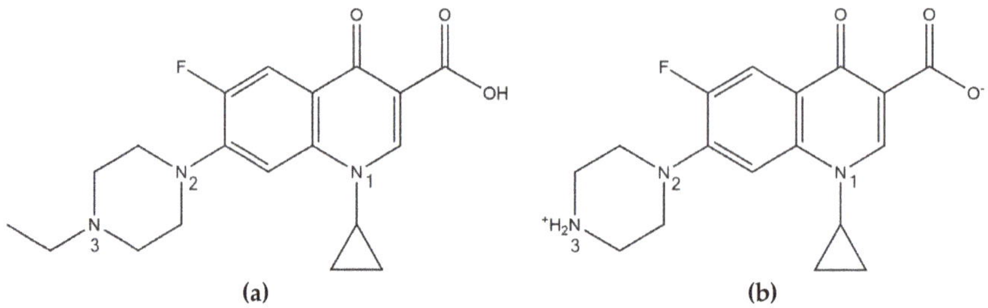

Figure 1. Chemical structure of (**a**) enrofloxacin and (**b**) ciprofloxacin.

CIP and ENRO are both poorly water-soluble drugs, with an intrinsic solubility of approximately 0.1 mg/mL and 0.4 mg/mL, respectively [3]. Both drugs are least soluble at pH 7.4 and exist predominantly in the zwitterionic form in neutral solutions [4]. Despite the theoretically higher hydrophilicity of CIP due to the absence of an aliphatic group in the N3 position, the strong crystal lattice of this drug reduces its aqueous solubility below that of ENRO. In addition, the extra ethyl group of ENRO increases its lipophilicity and permeability, resulting in greater absorption in rat in situ permeability studies than CIP [3]. However, the permeability of ENRO still falls within the limits of poorly permeable [5].

One of the most commonly used techniques to improve the solubility of ionizable drugs is salt formation. A number of crystalline ENRO salts have been produced by Karanam et al. and all were found to be significantly more water-soluble than the pure drug [2]. The piperazine N3 nitrogen of ENRO is positively charged in the salts containing acidic counterions, such as succinic acid, fumaric acid, and maleic acid, and forms an ionic bond with the carboxylate groups of the acids. The carboxylic acid of the drug, on the other hand, remains unionized. By contrast, ENRO exists in the anionic state in the ammonium salt, with a negatively charged carboxylate group and neutral piperazine group [2]. The solubility of ENRO was also increased significantly via formulation as the saccharinate salt. Like the equivalent CIP salt, an ionic interaction between the negatively charged saccharin molecule and positively charged N3 amino group of ENRO was detected in this compound [6].

The solubility of a drug may also be increased by converting it to the amorphous form. This involves the disruption of the crystal lattice, producing a disordered, high-energy version of the drug [7]. This approach is usually avoided during commercial drug development due to the intrinsic instability of amorphous solids. However, suitable excipients can be used to stabilize the amorphous form and prevent its crystallization. This stabilization is usually brought about via interactions between the components, such as hydrogen or ionic bonds, and/or through steric hindrance, e.g., by polymers with long chains [8]. Recently, the formation of various amorphous solid dispersions (ASDs) and amorphous salts of CIP was investigated by our group. Due to the poor solubility and thermal degradation of the drug, these were mainly prepared by ball milling [1,9]. Promising results were obtained with a number of CIP ASDs containing various acidic polymers. They were found to increase the glass transition temperature (T_g) and solubility of CIP, while the permeability and antibacterial activity of the drug was either unchanged or moderately improved [9]. As an ionic interaction between the drug and polymer was identified in each as these ASDs, they may also be referred to as amorphous polymeric salts (APSs) [9]. The majority of ASDs described in the literature are stabilized by nonionic interactions between the components, such as hydrogen bonds, and do not involve proton transfer between the drug and polymer. The apparent solubility of APSs may be even further improved in comparison to unionized ASDs, due to the amalgamation of both approaches, i.e., drug ionization and amorphization.

Amorphous salts of CIP containing succinic acid or amino acids as counterions have also been prepared. While these formulations were far more soluble than the CIP ASDs, they were less stable when exposed to high humidity and in most cases decreased the permeability of the drug [10,11].

Unlike many other poorly soluble drugs, there is very little in the literature regarding amorphous solid dispersions of ENRO, and no mention of the pure amorphous form of the drug. However, one study by Chun and Choi described the preparation of an enrofloxacin–Carbopol "complex" by mixing a solution of ENRO in 1% acetic acid with that of Carbopol in water, filtering and washing the precipitate, and then drying and milling the resultant powder [12]. The product was found to be X-ray amorphous but lacked a clear T_g. The authors hypothesized that the positively charged tertiary amine of the drug formed an ionic bond with the carboxylate anions of Carbopol. Consequently, when the dissolution rate of the complex was found to be lower than that of the pure drug, this was attributed to the strength of the drug–polymer interactions.

Due to the absence of research in this area, the main aim of this project was to prepare ASDs of ENRO and to examine their solid-state and pharmaceutical properties. As previously mentioned, the chemical structure of ENRO differs from that of CIP by the presence of an ethyl group on its N3 piperazine amino group. It was of interest to determine whether this has an impact on the interactions that the drug can form with various polymers and whether the biopharmaceutical properties of this veterinary drug can be improved. The solid-state characteristics and water uptake behavior of the successfully prepared dispersions were also examined and compared to those of equivalent CIP ASDs produced in an earlier study [9]. In addition, the solubility, dissolution, and antibacterial activity of the ENRO dispersions were investigated in order to determine the effect of physicochemical transformations on these biopharmaceutical properties of the drug.

2. Materials and Methods

2.1. Materials

Enrofloxacin (ENRO) was obtained from Glentham Life Sciences (Wiltshire, UK) and Ciprofloxacin (CIP) was purchased from Carbosynth Limited (Berkshire, UK). Polyvinylpyrrolidone K17 (PVP: Plasdone C-15) was sourced from ISP Technologies (New Jersey, USA), poly(vinyl alcohol) (PVA: 98% hydrolyzed, Mw 13000–23000) was obtained from Sigma-Aldrich (St. Louis, MO, USA), and Carbopol 981 was purchased from BF Goodrich (OH, USA). Methacrylic acid methyl methacrylate copolymer (Eudragit L100) and methacrylic acid ethyl acrylate copolymer (Eudragit L100-55) were kindly donated by Evonik Röhm GmbH (Darmstadt, Germany), while hydroxypropyl methylcellulose acetate succinate grades LG and MG (HPMCAS-LG and HPMCAS-MG) were provided by Shin-Etsu Chemical Co., Ltd. (Tokyo, Japan).

Fasted state simulated intestinal fluid (FaSSIF) was produced by adding 2.24 g SIF® Powder Original (biorelevant.com, Surrey, UK) to one liter of FaSSIF phosphate buffer, consisting of 19.5 mM NaOH (Riedel-de Haën, Seelze, Germany), 25 mM $NaH_2PO_4 \cdot H_2O$ (Merck, Darmstadt, Germany) and 106 mM NaCl (Sigma-Aldrich Ireland Ltd., Arklow, Ireland), adjusted to pH 6.5 with NaOH. Triethylamine was obtained from Sigma-Aldrich Ireland Ltd., (Arklow, Ireland). Brain–heart infusion (BHI) broth was obtained from bioMérieux (Marcy l'Étoile, France), while plates with Columbia agar supplemented with sheep blood were sourced from Oxoïd (Dardilly, France). All other chemicals and solvents were of analytical grade.

2.2. Methods

2.2.1. Sample Preparation

Solid dispersions were produced by dry ball milling ENRO and various polymers as described previously [9]. Briefly, the process was carried out at room temperature (22–25 °C) with a Retsch planetary ball mill PM 100 (Haan, Germany). The polymer concentration used was 40–60% (w/w), and a total of 2 g of powder was loaded to 50 mL stainless steel grinding jars containing three 20 mm stainless steel milling balls. Each mixture was milled for 1–6 h in total, in intervals of 15 min with 10 min breaks in between. Crystalline ENRO was quench cooled by heating the drug to the endset of melting (~235 °C) at 10 °C/min in a differential scanning calorimetry (DSC) machine, and then immediately removing the sample to allow it to cool quickly at room temperature. Physical mixtures (PMs) were prepared by mixing relevant concentrations of ENRO and the polymers in a pestle and mortar for a few minutes.

2.2.2. Powder X-ray Diffraction (PXRD)

PXRD was performed at room temperature using a benchtop Rigaku MiniflexII X-ray diffractometer (Tokyo, Japan) and a Haskris cooler (Illinois, USA) as described previously [1].

2.2.3. Solid-State Fourier Transform Infrared Spectroscopy (FTIR)

A Spectrum One FTIR spectrometer (PerkinElmer, Connecticut, USA) was utilized to obtain FTIR data [1,9]. The following parameters of the analysis, accumulating 10 scans in total, were employed: 450–4000 cm^{-1} was spectral range, 4 cm^{-1} was resolution, while the scan speed was 0.2 cm/s. A sample concentration of 1% (w/w) was obtained, diluting the powdered sample with KBr and making disks suitable for FTIR by applying pressure of approximately 10 bar for 1 min.

Deconvolution of the FTIR spectra was conducted to facilitate their comparison. OriginPro 7.5 software was used to subtract the baseline and carry out Gaussian peak fitting on the spectra. In each case, seven overlapping peaks were detected in the region under examination, whose combined area and shape were similar to those of the original bands [1].

2.2.4. Differential Scanning Calorimetry (DSC)

DSC analysis on 5–10 mg samples was done using a Mettler Toledo DSC (Schwerzenbach, Switzerland) under nitrogen purge and employing sealed 40 µL aluminum pans with three pin-holes in the lid [9]. To expose the glass transition temperature (T_g) of the samples, the powders were first subjected to a first heating cycle from 25 to 65 °C to remove the residual moisture, then the samples were cooled to 25 °C and re-heated to 250 °C at a rate of 10 °C/min.

2.2.5. Modulated Temperature Differential Scanning Calorimetry (MTDSC)

The T_gs of the ASDs were detected by MTDSC using a Q200 DSC instrument and TA Instruments DSC Refrigerated Cooling System (TA Instruments, New Castle, Delaware). Samples of 3–4 mg were heated in aluminum pans with sealed aluminum lids. Nitrogen was used as the purge gas at a flow rate of 20 mL/min. Samples were heated from 0 °C to 110–185 °C at 2 °C/min, with an amplitude of ± 0.318 °C and a modulation period of 60 s. Results were analyzed with the Universal Analysis 2000 software (TA Instruments). The midpoint of the transition was taken as the T_g. Sapphire was used to calibrate the heat capacity, while indium was used for the calibration of enthalpy and temperature. All measurements were carried out in triplicate.

2.2.6. Calculation of Theoretical Glass Transition (T_g) Values with Gordon–Taylor Equation

The theoretical T_gs of the ASDs were calculated using the Gordon–Taylor equation [13,14]:

$$T_g = \frac{w_1 T_{g1} + K w_2 T_{g2}}{w_1 + K w_2}, \quad (1)$$

where K is approximately equal to:

$$K \approx \frac{T_{g1} \rho_1}{T_{g2} \rho_2}. \quad (2)$$

w_1 and w_2 are the weight fractions of the components, T_{g1} and T_{g2} are the glass transition temperatures of ENRO and the polymer, and ρ_1 and ρ_2 are the true densities of the two constituents. The T_gs of the polymers were sourced from the literature: HPMCAS-LG, 119 °C [15]; HPMCAS-MG, 120 °C [15]; Eudragit L100, 130 °C [16]; Eudragit L100-55, 96 °C [16]; and Carbopol, 105 °C [17]. Further, the average true density data were obtained from the published resources: ENRO, 1.385 g/cm^3 [18]; HPMCAS-LG and HPMCAS-MG, 1.29 g/cm^3 [19]; Eudragit L100, 0.84 g/cm^3 [20]; Eudragit L100-55, 0.83 g/cm^3 [16]; and Carbopol, 1.4 g/cm^3 [21].

2.2.7. High-Speed Differential Scanning Calorimetry (HSDSC)

HSDSC was performed on crystalline ENRO, under helium purge, with a PerkinElmer Diamond DSC (Waltham, MA, USA) supported by a ULSP B.V. 130 cooling system (Ede, The Netherlands) as described previously [1]. Around 3–5 mg samples were first encapsulated in aluminum pans (18 µL) and heated from 25 to 300 °C at a rate of 300–500 °C/min.

2.2.8. Thermogravimetric Analysis (TGA)

TGA was done using a Mettler TG50 measuring module coupled to a Mettler Toledo MT5 balance (Schwerzenbach, Switzerland) [1]. The heating rate employed was 10 °C/min and samples (8–10 mg) were loaded into open aluminum pans.

2.2.9. Dynamic Vapor Sorption (DVS) and Mathematical Modeling Using Young–Nelson Equations

DVS studies were performed using an Advantage-1 automated gravimetric vapor sorption analyzer (Surface Measurement Systems Ltd., London, UK) at 25.0 ± 0.1 °C, between 0 and 90% RH, in steps of 10% RH, as described previously [9]. The complete sorption and desorption profile is shown as an isotherm. PXRD analysis was performed on all samples following DVS to identify any solid-state transformations.

In order to determine how water uptake occurs in the ASDs, the experimental sorption and desorption data were fitted to equations using the Young–Nelson model, as described previously [22,23]:

$$M_s = A(\beta + \theta) + B(\theta)RH, \quad (3)$$

$$M_d = A(\beta + \theta) + B(\theta)RH_{max}. \quad (4)$$

M_s and M_d are the amount of water sorbed and desorbed, respectively, at each relative humidity value. This is expressed as a fraction of the dry mass of the sample. A and B are constants which can be defined as follows:

$$A = \frac{\rho_w Vol_M}{W_m}, \quad (5)$$

$$B = \frac{\rho_w Vol_A}{W_m}. \quad (6)$$

ρ_w is the density of water, W_m is the weight of the dry sample, and Vol_M and Vol_A are the volume of adsorbed and absorbed water, respectively. In Equations (3) and (4), θ represents the fraction of sample surface that is covered by at least one layer of water molecules, and β is the mass of absorbed water at 100% RH. B(θ)RH is therefore the mass of absorbed water at a particular fraction of monolayer coverage, θ, and RH level. A(β + θ) is equal to the total amount of adsorbed water, while Aθ is the mass of water in an entire adsorbed monolayer, as a fraction of the dry mass of the material. Aβ is the mass of water adsorbed in a multilayer. θ and β may be further defined as follows [23]:

$$\theta = \frac{RH}{RH + E(1 - RH)}, \quad (7)$$

$$\beta = -\frac{ERH}{E - (E-1)RH} + \frac{E^2}{(E-1)} \ln a \frac{E - (E-1)RH}{E} a - (E+1) \ln(1 - RH). \quad (8)$$

E is an equilibrium constant between water in the monolayer and condensed water adsorbed externally to the monolayer:

$$E = e^{-\left[\frac{q_1 - q_L}{k_B T}\right]}. \tag{9}$$

q_1 is the heat of adsorption of water on the solid, q_L is the heat of condensation of water, both in Joules/mole, then T is the temperature in Kelvin and k_B is Boltzmann's constant (1.38×10^{-23} J/K).

The experimental data obtained from DVS studies of the ENRO ASDs, as well as equivalent CIP ASDs that were previously prepared [9], were fitted to Equations (3) and (4) by iterative multiple linear regression. The sum of the squares of the residuals between the experimental and calculated values was used as fitting criteria. The multiple correlation coefficient (r) was calculated using Microsoft Excel 2007. Using the calculated values of A, B, θ, and β, the profiles of water adsorbed in monolayer (Aθ) and multilayer (Aβ), and of absorbed water (Bθ) were determined [23].

2.2.10. Dynamic Solubility Study

A volume of 5 mL of FaSSIF was added to 40 mL glass vials and placed into jacketed beakers connected to a Lauda M12 waterbath at 37 °C (Lauda-Königshofen, Germany). A quantity of pure drug or ASD, in excess of the expected saturated solubility (25–200 mg, depending on the sample), was added to the vials containing the aliquot of FaSSIF and stirred at 1000 rpm. At different time points, over a 2 h period, samples were taken for the stirred suspensions and filtered with 0.45 µm PTFE membrane filters (VWR, USA). The filtered solutions were then diluted appropriately with a 2.9 g/L solution of phosphoric acid, previously adjusted to pH 2.3 with trimethylamine [2]. The concentration of ENRO in each of the diluted samples was determined by UV spectrophotometry as described below. The solid material left in the vials at the end of the studies was filtered and analyzed by PXRD.

2.2.11. Dissolution Study

Dissolution studies were carried out at 37 °C, using a paddle apparatus (Apparatus II) with a continuous rotation of 100 rpm. A quantity of sample corresponding to approximately 10% of the final drug concentration obtained in the solubility study was added to 300 mL of FaSSIF (ENRO: 25 mg, ENRO/Eudragit L100: 287.5 mg, ENRO/HPMCAS-LG: 967.5 mg and ENRO/HPMCAS-MG: 517.5 mg). One milliliter aliquots was taken at specific time points over the 2 h period of the study and replaced with 1 mL of FaSSIF. Each sample was filtered with a 0.45 µm PTFE membrane filter (VWR, USA) and diluted with a 2.9 g/L solution of phosphoric acid, previously adjusted to pH 2.3 with triethylamine. The concentration of ENRO in each of the diluted samples was then measured by UV spectrophotometry. The cumulative quantity of dissolved drug at each time point was calculated by taking account of the 1 mL aliquots taken for analysis. Each study was carried out in triplicate.

2.2.12. UV Spectrophotometry

UV analysis was performed using a Shimadzu UV-1700 PharmaSpec UV-vis spectrophotometer (Shimadzu Corp., Kyoto, Japan) using quartz cuvettes with a 1 cm optical path length. The reference was a 2.9 g/L solution of phosphoric acid, previously adjusted to pH 2.3 with triethylamine. This buffer was also used to produce a range of concentrations of pure ENRO, in order to construct a calibration curve. The λ_{max} of these solutions was found to be 277 nm; therefore, UV absorbance was measured at this wavelength.

2.2.13. Bacterial Studies

For these studies, *Staphylococcus aureus* ATCC 25923, *Escherichia coli* ATCC 25922, *Pseudomonas aeruginosa* ATCC 27853, and *Klebsiella pneumoniae* DSM 16609 were cultured on Columbia agar supplemented with sheep blood. The inoculum was prepared as previously described [9,24]. The density of the *S. aureus* suspension was adjusted so that it equaled that of the 1.1 McFarland standard, and then further diluted 100-fold with BHI medium. The *P. aeruginosa*, *E. coli*, and *K. pneumoniae* suspensions, on the other hand, were adjusted to equal that of the 0.5 McFarland standard, and then diluted 10-fold.

The minimum inhibitory concentrations (MICs) and minimum bactericidal concentrations (MBCs) of ENRO and the ASDs were determined using a broth microdilution method, as previously described [9,24].

3. Results and Discussion

3.1. Production of Amorphous Solid Dispersions/Amorphous Polymeric Salts

Ball milling was first carried out on crystalline "as received" ENRO to determine whether it is possible to amorphize the drug in this manner. However, following four hours of milling at room temperature, a disordered, semi-crystalline solid was obtained (Figure 2a). This was also the case with CIP [1]. The most intense peaks in the X-ray diffractogram of the unprocessed ENRO powder are visible at 7.4, 9.8, 14.9, and 25.8 2θ degrees. These peaks are also present in the diffractogram of ball milled ENRO; however, their intensity is reduced. Quench cooling ENRO, on the other hand, resulted in an X-ray amorphous material (Figure 2a).

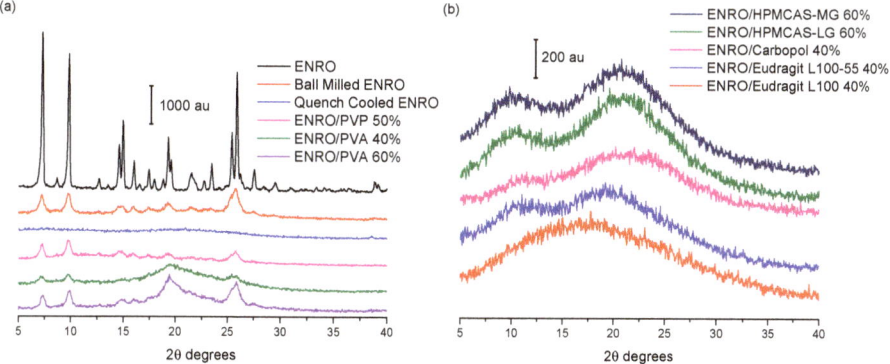

Figure 2. Powder X-ray diffraction (PXRD) diffractograms of (**a**) enrofloxacin (ENRO) and semi-crystalline solid dispersions, and (**b**) ENRO amorphous solid dispersions (ASDs).

In previous studies with CIP, X-ray amorphous solid dispersions were obtained when the drug was ball milled with Eudragit L100, Eudragit L100-55, Carbopol, HPMCAS-LG, and HPMCAS-MG. All of these polymers are acidic, and FTIR analysis confirmed the presence of an ionic bond between the positively charged piperazine amino group of CIP and the carboxylate groups of the polymers in the ASDs [9]. These acidic polymers also proved to be suitable co-formers for ENRO, with each resulting in an X-ray amorphous formulation (Figure 2b). As was the case with CIP, a polymer concentration of 60% (w/w) was required to fully amorphize mixtures of CIP and HPMCAS, whereas 40% (w/w) was adequate for Eudragit L100, Eudragit L100-55, and Carbopol. Although the product obtained with 40% (w/w) HPMCAS-LG was almost X-ray amorphous following 4 h of milling, very small peaks could still be detected by PXRD at 9.8 and 25.8 2θ degrees, corresponding to the most prominent peaks of anhydrous ENRO (Figure S1). A slightly more crystalline product was obtained with HPMCAS-MG under the same conditions, which decreased in intensity following a further 2 h of milling but did not disappear entirely. In contrast to CIP, which required a total of 6 h of milling and a reduced temperature of 2–5 °C to form ASDs with 60% (w/w) HPMCAS [9], 4 h of milling at room temperature was adequate for the corresponding ENRO ASDs (Figure S1). This indicates that the polymers may interact more readily with ENRO than CIP possibly due to the weaker crystal lattice of ENRO, which would facilitate its amorphization. These results show that the presence of an extra ethyl group in the structure of ENRO does not appear to negatively affect its ability to interact with these acidic polymers. To enable closer comparison with the equivalent CIP ASDs, the ENRO/HPMCAS ASDs containing 60% (*w/w*) polymer that were milled for 6 h were used for further studies.

In contrast to the acidic excipients, when CIP was milled with neutral polymers such as PVP and PVA at a concentration of 40–60% (*w/w*), a semi-crystalline product was obtained [9]. This was also the case with ENRO (Figure 2a). The fact that X-ray amorphous solid dispersions were only formed when ENRO was milled with acidic polymers containing carboxylic acid groups suggests that the drug is interacting with these substances via ionic bonds, as was the case with CIP. Likewise, in all of the ENRO salts produced by Karanam et al. containing an acidic counterion, proton transfer from the acid to the piperazine tertiary amine (N3) of the drug occurred, resulting in an ionic interaction between the two moieties [2]. A similar reaction may take place between the N3 of ENRO and the polymers in these ASDs.

3.2. Solid-State Fourier Transform Infrared Spectroscopy

The results of FTIR analysis of the ASDs, PMs, and starting materials are shown in Figure 3a–d. A sharp peak is located at 1737 cm^{-1} in the spectrum of crystalline ENRO due to the carbonyl stretch of its unionized carboxylic acid group. While the process of ball milling introduced some disorder to the crystal lattice of ENRO, the FTIR spectrum of the ball milled drug was almost identical to the crystalline ENRO starting material. The greater molecular disorder of quench cooled ENRO, on the other hand, is evident in the broader and less intense peaks of its spectrum (Figure 3d). Slight peak shifts were also seen with this sample, in particular, the carboxylic acid C=O stretch, which shifted to 1728 cm^{-1}. This can be attributed to changes in the drug's intermolecular interactions upon amorphization, such as hydrogen bonding [25]. Interestingly, the COOH carbonyl stretch of the drug also shifted to lower wavenumbers in the spectrum of the crystalline ENRO saccharinate salt, in which the piperazine N3 amino group of the drug is positively charged [6]. This carbonyl peak underwent a similar shift with all of the ASDs. Therefore, while the carboxylic acid of ENRO remains unionized in the ASDs, changes in the hydrogen bonding of this group clearly occur upon amorphization. This shift may also be related to changes in the ionization state of the drug.

The main differences between the spectra of ENRO and the ASDs may be seen in the 1650–1450 cm^{-1} region. In the case of crystalline ENRO, the carbonyl stretch of its ketone group appears as a sharp, strongly absorbing peak at 1628 cm^{-1}. The medium intensity shoulder at 1611 cm^{-1} may be assigned to C=C stretching vibrations of the drug's aromatic ring. While these peaks are not significantly shifted in the spectra of the ASDs, differences in their relative absorbance were observed. In crystalline ENRO, the absorbance of the ketone peak is approximately 1.8 times greater than that of the aromatic peak. This ratio decreases to 1.5–1.7 for each of the ASDs. However, a similar decrease in the relative absorbance of these peaks was also seen with quench cooled ENRO and is therefore likely due to changes in the interactions of these groups upon amorphization.

The peaks at 1508 and 1469 cm^{-1} in the spectrum of ENRO may be attributed to C=C stretching of the aromatic ring, and C–C stretching of the drug's piperazine group, respectively [26]. The shape of these peaks was altered notably in the ASDs, and the presence of multiple overlapping peaks became evident. In order to separate the individual peaks in this region and to quantify their relative absorbance, deconvolution of the spectra, with Gaussian peak fitting, was carried out. The resulting spectra are shown in Figure S2. Deconvolution allowed the detection of a further peak at approximately 1453 cm^{-1} in the spectrum of ENRO, which may be tentatively assigned to the C–H bending vibrations of the ethyl group. This peak is also present in the spectra of ball milled and quench cooled ENRO, and all of the ASDs, along with an additional peak at approximately 1494 cm^{-1}. Although a slight broadening is visible at this wavenumber in the spectrum of crystalline ENRO, it is not as distinct as with the other samples. Clear differences in the relative absorbance of these peaks may also be seen between the pure drug and ASDs. For instance, in crystalline ENRO, the area of the peak at 1508 cm^{-1} is approximately two times smaller than the combined area of the peaks at 1469–1453 cm^{-1}. While a similar ratio was obtained with the equivalent peaks in ball milled ENRO, with the quench cooled form of the drug, it decreased to 1.9. With the ASDs, on the other hand, this ratio decreased further to 1.3–1.55. Similarly, in the spectra of crystalline and ball milled ENRO, the absorbance of the peak at 1469 cm^{-1} is clearly greater than that at 1453 cm^{-1}. By contrast, in each of the ASDs, as well as quench cooled ENRO, the maximum absorbance of these peaks did not differ greatly. Similar changes in this region were seen in the spectra of the partially crystalline ENRO/PVA solid dispersion, whereas the less disordered ENRO/PVP more closely resembled the crystalline ENRO starting material (Figure S3).

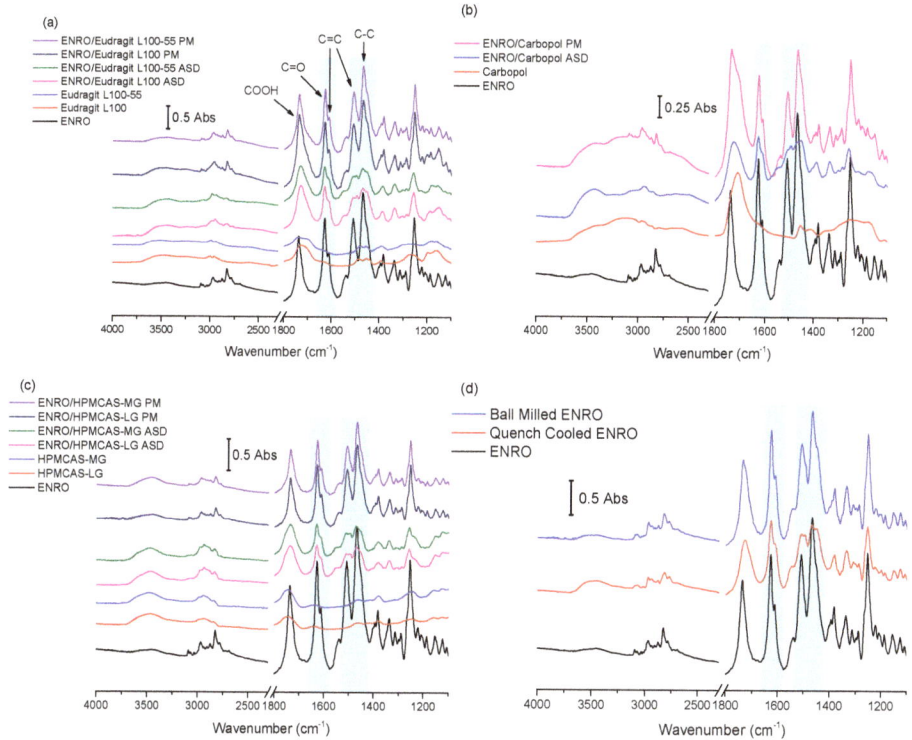

Figure 3. FTIR spectra of ASDs and physical mixtures (PM) containing (**a**) Eudragit L100 and Eudragit L100-55 40% (*w*/*w*) (**b**) Carbopol 40% (*w*/*w*), (**c**) hydroxypropyl methylcellulose acetate succinate grades LG and MG (HPMCAS-LG and HPMCAS-MG) 60% (*w*/*w*), and (**d**) ball milled and quench cooled ENRO. The areas of the spectra that undergo significant changes upon amorphization are highlighted in violet.

As previously mentioned, the terminal tertiary amine of ENRO (N3) may be protonated in these ASDs, forming ionic bonds with the carboxylate groups of the polymers. If this is the case, the main differences in the FTIR spectra of the ASDs compared to the starting materials or PMs can be attributed to the change in ionization state of the drug, and the presence of an additional $^+$N–H bond. Unfortunately, the $^+$N–H stretch is difficult to assign with certainty, as it will produce a weak band in the 3000–2600 cm^{-1} region that possibly overlaps with others, such as that of the C–H stretch of the neighbouring aliphatic group [26]. Similarly, the $^+$N–H bend of a tertiary amine salt generally appears as a very weak band in the 1610–1500 cm^{-1} region [27] and therefore is likely to be obscured by more intense peaks in the spectra of the ASDs. However, as described above, a number of differences in the 1450–1550 cm^{-1} region of the spectra of ENRO and the ASDs were observed. Therefore, it is possible that the presence of the peak corresponding to the $^+$N–H bend contributed to the variations in this area of the spectra. In addition, as the peaks in this region correspond to groups surrounding the terminal amino group of ENRO, it is likely that they would be altered upon the protonation of N3.

The hypothesis that ENRO is protonated in these ASDs is supported by the FTIR analysis of ENRO salts conducted by Karanam et al. [2]. In the spectra of each of the salts containing an acidic counterion, a decrease in the absorbance of the peak around 1469 cm^{-1} relative to that at 1508 cm^{-1} can be seen, in common with the ENRO ASDs. Single crystal X-ray diffraction confirmed that the N3 of the drug was protonated in these salts and formed an ionic bond with the carboxylate groups of the acids. Therefore, it is likely that ENRO is in the same cationic state in these ASDs and interacts with the acidic groups of the polymers to form amorphous polymeric salts (APSs). The fact that the spectrum of quench cooled ENRO is similar to that of the ASDs may be due to the partial conversion of the drug to the zwitterion.

3.3. Thermal Analysis

The conventional DSC thermograms of ENRO and the ASDs are shown in Figure 4. The melting point onset of crystalline ENRO, as well as the ball milled and quench cooled drug, was approximately 225 °C. Its lower melting point in comparison to CIP (approximately 272 °C) [1] can be explained by the less extensive intermolecular bonds in ENRO. In contrast to the pure drug, the thermograms of the ASDs were missing a clear melting point. Similarly, the ASDs did not show distinct crystallization exotherms during DSC analysis, although the small, broad peaks visible at approximately 157 °C and 148 °C in the thermograms of ENRO/HPMCAS-LG and ENRO/HPMCAS-MG, respectively, may be due to some crystallization. The indistinct nature of the thermograms can be attributed to the amorphous nature of these formulations and their stability upon heating [28]. By contrast, ball milled and quench cooled ENRO had clear crystallization peaks at approximately 73 °C and 106 °C, respectively, confirming their lower resistance to crystallization. The particularly low crystallization temperature of ball milled ENRO is to be expected, as the residual crystallinity present in this sample would enable crystal growth to occur more quickly upon heating.

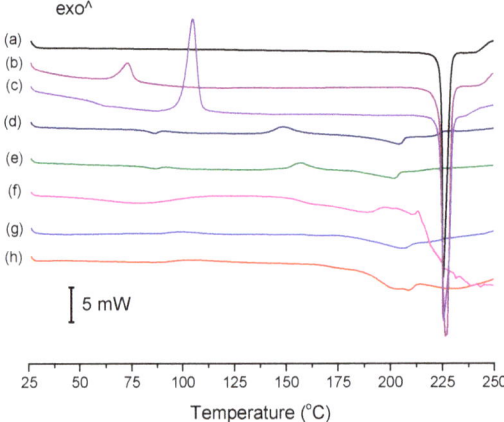

Figure 4. Differential scanning calorimetry (DSC) thermograms of (a) crystalline ENRO, (b) ball milled ENRO, (c) quench cooled ENRO, (d) ENRO/HPMCAS-MG, (e) ENRO/HPMCAS-LG, (f) ENRO/Carbopol, (g) ENRO/Eudragit L100-55, and (h) ENRO/Eudragit L100. The thermograms of the ASDs are those obtained from the second heating cycle, following initial heating to 65 °C to allow for residual water removal.

The T_g of quench cooled ENRO was detected at 58.9 °C, which is significantly lower than that of CIP (86.7 °C). Again, this may be attributed to the weaker intermolecular interactions present in ENRO. As a distinct T_g could not be found for all of the ENRO ASDs using conventional DSC; they were therefore analyzed by MTDSC. The resultant T_gs are listed in Table 1. In each case, a single T_g was detected. This suggests that the drug is miscible with each of these polymers, and that they form a single homogeneous phase [29]. Due to its low amorphous content and high crystallization rate, no T_g could be determined for ball milled ENRO with either DSC technique.

The Gordon–Taylor (G–T) equation was used to calculate the expected T_gs of the ASDs, given their weight percentage of drug and polymer. From Table 1, it can be seen that the experimental T_gs of the ASDs containing Carbopol, Eudragit L100, and Eudragit L100-55 were substantially higher than the theoretically derived values. Such large positive deviations from the predicted T_gs suggest that strong interactions exist between the components and are particularly indicative of polymeric salt formation [30]. By contrast, the experimental and G–T T_gs of the HPMCAS-containing ASDs differed by only a few degrees. This suggests that these polymers are fully miscible with ENRO but do not form strong interactions with the drug or that the heteromolecular drug–polymer interactions may be of a similar strength to the homomolecular interactions present in the individual raw materials [31].

Similar results have been obtained with the equivalent CIP ASDs, whereby the experimental T_gs of those containing Eudragit L100 and L100-55 deviated from the predicted values by a much greater degree than those containing either grade of HPMCAS. The apparently weaker interactions present in the latter ASDs were attributed to the lower proportion of carboxylic acid groups present in HPMCAS compared to the other polymers [9]. This would explain why a polymer concentration of 60% (w/w) was required to produce X-ray amorphous solid dispersions with these polymers, whereas a concentration of 40% (w/w) was sufficient with the others.

Table 1. Glass transition temperatures (T_g) of ENRO and ENRO ASDs.

Sample	Experimental T_g (°C)	G-T T_g (°C)
ENRO	58.9	N/A
ENRO/Eudragit L100	109.9 ± 1.6	82.5
ENRO/ Eudragit L100-55	103.2 ± 0.2	74.0
ENRO/Carbopol	155.6 ± 0.2	71.4
ENRO/HPMCAS-LG	86.8 ± 0.4	85.6
ENRO/HPMCAS-MG	83.3 ± 0.4	85.9

As previously mentioned, CIP exists as a zwitterion in the solid state, with a positively charged piperazine amino group and negatively charged carboxylate group. However, it has been shown to convert to the unionized form upon melting, due to intramolecular proton transfer. This was visualized as a small endothermic peak in the DSC thermogram of the drug, just prior to the melting endotherm. However, this low energy event was only visible when CIP was heated at 500 °C/min [1]. HSDSC analysis was therefore carried out on crystalline ENRO in order to determine if it also undergoes proton transfer at high temperatures, in this case from the unionized form to the zwitterion. However, even when heated at the maximum heating rate of 500 °C/min, the drug did not show any evidence of solid-state transformation (Figure S4). After heating ENRO to the endset of melting and allowing it to cool slowly, PXRD and FTIR analysis confirmed that the drug remained in the unionized anhydrous state; thus, the ethyl moiety attached to N3 prevented the proton transfer.

While crystalline ENRO is pale yellow, quench cooled ENRO is a more vibrant golden color, and when heated to 250 °C, the drug becomes dark orange/rusty. CIP also turns from off-white to a yellow color prior to melting; however, when heated past its melting point, it becomes brown due to substantial degradation. From the TGA curves obtained with ENRO and the ASDs (Figure 5), crystalline and ball milled ENRO do not appear to undergo substantial thermal degradation, decreasing in mass by only 3.4% over the course of the TGA analysis. CIP, on the other hand, is much more prone to thermal degradation, with a mass loss of 12.8% and 17.3% being obtained with the crystalline and ball milled forms of the drug, respectively [1]. An initial mass loss was observed below 70 °C with all of the amorphous formulations due to water evaporation. This is to be expected with ASDs, as the hygroscopic nature of amorphous drugs and polymers results in the absorption of atmospheric water vapor. The amorphous samples also degraded to a greater degree than the pure drug, in particular, the Carbopol ASD. Amorphous solids are typically more reactive than their crystalline counterparts, as their higher molecular mobility can enable such degradation reactions to occur [32]. Alternatively, this mass loss may simply be due to degradation of the polymers.

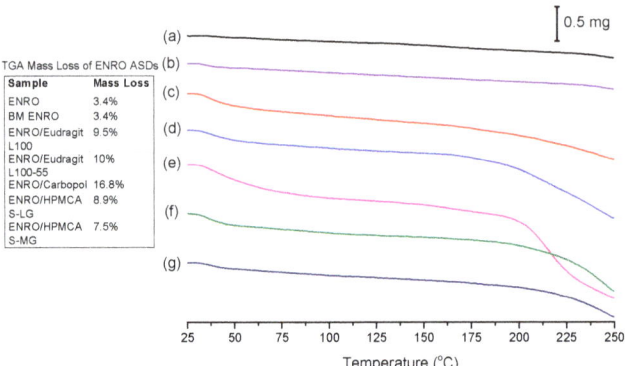

Figure 5. TGA analysis of (a) crystalline ENRO, (b) ball milled (BM) ENRO, (c) ENRO/Eudragit L100, (d) ENRO/Eudragit L100-55, (e) ENRO/Carbopol, (f) ENRO/HPMCAS-LG, and (g) ENRO/HPMCAS-MG.

3.4. Water Sorption Studies

The stability of the ASDs when exposed to various humidity levels was examined by DVS. At the end of the sorption cycle, at 90% RH, ENRO absorbed only 0.13% (w/w) water. This increased to 2.9% for the ball milled drug, due to the increase in disordered material (Figure 6a). CIP also absorbed low levels of water during DVS analysis, increasing in mass by only 0.6% (w/w) [33]. PXRD analysis of the drugs at the end of the sorption studies revealed that both ENRO remained in the same solid state, with PXRD patterns matching those of the starting materials (Figure S5) [33]. In contrast to the crystalline drug, the ENRO ASDs were far more hygroscopic, absorbing 16–19% of their mass in water. Very similar levels of water uptake were observed with the CIP ASDs [9]. The higher hygroscopicity of the amorphous formulations can be explained by the random orientation of their molecules. This leads to a larger free volume and enables the penetration of water into the samples [34]. In addition, polymers are often more hygroscopic than the amorphous form of a drug, which increases the tendency of an ASD to take up moisture [35].

As can be seen in Figure 6b, the isotherms obtained with the ASDs containing Eudragit L100, Eudragit L100-55 and Carbopol were very similar in shape, with significant hysteresis. Hysteresis is commonly encountered with amorphous or porous solids, as water can absorb into the interior of the material [36]. If water diffuses into the sample bulk more quickly than it can return to the surface, then, at the same RH level, a greater amount of moisture will be present during desorption than sorption, resulting in the appearance of hysteresis.

Unlike the other ASDs, the isotherms of both ENRO/HPMCAS ASDs were convex in shape with a small amount of hysteresis, suggesting that water was mainly adsorbed to the outer surfaces of these samples (Figure 6c). Therefore, the water uptake behavior of the ENRO ASDs differs depending on the polymer used. This was further examined by fitting the sorption and desorption data to the Young–Nelson equations. According to the Young–Nelson model, water can be taken up by a sample in three different ways: adsorbed as a monomolecular layer, adsorbed as a multilayer, or absorbed into the interior of the solid [22]. The parameters calculated using the Young–Nelson equations are listed in Table S1, and the isotherms obtained using this approach are shown in Figure 7 and Figure S6. The corresponding CIP ASDs were also examined for comparison (Figure S7).

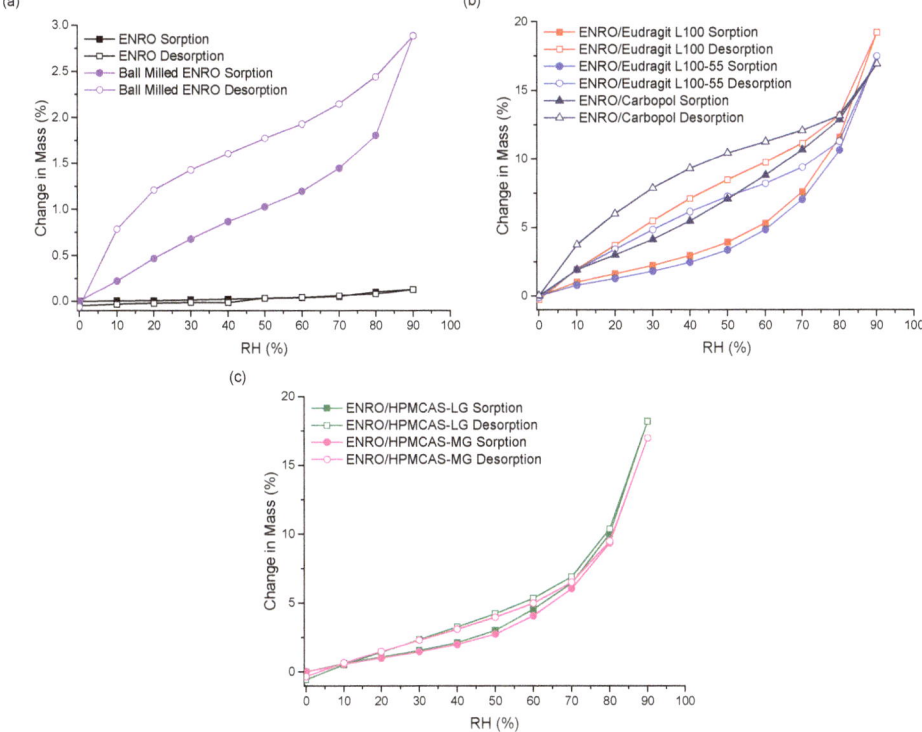

Figure 6. DVS isotherms of (**a**) crystalline and ball milled ENRO, (**b**) ENRO ASDs containing 40% (*w/w*) Eudragit L100, Eudragit L100-55, and Carbopol, and (**c**) ENRO ASDs containing 60% (*w/w*) HPMCAS-LG and HPMCAS-MG.

As predicted from the DVS isotherms, the major water uptake mechanism of the ENRO ASDs containing Eudragit L100, Carbopol, and Eudragit L100-55 was water absorption (Figure 7a,b and Figure S6a). The small degree of absorption that occurred with the ENRO/HPMCAS ASDs confirms that they are somewhat porous, but less so than the other ASDs, as suggested by the minor hysteresis in their DVS isotherms. Unlike the other samples, the majority of water taken up by ENRO/HPMCAS ASDs was bound to their exterior surfaces as a multilayer. Multilayer formation begins at low RH levels and appears to occur simultaneously with monolayer adsorption (Figure 7c and Figure S6b). By contrast, the major water uptake mechanism for the CIP ASDs containing HPMCAS was absorption (Figure S7). This suggests that the CIP/HPMCAS ASDs are more porous than the corresponding ENRO ASDs, or the polymers may be capable of swelling to a greater degree in the former formulations. As with the ENRO ASDs, water is primarily absorbed into the interior of the CIP ASDs containing Eudragit L100, Eudragit L100-55, and Carbopol. However, the water distribution patterns obtained with the ENRO and CIP ASDs containing Carbopol differed somewhat from the others. The monolayer adsorption of these samples increased more gradually over the course of the study and was also more extensive. This may be due to the presence of more hydrophilic groups on the surface of these ASDs, which can interact with water molecules [23].

With both sets of ASDs, the highest value of E was obtained with those containing HPMCAS-LG, followed by HPMCAS-MG (Table S1). However, this constant was more than 10 times larger for the ENRO/HPMCAS samples than those containing CIP. This indicates that water molecules form much stronger and extensive interactions with the surface of these samples [37] and explains why water appears to be mainly adsorbed to the surface of these ASDs in a multilayer. The value of the regression coefficient, r, was ≥0.98 for all of the ASDs, showing that there was a good fit between the experimental and estimated values of the different parameters (Table S1). Therefore, application of the Young–Nelson model is a suitable approach for comparing the water uptake of these samples.

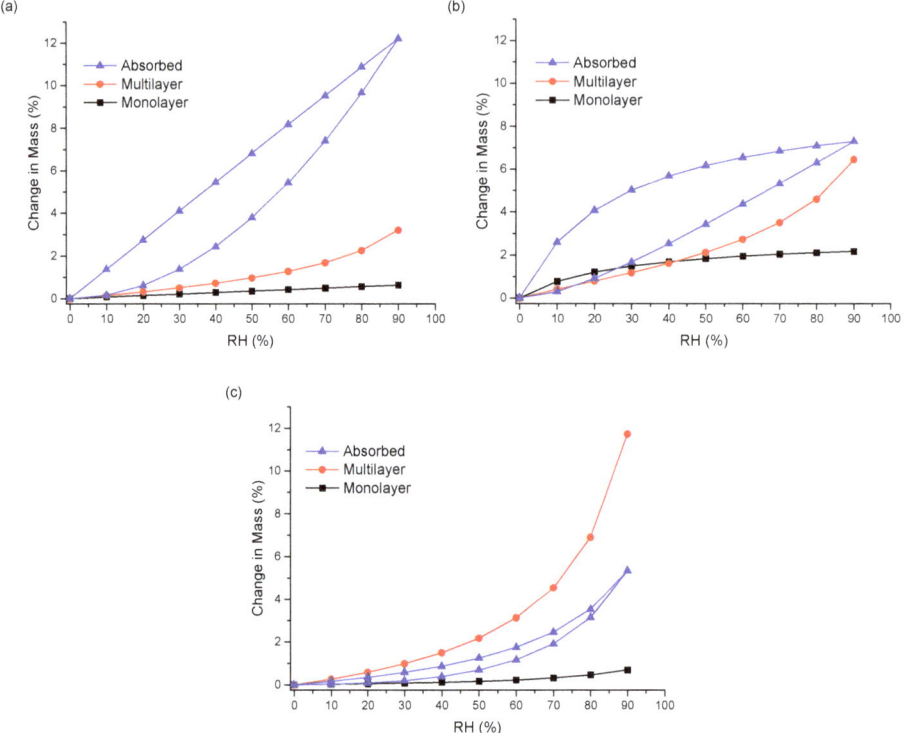

Figure 7. Water distribution patterns according to the Young–Nelson model in ENRO ASDs containing (**a**) Eudragit L100 40% (*w/w*), (**b**) Carbopol 40% (*w/w*), and (**c**) HPMCAS-LG 60% (*w/w*).

The permeation of water molecules into the interior of an amorphous solid can increase its free volume, resulting in a decrease in T_g [38]. Water sorption is also known to increase the molecular mobility and thus crystallization rate of amorphous substances, and to decrease the crystallization onset temperature [39]. However, despite the plasticizing effects of sorbed water, all five of the ENRO ASDs remained X-ray amorphous following DVS analysis (Figure S5). This was also the case for the corresponding CIP polymeric ASDs [9]. The high stability of these ASDs may be due to stabilizing drug–polymer interactions, the presence of which was suggested by the results of FTIR and DSC analysis. Polymers are also known to have anti-plasticizing effects and to reduce the molecular mobility of amorphous formulations, while steric hindrance from polymer chains can prevent the nucleation and crystal growth of drug molecules [8,40]. In contrast to the polymeric ASDs, amorphous CIP salts containing succinic acid or amino acids as counterions were unstable in humid environments and crystallized during DVS studies [33].

3.5. Solubility and Dissolution Studies

Due to issues with clumping and viscosity, solubility studies could not be carried out accurately on the ASDs containing Eudragit L100-55 and Carbopol, and these samples were therefore excluded from further studies. The superior solubility of the remaining ASDs in FaSSIF in comparison to crystalline ENRO is clear from Figure 8a. With the pure drug, a peak in concentration was seen at 30–60 s, which then quickly fell to a constant level of approximately 0.7 mg/mL. A steep initial increase in drug concentration was also seen with the ASDs containing HPMCAS-LG and HPMCAS-MG, which peaked after 5 and 2 min, respectively. While this supersaturation then fell after 10–15 min, the concentration was still significantly higher than that obtained with crystalline ENRO. This solubility enhancement was sustained for the remainder of the study, with final concentrations of 12.2 mg/mL and 5.6 mg/mL being obtained with ENRO/HPMCAS-LG and ENRO/HPMCAS-MG, respectively. In contrast to the other samples, a more gradual increase in drug concentration was seen with ENRO/Eudragit L100, followed by a plateau after 20 min. This sample was also less soluble than those containing HPMCAS, reaching a concentration of 4.6 mg/mL after 2 h.

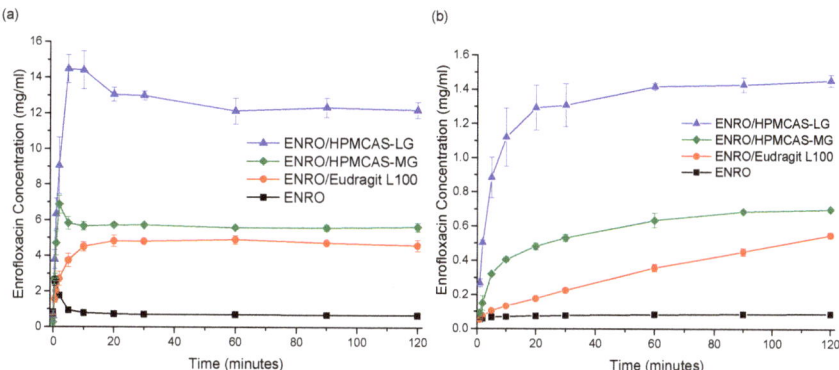

Figure 8. (a) Solubility and (b) dissolution studies in FaSSIF at 37 °C. The average of three experiments is plotted, ± the standard deviation.

The dissolution behavior observed with these ASDs is similar to that described by the "spring" and "parachute" model [41]. In this model, ASDs are described as "springs", as their high energy and lack of a crystal lattice results in rapid drug dissolution and supersaturation. However, this supersaturated state is thermodynamically unstable, and crystallization of a lower energy, less soluble form of the drug soon follows. Fortunately, excipients such as polymers may be used to inhibit or delay the precipitation of dissolved drug and thus act as "parachutes" [41]. Polymers can prevent nucleation and crystal growth via interactions with the drug, steric hindrance, and increased viscosity [42]. Although the concentration obtained with the ENRO/HPMCAS ASDs did decrease somewhat over the course of the study, the polymers present in these ASDs most likely prevented extensive crystallization of the drug in solution, enabling supersaturation to be maintained for at least 2 h. By avoiding the rapid generation of supersaturation, less nucleation and crystallization would be expected to occur with the ENRO/Eudragit L100 ASD. This was confirmed by PXRD analysis of the excess solid recovered at the end of the solubility studies. In each case, enrofloxacin hexahydrate [2] was detected; however, with ENRO/Eudragit L100, the sample was far less crystalline (Figure S8).

Similarly enhanced concentrations were obtained with the ENRO ASDs in dissolution studies in comparison to crystalline ENRO. Following 2 h, the highest concentration was achieved with ENRO/HPMCAS-LG, at 1.45 ± 0.03 mg/mL (44.8 ± 1.2% of ENRO released), followed by ENRO/HPMCAS-MG (0.70 ± 0.01 mg/mL, 40.5 ± 0.6% of ENRO released) and ENRO/Eudragit L100 (0.55 ± 0.02 mg/mL, 57.4 ± 1.8% of ENRO released). Crystalline ENRO, on the other hand, only attained 0.09 ± 0.00 mg/mL (104.7 ± 4.5% of ENRO released) over the course of the study (Figure 8b). Apart from concentration, the ASDs also differed in the shape of their dissolution profiles. With ENRO/HPMCAS-LG, the drug concentration increased quite rapidly at the start of the study and then remained fairly constant for the remainder. While a similar profile was obtained with ENRO/HPMCAS-MG, the initial drug release was more gradual than with the LG grade of polymer. As was the case in the solubility study, the final concentration obtained with ENRO/HPMCAS-MG was approximately half that of ENRO/HPMCAS-LG. However, ASDs containing different grades of HPMCAS are known to demonstrate different rates and extents of drug release, due to differences in their succinoyl and acetyl content [43]. This may affect the pH of the diffusion layer surrounding the ASD particles, or the strength of drug–polymer interactions.

A steady, linear increase in drug concentration was observed with ENRO/Eudragit L100. As no leveling off occurred during the study, it is possible that the drug concentration would continue to rise during longer-term studies, similar to an extended release formulation. The gradual dissolution of ENRO from this ASD may be due to strong drug–polymer interactions, which could delay the dissociation and dissolution of the drug [44]. Such interactions would also explain the higher than predicted T_g of this formulation and the absence of crystallization during DSC analysis, unlike the ASDs containing HPMCAS. Alternatively, this polymer may be less soluble than HPMCAS, which would reduce the diffusion of water into the ASD and, thus, its dissolution rate.

Visible differences in the behavior of the ENRO ASD powders were also evident during dissolution studies. Both ENRO/Eudragit L100 and ENRO/HPMCAS-MG formed clumps when added to the dissolution vessels. While these eventually dissolved in the case of ENRO/HPMCAS-MG, with ENRO/Eudragit L100, they remained largely intact for the duration of the study. This would have hindered the release of the drug and reduced the surface area exposed to the dissolution medium. By contrast, no clumping was observed with ENRO/HPMCAS-LG, which enabled faster dissolution and higher concentrations of ENRO to be achieved.

From the results of this study and that of a previous investigation involving CIP, it can be concluded that ENRO is the more soluble of the two fluoroquinolones in FaSSIF. CIP was found to have a solubility of only 0.14 mg/mL in this medium [9], which is five times lower than that of ENRO. Higher drug concentrations were also obtained with the ENRO ASDs than the equivalent CIP ASDs. Similarly, ENRO has been reported to be more soluble than CIP in pH 7.4 phosphate buffer [5]. As previously mentioned, the greater solubility of ENRO may be explained by its weaker crystal lattice, which would facilitate the release of drug molecules into solution.

3.6. Bacterial Studies

The minimum inhibitory concentrations (MICs) and minimum bactericidal concentrations (MBCs) of ENRO and the ASDs are listed in Table 2. Values that differ significantly from those of ENRO are shown in bold. In each case, the MBC should be greater than the MIC, as a larger quantity of drug is required to bring about bacterial death rather than growth inhibition. If the ratio of MBC to MIC is ≤4, this indicates that a drug is bactericidal [45], which was the case for ENRO and the ASDs in all species of bacteria in this study. A MIC of ≤0.5 µg/mL may be considered as susceptible to ENRO, while ≥2 µg/mL indicates bacterial resistance, and 1 µg/mL is intermediate [46]. Therefore, from the results of this study, it can be concluded that *E. coli*, *S. aureus*, and *K. pneumoniae* are susceptible to ENRO, while *P. aeruginosa* is not. As was the case with CIP, *E. coli* was found to be the most susceptible of these bacteria to ENRO, having a MIC of 0.004–0.0016 µg/mL. Quite low MIC levels were also obtained in *K. pneumoniae* (0.032–0.125 µg/mL), followed by *S. aureus* (MIC 0.125–0.25 µg/mL). By contrast, much higher MIC and MBC values of 4–8 µg/mL were obtained with *P. aeruginosa*. However, the outer membrane of this bacteria is known to be far less permeable than that of *E. coli*, while fluoroquinolones are also believed to be substrates for an active efflux system within *P. aeruginosa* [47]. In each case, the MIC values obtained for ENRO in these four species agree well with those reported previously [46].

As can be seen from Table 2, the formulation of ENRO as an ASD did not significantly affect its antibacterial activity in any species of bacteria, while the MIC and MBC obtained with ENRO/HPMCAS-MG was significantly lower in *E. coli* and *K. pneumoniae* than in the pure drug. Similar results were previously obtained with CIP ASDs, whereby the MIC and MBC of CIP/HPMCAS-MG was significantly lower than crystalline CIP in all four of these species, while the MIC of CIP/HPMCAS-LG was also significantly reduced in *E. coli*, and its MBC was lower in both *E. coli* and *S. aureus*. These ASDs were also found to increase the passive transmembrane permeability of CIP [9]. Therefore, it is possible that the formulation of ENRO as an ASD with HPMCAS-MG also improved its permeability, enabling more of the drug to be transported through the bacterial cell membranes via passive diffusion.

Table 2. Minimum inhibitory concentration and minimum bactericidal concentration of enrofloxacin and ASDs in various bacteria [a].

Sample	S. aureus	P. aeruginosa	E. coli	K. pneumoniae
Minimum Inhibitory Concentration (µg/mL)				
ENRO	0.25	4	0.016	0.125
ENRO/Eudragit L100	0.125–0.25	4	0.008–0.016	0.063–0.125
ENRO/HPMCAS-LG	0.125–0.25	4–8	0.008–0.016	0.063–0.125
ENRO/HPMCAS-MG	0.125	4	**0.004–0.008**	**0.032–0.063**
Minimum Bactericidal Concentration (µg/mL)				
ENRO	0.25	4	0.032	0.25
ENRO/Eudragit L100	0.25	8	0.016	0.125
ENRO/HPMCAS-LG	0.125	8	0.016	0.125
ENRO/HPMCAS-MG	0.125	4	**0.008**	**0.063**

[a] The values shown in bold differ significantly from those of pure crystalline ENRO.

4. Conclusions

In this study, ball milling was successfully used to produce several ASDs of ENRO. Despite its extra ethyl group, ENRO behaved similarly to CIP in terms of polymer compatibility, with each drug only forming X-ray amorphous ASDs with acidic polymers. The results of FTIR analysis indicate that the terminal tertiary amine of ENRO is protonated in these ASDs and forms an ionic bond with the carboxylate groups of the polymers. The high T_gs of the ASDs and their resistance to crystallization during DSC analysis reinforces the suggestion that strong interactions exist between the components and formation of amorphous polymeric salts. Although the ASDs were hygroscopic, they remained X-ray amorphous during water sorption studies due to the stabilizing effects of the polymers. The ASDs also generated significantly higher drug concentrations than crystalline ENRO during solubility and dissolution testing, and these levels were sustained for the duration of the studies. As the prolongation of supersaturation is believed to increase drug absorption, the in vivo absorption of these formulations is likely to be superior to that of the pure drug. In addition, the antimicrobial activity of ENRO was not decreased by ASD formation, while it was improved by ENRO/HPMCAS-MG in E. coli and S. aureus. This study has therefore demonstrated that the formulation of ENRO as a polymeric ASD, or more accurately, an APS, can improve a number of the drug's biopharmaceutical properties, making this an attractive formulation option.

Supplementary Materials: The following are available online at http://www.mdpi.com/1999-4923/11/6/268/s1, Figure S1: PXRD analysis of solid dispersions formed by milling ENRO with 40–60% (w/w) HPMCAS-LG and HPMCAS-MG for 4 h at room temperature, Figure S2: FTIR peak deconvolution of (a) crystalline ENRO, (b) ball milled ENRO, (c) quench cooled ENRO, (d) ENRO/Eudragit L100, (e) ENRO/Eudragit L100-55, (f) ENRO/Carbopol, (g) ENRO/HPMCAS-LG, and (h) ENRO/HPMCAS-MG. Dotted black line: recorded spectrum; solid blue lines: deconvoluted individual Gauss peaks; and solid red line: sum of the component peaks, Figure S3: FTIR spectra of ENRO semi-crystalline solid dispersions (SD) and physical mixtures (PM) containing (a) 50% (w/w) PVP, and (b) 40% (w/w) PVA, Figure S4: HSDSC analysis of crystalline ENRO using various heating rates, Figure S5: PXRD analysis of ENRO and ENRO ASDs following DVS studies, Figure S6: Water distribution patterns according to the Young–Nelson model in ENRO ASDs containing (a) Eudragit L100-55 40% (w/w) and (b) HPMCAS-MG 60% (w/w), Figure S7: Water distribution patterns according to the Young–Nelson model in CIP ASDs containing (a) Eudragit L100 40% (w/w), (b) Eudragit L100-55 40% (w/w), (c) Carbopol 40% (w/w), (d) HPMCAS-LG 60% (w/w), and (e) HPMCAS-MG 60% (w/w), Figure S8: PXRD analysis of ENRO and ENRO ASDs following solubility studies in FaSSIF, Table S1: Parameters estimated from the Young–Nelson Model for the CIP and ENRO ASDs.

Author Contributions: Conceptualization, H.M. and L.T.; Formal analysis, H.M. and A.U.; Funding acquisition, L.T.; Methodology, H.M., A.U., and L.T.; Supervision, L.T.; Visualization, H.M.; Writing—original draft, H.M.; Writing—review and editing, H.M., A.U., and L.T.

Funding: This research was funded by the Synthesis and Solid State Pharmaceutical Centre (SSPC), financed by a research grant from Science Foundation Ireland (SFI) and co-funded under the European Regional Development Fund (Grant Number 12/RC/2275).

Conflicts of Interest: The authors declare no conflict of interest.

References

1. Mesallati, H.; Mugheirbi, N.A.; Tajber, L. Two Faces of Ciprofloxacin: Investigation of Proton Transfer in Solid State Transformations. *Cryst. Growth Des.* **2016**, *16*, 6574–6585. [CrossRef]
2. Karanam, M.; Choudhury, A.R. Structural landscape of pure enrofloxacin and its novel salts: Enhanced solubility for better pharmaceutical applicability. *Cryst. Growth Des.* **2013**, *13*, 1626–1637. [CrossRef]
3. Escribano, E.; Calpena, A.C.; Garrigues, T.M.; Freixas, J.; Domenech, J.; Moreno, J. Structure-absorption relationships of a series of 6-fluoroquinolones. *Antimicrob. Agents Chemother.* **1997**, *41*, 1996–2000. [CrossRef] [PubMed]
4. Lizondo, M.; Pons, M.; Gallardo, M.; Estelrich, J. Physicochemical properties of enrofloxacin. *J. Pharm. Biomed. Anal.* **1997**, *15*, 1845–1849. [CrossRef]
5. Blokhina, S.V.; Sharapova, A.V.; Ol'khovich, M.V.; Volkova, T.V.; Perlovich, G.L. Solubility, lipophilicity and membrane permeability of some fluoroquinolone antimicrobials. *Eur. J. Pharm. Sci.* **2016**, *93*, 29–37. [CrossRef]
6. Romañuk, C.B.; Manzo, R.H.; Linck, Y.G.; Chattah, A.K.; Monti, G.A.; Olivera, M.E. Characterization of the solubility and solid-state properties of saccharin salts of fluoroquinolones. *J. Pharm. Sci.* **2009**, *98*, 3788–3801. [CrossRef] [PubMed]
7. Brough, C.; Williams, R.O. Amorphous solid dispersions and nano-crystal technologies for poorly water-soluble drug delivery. *Int. J. Pharm.* **2013**, *453*, 157–166. [CrossRef]
8. Yang, J.; Grey, K.; Doney, J. An improved kinetics approach to describe the physical stability of amorphous solid dispersions. *Int. J. Pharm.* **2010**, *384*, 24–31. [CrossRef]
9. Mesallati, H.; Umerska, A.; Paluch, K.J.; Tajber, L. Amorphous Polymeric Drug Salts as Ionic Solid Dispersion Forms of Ciprofloxacin. *Mol. Pharm.* **2017**, *14*, 2209–2223. [CrossRef]
10. Mesallati, H.; Conroy, D.; Hudson, S.; Tajber, L. Preparation and characterization of amorphous ciprofloxacin-amino acid salts. *Eur. J. Pharm. Biopharm.* **2017**, *121*, 73–89. [CrossRef]
11. Mesallati, H.; Tajber, L. Polymer/Amorphous Salt Solid Dispersions of Ciprofloxacin. *Pharm. Res.* **2017**, *34*, 2425–2439. [CrossRef] [PubMed]
12. Chun, M.-K.; Choi, H.-K. Preparation and characterization of enrofloxacin/carbopol complex in aqueous solution. *Arch. Pharm. Res.* **2004**, *27*, 670–675. [CrossRef] [PubMed]
13. Gordon, M.; Taylor, J.S. Ideal copolymers and the second-order transitions of synthetic rubbers. i. non-crystalline copolymers. *J. Appl. Chem.* **2007**, *2*, 493–500. [CrossRef]
14. Lu, Q.; Zografi, G. Phase behavior of binary and ternary amorphous mixtures containing indomethacin, citric acid, and PVP. *Pharm. Res.* **1998**, *15*, 1202–1206. [CrossRef] [PubMed]
15. Ashland AquaSolve Hydroxypropylmethylcellulose Acetate Succinate. Available online: https://www.ashland.com/file_source/Ashland/Industries/Pharmaceutical/Links/PC-12624.6_AquaSolve_HPMCAS_Physical_Chemical_Properties.pdf (accessed on 18 May 2019).
16. Evonik Industries EUDRAGIT® Functional Polymers. Available online: https://healthcare.evonik.com/product/health-care/downloads/evonik-eudragit-brochure.pdf (accessed on 18 May 2019).
17. Surikutchi, B.T.; Patil, S.P.; Shete, G.; Patel, S.; Bansal, A.K. Drug-excipient behavior in polymeric amorphous solid dispersions. *J. Excipients Food Chem.* **2013**, *4*, 70–94.
18. LookChem Enrofloxacin. Available online: http://www.lookchem.com/Enrofloxacin/ (accessed on 18 May 2019).
19. Xiang, T.X.; Anderson, B.D. Molecular dynamics simulation of amorphous hydroxypropyl-methylcellulose acetate succinate (HPMCAS): Polymer model development, water distribution, and plasticization. *Mol. Pharm.* **2014**, *11*, 2400–2411. [CrossRef] [PubMed]

20. *Evonik EUDRAGIT L 100 and EUDRAGIT S 100*; Technical Information; Evonik Inductries AG: Essen, Germany, 2011.
21. *Pharmaceutical Polymers Typical Properties and Specifications*; Lubrizol Advanced Materials, Inc.: Cleveland, OH, USA, 2013.
22. Young, J.; Nelson, G.L. Theory of hysteresis between sorption and desorption isotherms in biological materials. *Trans. Am. Soc. Agric. Eng* **1967**, *10*, 260–263. [CrossRef]
23. Bravo-Osuna, I.; Ferrero, C.; Jiménez-Castellanos, M.R. Water sorption-desorption behaviour of methyl methacrylate-starch copolymers: Effect of hydrophobic graft and drying method. *Eur. J. Pharm. Biopharm.* **2005**, *59*, 537–548. [CrossRef] [PubMed]
24. Umerska, A.; Cassisa, V.; Matougui, N.; Joly-Guillou, M.L.; Eveillard, M.; Saulnier, P. Antibacterial action of lipid nanocapsules containing fatty acids or monoglycerides as co-surfactants. *Eur. J. Pharm. Biopharm.* **2016**, *108*, 100–110. [CrossRef] [PubMed]
25. Löbmann, K.; Laitinen, R.; Grohganz, H.; Strachan, C.; Rades, T.; Gordon, K.C. A theoretical and spectroscopic study of co-amorphous naproxen and indomethacin. *Int. J. Pharm.* **2013**, *453*, 80–87. [CrossRef] [PubMed]
26. Gunasekaran, S.; Anita, B. Spectral investigation and normal coordinate analysis of piperazine. *Indian J. Pure Appl. Phys.* **2008**, *46*, 833–838.
27. Sahil, K.; Prashant, B.; Akanksha, M.; Premjeet, S.; Devashish, R. Interpretation of Infra Red Spectra. *Int. J. Pharm. Chem. Sci.* **2012**, *1*, 174–200.
28. Mahlin, D.; Bergström, C.A.S. Early drug development predictions of glass-forming ability and physical stability of drugs. *Eur. J. Pharm. Sci.* **2013**, *49*, 323–332. [CrossRef] [PubMed]
29. Brostow, W.; Chiu, R.; Kalogeras, I.M.; Vassilikou-Dova, A. Prediction of glass transition temperatures: Binary blends and copolymers. *Mater. Lett.* **2008**, *62*, 3152–3155. [CrossRef]
30. Jensen, K.T.; Löbmann, K.; Rades, T.; Grohganz, H. Improving co-amorphous drug formulations by the addition of the highly water soluble amino acid, Proline. *Pharmaceutics* **2014**, *6*, 416–435. [CrossRef] [PubMed]
31. Konno, H.; Taylor, L.S. Influence of different polymers on the crystallization tendency of molecularly dispersed amorphous felodipine. *J. Pharm. Sci.* **2006**, *95*, 2692–2705. [CrossRef] [PubMed]
32. Lubach, J.W.; Xu, D.; Segmuller, B.E.; Munson, E.J. Investigation of the effects of pharmaceutical processing upon solid-state NMR relaxation times and implications to solid-state formulation stability. *J. Pharm. Sci.* **2007**, *96*, 777–787. [CrossRef] [PubMed]
33. Paluch, K.J.; McCabe, T.; Müller-Bunz, H.; Corrigan, O.I.; Healy, A.M.; Tajber, L. Formation and physicochemical properties of crystalline and amorphous salts with different stoichiometries formed between ciprofloxacin and succinic acid. *Mol. Pharm.* **2013**, *10*, 3640–3654. [CrossRef]
34. Andronis, V.; Yoshioka, M.; Zografi, G. Effects of sorbed water on the crystallization of indomethacin from the amorphous state. *J. Pharm. Sci.* **1997**, *86*, 346–357. [CrossRef] [PubMed]
35. Marsac, P.J.; Konno, H.; Rumondor, A.C.F.; Taylor, L.S. Recrystallization of nifedipine and felodipine from amorphous molecular level solid dispersions containing poly(vinylpyrrolidone) and sorbed water. *Pharm. Res.* **2008**, *25*, 647–656. [CrossRef] [PubMed]
36. Sheokand, S.; Modi, S.R.; Bansal, A.K. Dynamic vapor sorption as a tool for characterization and quantification of amorphous content in predominantly crystalline materials. *J. Pharm. Sci.* **2014**, *103*, 3364–3376. [CrossRef] [PubMed]
37. Miao, P.; Naderi, M.; Acharya, M.; Burnett, D.; Williams, D.; Ng, T.H.; Song, J. *Characterisation of Wheat Straw for Biofuel Application*. DVS Appl. Note 57. Available online: https://www.surfacemeasurementsystems.com/downloads/dvs-application-notes (accessed on 8 June 2019).
38. Zografi, G. States of water associated with solids. *Drug Dev. Ind. Pharm.* **1988**, *14*, 1905–1926. [CrossRef]
39. Mehta, M.; Kothari, K.; Ragoonanan, V.; Suryanarayanan, R. Effect of Water on Molecular Mobility and Physical Stability of Amorphous Pharmaceuticals. *Mol. Pharm.* **2016**, *13*, 1339–1346. [CrossRef] [PubMed]
40. Knapik, J.; Wojnarowska, Z.; Grzybowska, K.; Tajber, L.; Mesallati, H.; Paluch, K.J.; Paluch, M. Molecular Dynamics and Physical Stability of Amorphous Nimesulide Drug and Its Binary Drug-Polymer Systems. *Mol. Pharm.* **2016**, *13*, 1937–1946. [CrossRef]
41. Guzmán, H.R.; Tawa, M.; Zhang, Z.; Ratanabanangkoon, P.; Shaw, P.; Gardner, C.R.; Chen, H.; Moreau, J.P.; Almarsson, Ö.; Remenar, J.F. Combined use of crystalline salt forms and precipitation inhibitors to improve oral absorption of celecoxib from solid oral formulations. *J. Pharm. Sci.* **2007**, *96*, 2686–2702. [CrossRef]

42. Chen, Y.; Liu, C.; Chen, Z.; Su, C.; Hageman, M.; Hussain, M.; Haskell, R.; Stefanski, K.; Qian, F. Drug-polymer-water interaction and its implication for the dissolution performance of amorphous solid dispersions. *Mol. Pharm.* **2015**, *12*, 576–589. [CrossRef]
43. Tanno, F.; Nishiyama, Y.; Kokubo, H.; Obara, S. Evaluation of Hypromellose Acetate Succinate (HPMCAS) as a Carrier in Solid Dispersions. *Drug Dev. Ind. Pharm.* **2004**, *30*, 9–17. [CrossRef]
44. Van den Mooter, G.; Wuyts, M.; Blaton, N.; Busson, R.; Grobet, P.; Augustijns, P.; Kinget, R. Physical stabilisation of amorphous ketoconazole in solid dispersions with polyvinylpyrrolidone K25. *Eur. J. Pharm. Sci.* **2000**, *12*, 261–269. [CrossRef]
45. Pankey, G.A.; Sabath, L.D. Clinical Relevance of Bacteriostatic versus Bactericidal Mechanisms of Action in the Treatment of Gram-Positive Bacterial Infections. *Clin. Infect. Dis.* **2004**, *38*, 864–870. [CrossRef] [PubMed]
46. Bayer HealthCare LLC Animal Health Division Baytril®—Enrofloxacin Injection, Solution. Available online: https://dailymed.nlm.nih.gov/dailymed/drugInfo.cfm?setid=7deb5c76-90c2-471c-9e68-bcc088af5cac (accessed on 18 May 2019).
47. Hancock, R.E.W. Resistance Mechanisms in Pseudomonas aeruginosa and Other Nonfermentative Gram-Negative Bacteria. *Clin. Infect. Dis.* **1998**, *27*, S93–S99. [CrossRef] [PubMed]

© 2019 by the authors. Licensee MDPI, Basel, Switzerland. This article is an open access article distributed under the terms and conditions of the Creative Commons Attribution (CC BY) license (http://creativecommons.org/licenses/by/4.0/).

Article

Delivery of Poorly Soluble Drugs via Mesoporous Silica: Impact of Drug Overloading on Release and Thermal Profiles

Tuan-Tu Le, Abdul Khaliq Elzhry Elyafi, Afzal R. Mohammed and Ali Al-Khattawi *

Aston Pharmacy School, School of Life and Health Sciences, Aston University, Birmingham B4 7ET, UK; let3@aston.ac.uk (T.-T.L.); elzhryea@aston.ac.uk (A.K.E.E.); a.u.r.mohammed@aston.ac.uk (A.R.M.)
* Correspondence: a.al-khattawi@aston.ac.uk; Tel.: +44-(0)-121-204-4735

Received: 5 April 2019; Accepted: 3 June 2019; Published: 10 June 2019

Abstract: Among the many methods available for solubility enhancement, mesoporous carriers are generating significant industrial interest. Owing to the spatial confinement of drug molecules within the mesopore network, low solubility crystalline drugs can be converted into their amorphous counterparts, which exhibit higher solubility. This work aims to understand the impact of drug overloading, i.e., above theoretical monolayer surface coverage, within mesoporous silica on the release behaviour and the thermal properties of loaded drugs. The study also looks at the inclusion of hypromellose acetate succinate (HPMCAS) to improve amorphisation. Various techniques including DSC, TGA, SEM, assay and dissolution were employed to investigate critical formulation factors of drug-loaded mesoporous silica prepared at drug loads of 100–300% of monolayer surface coverage, i.e., monolayer, double layer and triple layer coverage. A significant improvement in the dissolution of both Felodipine and Furosemide was obtained (96.4% and 96.2%, respectively). However, incomplete drug release was also observed at low drug load in both drugs, possibly due to a reversible adsorption to mesoporous silica. The addition of a polymeric precipitation inhibitor HPMCAS to mesoporous silica did not promote amorphisation. In fact, a partial coating of HPMCAS was observed on the exterior surface of mesoporous silica particles, which resulted in slower release for both drugs.

Keywords: mesoporous; poorly soluble drugs; solubility enhancement; solid dispersion; amorphisation; spray drying

1. Introduction

The most important properties of promising drug candidates for oral dosage form are aqueous solubility and intestinal permeability. Over 40% of drugs on the market are BCS class II and IV, which have low solubility. Furthermore, new chemical entities are even less soluble compared to marketed products with a projection of up to 70–90% of drug candidates in the pipeline suffer from low solubility [1]. Following oral administration, drugs must dissolve in gastrointestinal fluids in order to be absorbed into the systemic circulation and exert a therapeutic action. The formulation development of low solubility drugs (BCS class II and IV) faces a great challenge as these drugs are poorly absorbed and usually exhibit subsequent low or variable oral bioavailability [2].

The problem of low solubility can be addressed by using solubilisation techniques, namely solid dispersion systems, size reduction, salt formation, prodrug, liposomes, etc. Among those techniques, solid dispersions are preferred by the industry due to their practicality and low cost. This technique is mainly based on a so-called "amorphisation", whereby the crystalline drugs are converted into their high energy amorphous form, which exhibits a superior solubility in comparison with that of the original ones [3]. Mesoporous materials, e.g., mesoporous silica, a subclass of solid dispersion systems, are considered highly effective for drug amorphisation due to their ability to achieve the spatial confinement of drug molecules within their nanometre-scale pore structure [4]. Shen et al. [5] suggested that drugs would exist in an amorphous state within the mesoporous silica if the pore size were smaller than 12 times the drug's molecular size. Mesoporous carriers also offer formulation flexibility due to tunable pore size and surface area [6]. Furthermore, this approach has shown applicability for both existing poorly soluble drugs and drug candidates in the pipeline with various chemical structures. Utilising mesoporous silica to enhance the bioavailability of poorly soluble drugs has been successfully demonstrated in various clinical studies conducted on rabbits, dogs, and mice [7].

For loading a drug into mesoporous silica, there are several techniques, which can be categorised into two main approaches: solvent-free methods and solvent-based methods. Solvent-free methods are comprised of physical mixing followed by heating to melt the drug, co-milling between the drug and mesoporous materials, and using supercritical carbon dioxide. Although solvent-free loading methods offer apparent advantages, e.g., no requirement for checking the residual solvent in drug products and low environmental impact, these methods are still under investigation to exhibit better performance in terms of the loading efficiency and stability of thermolabile drugs. On the other hand, solvent-based approaches offer a practical and straightforward solution for drug amorphisation within mesoporous silica. Simply put, a drug is dissolved in a suitable solvent, e.g., ethanol, then mixed/impregnated with mesoporous silica. The solvent can be removed by appropriate drying techniques at the end of the process. There are various factors that influence drug loading into mesoporous silica, such as the type of solvent, drug load, accessible surface area, and the pore volume of the mesoporous silica. In general, solvent-based loading techniques, especially spray drying, produce drug-loaded mesoporous silica with a high loading efficiency compared to solvent-free techniques [6]. However, higher drug loadings above 30% (w/w) could lead to incomplete amorphisation, i.e., a small amount of crystalline drug will remain on the exterior surface of the generated particles [5]. Drug molecules can theoretically adsorb onto the silica surface of mesopores as a monolayer or multilayers, depending on the drug's molecular dimension, accessible surface area, and pore size [8]. Dening and Taylor [9] studied ritonavir–loaded–mesoporous silica at various drug loads of a 25–150% monolayer surface coverage and found that drug release decreased significantly as the drug load increased. Therefore, it would be prudent to systematically investigate the impact of drug loading beyond monolayer surface coverage (overloading) on the thermal behaviour of drug within mesoporous silica. Furthermore, the addition of a precipitation inhibitor to the mesoporous silica has been previously investigated to overcome this recrystallisation challenge. Lainé et al. [10] found that hypromellose acetate succinate (HPMCAS), when combined with mesoporous silica, promoted a complete amorphisation of Celecoxib. However, the advantage of such ternary system combining mesoporous silica with drug and precipitation inhibitors still requires further study, to confirm whether poorly soluble drugs with different chemical natures and crystallisation tendencies can benefit from it.

Felodipine and Furosemide are BCS class II and class IV drugs, respectively. Felodipine is mainly absorbed in the small intestine [11], i.e., and alkaline pH environment, while the stomach is the favoured absorption site for Furosemide [12], i.e., acidic pH environment. Hence, Felodipine and Furosemide were selected as model drugs in this study to represent two scenarios after oral administration in drug release from mesoporous silica. The aim of this study was to investigate the impact of drug overloading within mesoporous silica at 100–300% of the theoretical monolayer's surface coverage on the release behaviour and thermal properties of loaded drugs. This included investigating the release profiles of the model drugs after loading within mesoporous silica alone and in combination with HPMCAS. Thermal profiles and particle morphology were also studied to confirm the amorphous/crystalline nature of the drug within the carrier system and the presence or lack of surface crystals. These studies will help elucidate the link between theoretical drug load and important formulation properties, such as loading efficiency, release behaviour, and the nature of the drug within mesoporous carriers.

2. Materials and Methods

2.1. Materials

Felodipine (FELO) and Furosemide (FURO) were obtained from Discovery Fine Chemicals (Dorset, UK) and Chemical Point (Surrey, UK), respectively. Mesoporous silica Syloid® XDP 3050 (specific surface area of 310 m²/g, average pore size of 22.4 nm, pore volume of 1.74 cm³/g) was kindly provided by W.R. Grace and Co. (Worms, Germany). Aqoat® (HPMCAS) was a generous gift from Harke Pharma (Muelheim an der Ruhr, Germany). Sodium phosphate monobasic, sodium phosphate dibasic, sodium chloride, and sodium lauryl sulfate (SLS) were purchased from Sigma-Aldrich (Dorset, UK). Hydrochloric acid 37%, acetone and ethanol were purchased from Fisher Scientific (Loughborough, UK). Deionised water was produced by Milli-Q Integral system (Hertfordshire, UK).

2.2. Methods

2.2.1. Preparation of Drug-Loaded Mesoporous Silica Particles

Theoretical drug load was calculated based on an assumption that drug molecules would adsorb to the surface area of mesoporous silica particles in a packing geometry that increases the bonding between drug molecules and silica surface, i.e., to maximise the contact surface [9]. The following equation described by Dening and Taylor [9] was used to calculate the drug load (%, w_{drug}/w_{Syloid}) to theoretically obtain monolayer adsorption (equivalent to 100% surface coverage) in mesoporous silica:

$$\text{Theoretical drug load at monolayer adsorption} \left(\%, \frac{g}{g}\right) = \frac{SSA \times M_w \times 10^{20}}{SA_M \times N_A}. \quad (1)$$

SSA: Specific surface area of mesoporous silica (m²/g), e.g., 310 m²/g for Syloid XDP 3050 (in-house data measured by gas adsorption porosimetry).

M_w: Molecular weight of model drug (g/mol).

SA_M: Maximum projected contact surface area of single molecule (Å²): calculated using the two largest molecular dimensions of drug molecule (Figure 1).

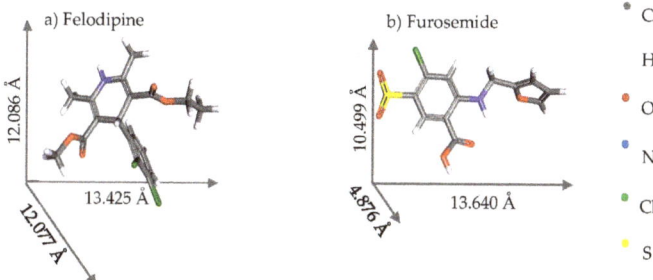

Figure 1. Molecular structures [13,14] and estimated molecular dimensions of Felodipine and Furosemide generated from the Cambridge Crystallographic Data Centre.

N_A: Avogadro's number (6.022 × 10²³).

Syloid was added to either FELO or FURO in ethanol (10 mg/mL) to form suspensions at various theoretical drug loads (%, w/w): 10.8, 21.6, and 32.4% for FURO; 12.6, 25.2, and 37.8% for FELO to represent monolayer adsorption (100% surface coverage), double layer adsorption (200% surface coverage), and triple layer adsorption (300% surface coverage), respectively. The ternary Drug–Syloid–HPMCAS formulations were prepared in a similar fashion with the ratio of drug to HPMCAS at 10:1. Syloid was added to solutions of HPMCAS and either FELO or FURO in ethanol-acetone 50:50 (10 mg/mL). All of the suspensions were gently stirred for 12 h, then spray-dried at inlet temperature of 100 °C using a mini spray dryer Buchi B-290 and inert loop Buchi B-295 (Flawil, Switzerland) in closed mode with a nitrogen flow rate of 600 L/min, with a feed rate of 5 mL/min and a drying gas flow rate of 30 m³/h. Spray-dried FELO or FURO (prepared in the same procedure without the incorporation of either Syloid or HPMCAS), physical mixtures between Syloid and either FELO, or FURO with a ratio of 1:1 were used as control samples.

2.2.2. Drug Loading Quantification

Drug loading within mesoporous silica samples was determined according to a TGA-based method described elsewhere [15–17]. Approximately 10 mg of drug–loaded mesoporous silica samples were placed into a platinum pan, then transferred to TGA instrument Perkin-Elmer Pyris 1 (Buckinghamshire, UK). The sample was treated with following temperature programme under a nitrogen flow of 20 mL/min: (1) heated from 50 to 120 °C and held for 5 min at 120 °C; (2) heated from 120 to 800 °C at a heating rate of 20 °C/min; and (3) held for 30 min at 800 °C. The drug load of FURO or FELO inside mesoporous silica was determined by the difference in weight loss between samples and blanks at a temperature range of 200–800 °C, whereby FELO or FURO were incinerated. The drug load was normalised to the surface area of the mesoporous silica to facilitate the comparison between studies using various types of mesoporous silica. Normalised surface area-based drug load and loading efficiency was calculated based on the two following equations.

$$\text{Surface area based drug load} \left(m^2/g\right) = \frac{\text{Actual drug load}}{\text{Specific surface area of mesoporou silica}} \quad (2)$$

$$\text{Loading efficiency } (\%) = \frac{\text{Actual drug load}}{\text{Theoretical drug load}} \quad (3)$$

2.2.3. In Vitro Drug Release Studies

Dissolution testing was performed by using a USP I apparatus (rotating basket, 50 rpm) in an Erweka DT 126 dissolution tester (Heusenstamm, Germany). Each sample containing 20 mg of FELO was filled into a HPMC hard-shell capsule and tested in 500 mL of pH 6.5 medium with 0.25% SLS at 37 °C (adapted from USP 36 monograph with a reduction of SLS concentration from 1.0 to 0.25%). Samples were withdrawn during a 120 min period at the following timepoints: 15, 30, 60, 90, and 120 min. The concentrations of dissolved FELO were determined according to a HPLC method described in United States Pharmacopoeia [18] with mobile phase of pH 3 phosphate buffer:acetonitrile:methanol (30:45:25), column C18 (15 cm × 4.6 mm, 5 µm), flow rate of 1 mL/min, injection volume of 40 µL, and UV detector at 362 nm in an Agilent 1200 HPLC system (Santa Clara, CA, USA).

For FURO, samples containing 40 mg was filled into a HPMC hard-shell capsule and tested in 900 mL of a pH 3.0 medium with 0.25% SLS at 37 °C. The medium was prepared by dissolving 2 g of sodium chloride and 2.5 g of SLS in 400 mL of deionised water, then adding 0.1 mL of hydrochloric acid 37% and diluting with deionised water to 1000.0 mL [17]. The concentrations of dissolved FURO were determined using a HPLC method described elsewhere [19] with mobile phase of a phosphate buffer of pH 3: acetonitrile 60:40, C18 column, column temperature: 35 °C, flow rate of 1 mL/min, injection volume of 10 µL, and UV detection at 234 nm.

2.2.4. Differential Scanning Calorimetry (DSC)

The thermal properties of samples were characterised by DSC instrument TA Q200 (New Castle, DE, USA). Each sample was accurately weighed (equivalently to 1 mg of FELO or FURO) into Tzero low-mass aluminium pan (sensitivity for a minimum sample size of 0.5 mg), and heated in the range of 50–250 °C (for FELO) or 100–300 °C (for FURO) at a scanning rate of 10 °C/min under nitrogen airflow of 50 mL/min. TA universal analysis 2000 software (version 4.5) was employed to analyse the resulting DSC thermograms.

2.2.5. Scanning Electron Microscopy (SEM)

The surface of the drug-loaded mesoporous silica particles was examined by a Philips XL30 ESEM FEG (Hillsboro, OR, USA) operating at 10 kV under a high vacuum. Prior to SEM imaging, samples were coated with gold by a sputter coater. Approximately 1 mg of each sample was placed onto a double-sided adhesive strip on a sample holder. SEM images were taken at 2000× magnification.

2.2.6. Statistical Analysis

Statistical analysis was carried out using GraphPad Prism 7.03 software. Statistically significant difference was considered at a *p* value < 0.05. All results are presented as mean ± standard deviation where applicable.

3. Results and Discussion

3.1. Thermal Profiles and Morphology of Drug-Loaded Mesoporous Silica

DSC analysis of Felodipine and Furosemide samples were presented in Figures 2 and 3. Results revealed that Felodipine was completely converted to amorphous form inside mesoporous silica at all drug loads. This can be confirmed through the lack of melting peak of crystalline Felodipine (146.0 °C) in DSC thermograms of any Felodipine-Syloid formulations. In contrast, raw material and spray dried Felodipine (without mesoporous silica), exhibited sharp endothermic peaks at 146.0 ± 0.6 °C, 144.3 ± 1.8 °C, respectively, confirming their crystalline state [20], as can be seen in Figure 4b. The DSC data also indicated there was no interaction between Felodipine and Syloid in their physical mixture as there was no change in temperature (146.0 ± 0.5 °C) and the shape of the Felodipine melting peak.

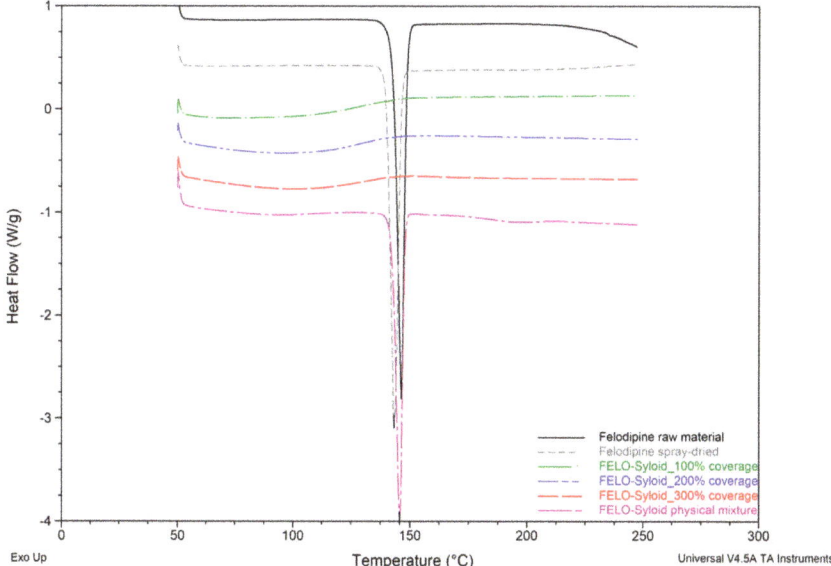

Figure 2. Differential Scanning Calorimetry (DSC) thermograms of Felodipine raw and spray-dried materials, Felodipine (FELO)-Syloid formulations at various drug loads, FELO-Syloid physical mixture.

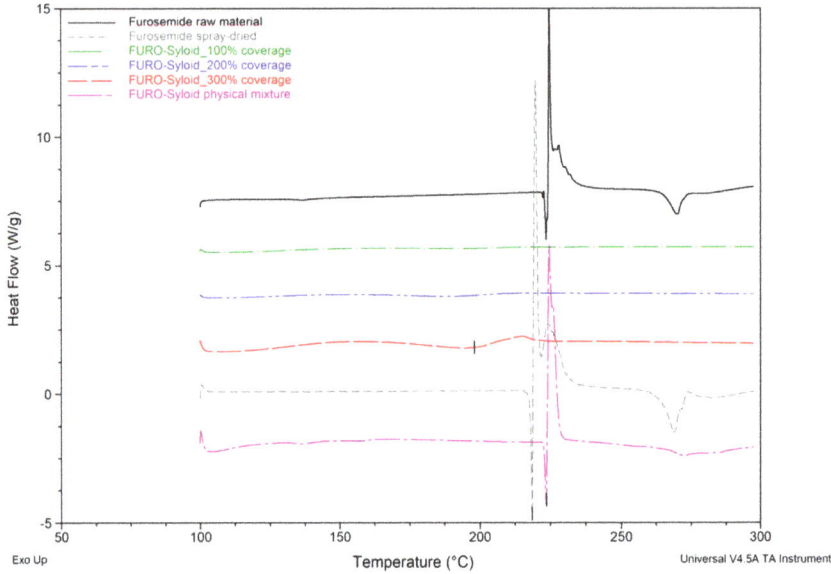

Figure 3. DSC thermograms of Furosemide raw and spray-dried materials, Furosemide (FURO)-Syloid formulations at various drug loads, FURO-Syloid physical mixture.

Sharp endothermic peaks at 222.8 ± 0.8 °C and 223.9 ± 0.3 °C were observed in DSC curves of Furosemide raw material and FURO-Syloid physical mixture respectively (Figure 3), which is in agreement with the crystalline Furosemide in previous study [21,22]. This was further verified by SEM image (Figure 4e). The DSC data of physical mixtures between Syloid and Furosemide or Felodipine suggested that the drug still remains crystalline if deposited externally onto mesoporous silica particles, i.e., the physical mixture had no effect on amorphisation. Spray-dried Furosemide exists in crystalline form after the spray drying process, as confirmed through endothermic peak at 218.9 ± 0.7 °C. DSC analysis also revealed that Furosemide loaded within Syloid was completely amorphised at drug loads of 100% and 200% surface coverage as no endothermic peak was detected. However, at 300% coverage a broad endothermic peak was detected at a temperature of 198.7 ± 4.3 °C, indicating a small amount of crystalline Furosemide. In addition, this endothermic peak is shifted slightly to a lower temperature (198.7 °C) compared to that of raw material (222.8 °C), possibly due to the presence of nanocrystals. This result is consistent with a previous observation of Ibuprofen-loaded mesoporous silica [15], whereby researchers suggested that a nanocrystal form would cause a melting point shift. The formation of Furosemide nanocrystals at the highest drug load of 300% surface coverage can be observed in SEM image (Figure 4f). After drug loading, the surface of silica becomes rough as can be seen in FELO-Syloid (Figure 4c), particularly in FURO-Syloid with many surface crystallites in comparison with original surface of mesoporous silica, which is relatively smoother (Figure 4a).

TGA results (Figure 5) were used to determine the drug load and loading efficiency and complement the thermal events observed in the DSC data. TGA curves of the original Syloid as well as the drug-loaded Syloid revealed a small weight loss at 130.9 ± 0.6 °C, which can be attributed to bound water loss. There was no further significant weight loss in the original Syloid from 200 to 800 °C. TGA weight% versus temperature and 1st derivative curves of Felodipine-Syloid showed a substantial weight loss caused by the drug decomposition in the range 200–800 °C. Felodipine has one-step decomposition starting from 277.7 ± 0.8 °C and reach the maximum decomposition rate at 338.3 ± 5.7 °C, while Furosemide has a multi-step decomposition, which takes place between 200 and 800 °C with an onset temperature of 227.5 ± 3.4 °C. Thermal decomposition of both Felodipine and

Furosemide are similar to the reported studies [22,23]. There was no notable difference in the shape of the TGA curves, as well as 1st derivative of the weight between the drug-loaded Syloid and ternary systems. This may have been caused by an overlap between the decomposition of drugs and that of the HPMCAS which happens in the range of 270 to 300 °C.

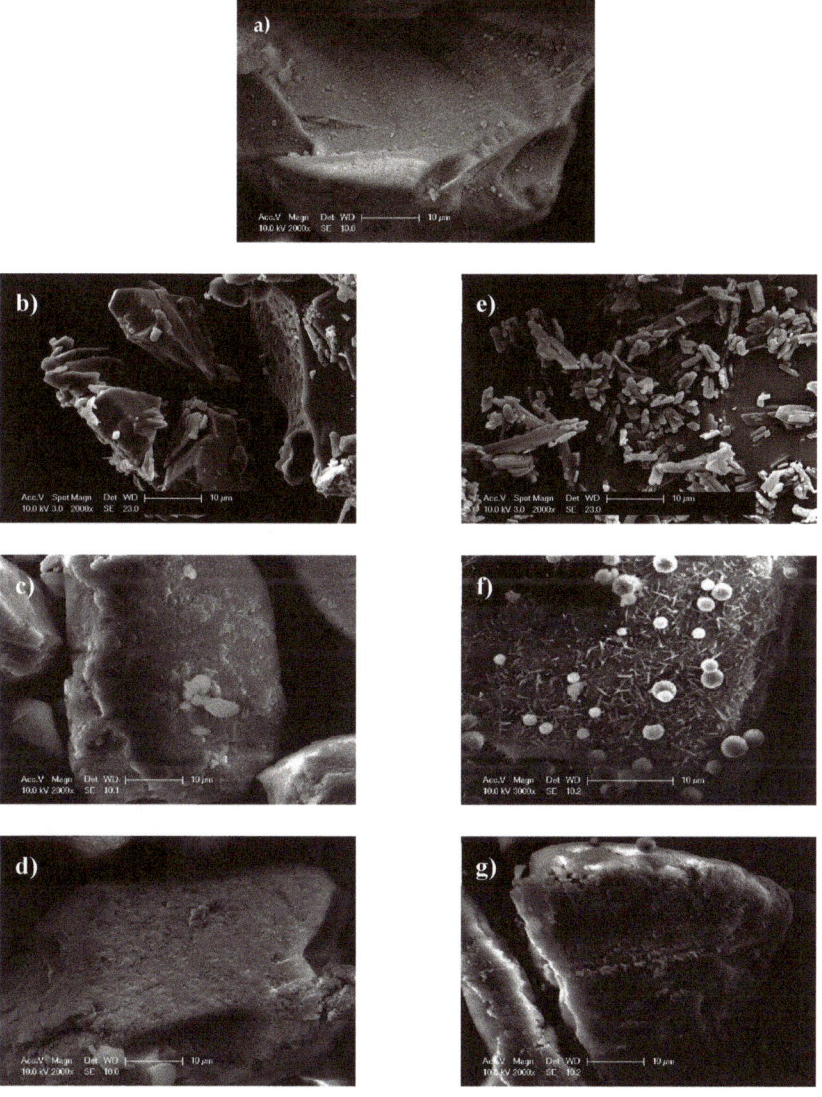

Figure 4. Particle surfaces of (**a**) mesoporous silica Syloid, (**b**) Felodipine raw material, (**c**) FELO-Syloid, (**d**) FELO-Syloid-hypromellose acetate succinate (HPMCAS), (**e**) Furosemide raw material, (**f**) FURO-Syloid, and (**g**) FURO-Syloid-HPMCAS. SEM images were taken for samples at a drug load of 300% surface coverage.

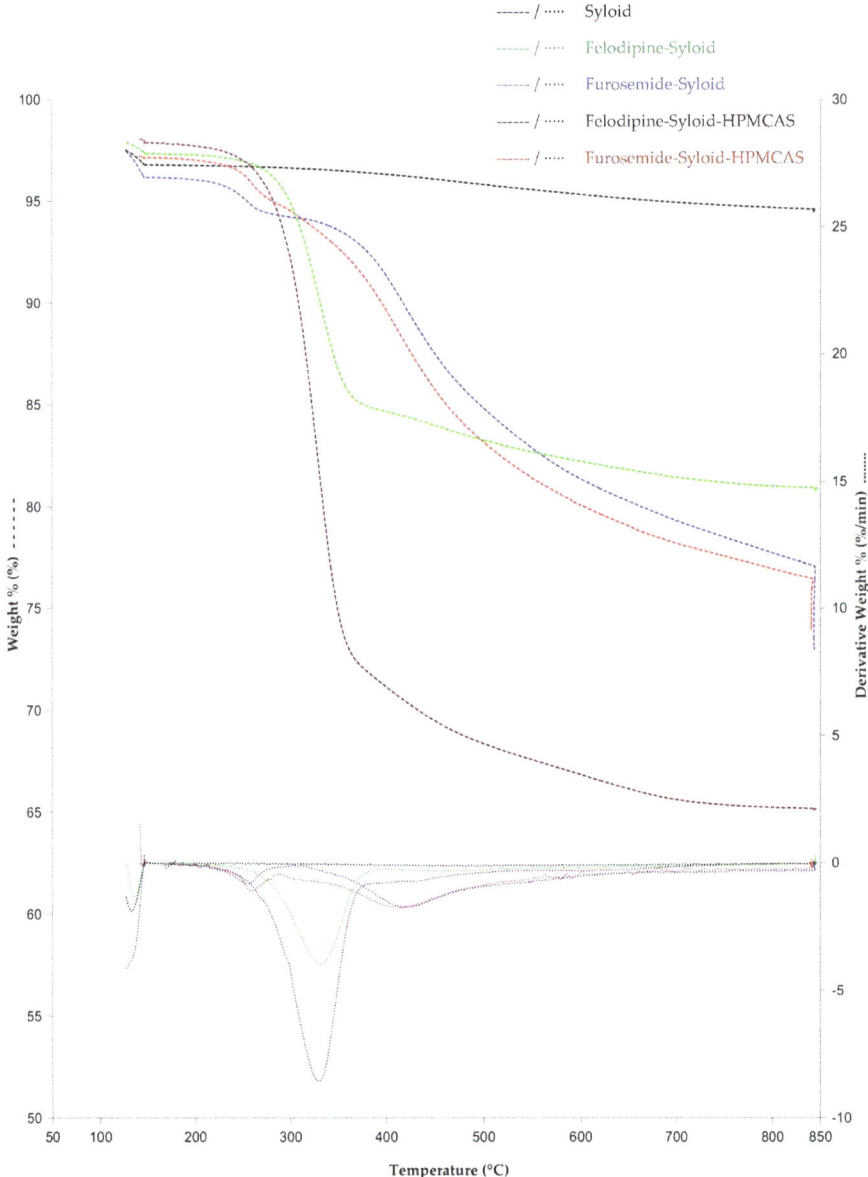

Figure 5. TGA weight% versus temperature and derivative weight% curves of Syloid, Felodipine-Syloid, Furosemide-Syloid, Felodipine-Syloid-HPMCAS, and Furosemide-Syloid-HPMCAS at a drug load of 300% surface coverage. Temperature programme under a nitrogen flow of 20 mL/min: (1) heat from 50 °C to 120 °C and held for 5 min at 120 °C; (2) heat from 120 °C to 800 °C at a heating rate of 20 °C/min; (3) held for 30 min at 800 °C.

Actual drug loads and loading efficiencies of Felodipine and Furosemide in mesoporous silica are presented in Table 1. As expected, the actual drug load increased as the theoretical surface coverage increased. Loading into mesoporous silica could be explained by physical adsorption, which depends on the extent of the available surface area of adsorbent and the amount of adsorbate. The loading process carried out at a low drug load, i.e., a 100% surface coverage, resulted in maximum loading efficiency (99.2% for Furosemide and 101.9% for Felodipine) and complete amorphisation. However, at higher drug load at 300% surface coverage, the free surface area of the adsorbent (mesoporous silica) became lower as more drug was added, leading to a lower loading efficiency of 57.4% and 73.8% for Furosemide and Felodipine, respectively.

Table 1. Drug loads and loading efficiency of Felodipine and Furosemide in mesoporous silica Syloid XDP 3050 (surface area of 310 m^2/g).

% of Theoretical Surface Coverage	Felodipine-Syloid				Furosemide-Syloid			
	Theoretical Drug Load (%, g/g)	Actual Drug Load (%, g/g)	Normalised Actual Drug Load (×10^{-3} g/m^2)	Loading Efficiency (%)	Theoretical Drug Load (%, g/g)	Actual Drug Load (%, g/g)	Normalised Actual Drug Load (×10^{-3} g/m^2)	Loading Efficiency (%)
100%	12.6	12.5 ± 0.3	0.403 ± 0.010	99.2 ± 2.4	10.8	11.0 ± 0.9	0.355 ± 0.029	101.9 ± 8.3
200%	25.2	15.6 ± 0.2	0.503 ± 0.006	61.9 ± 0.8	21.6	16.6 ± 0.3	0.535 ± 0.010 *	76.9 ± 1.4
300%	37.8	21.6 ± 0.3	0.697 ± 0.010 *	57.4 ± 0.8	32.4	23.9 ± 0.5	0.771 ± 0.016	73.8 ± 1.5

(*): Drug load at which a complete amorphisation was successfully obtained.

Previously, Ambrogi et al. [17] studied Furosemide-loaded mesoporous silica prepared by rotary evaporation and were able to attain an amorphous state Furosemide with SBA-15 silica at a drug content of up to 30.0% ± 0.2% (0.3 g of Furosemide per 0.7 g of SBA-15 mesoporous silica). Furosemide-Syloid in our study exhibited complete amorphisation with a drug load of up to 16.6 ± 0.3%, i.e., 0.166 g of Furosemide per 1 g of Syloid. However, it should be noted that the SBA-15 silica used by Ambrogi et al. [17] had a surface area which was 2.5 times larger than that of Syloid (791 m^2/g vs. 310 m^2/g, respectively). Hence, despite the apparent difference in weight-per-weight drug load, after normalisation based on surface area, there is no difference in weight-per-surface area drug load between the two studies (0.542 ± 0.004 × 10^{-3} g/m^2 in comparison with 0.535 ± 0.010 × 10^{-3} g/m^2).

For Felodipine-loaded mesoporous silica, there were two previous studies using solvent impregnation methods, in which silica had a specific surface area of 584 m^2/g and 1051 m^2/g with a drug load of 25% [24] and 18.3% [25], respectively. The amorphous Felodipine inside Syloid was obtained at a drug load of 21.7%. Similar to Furosemide, the weight-per-weight drug load of Felodipine-mesoporous silica was converted to weight-per-surface area to enable a like-for-like comparison between different studies. The Felodipine-Syloid in our study produced a higher weight-per-surface area drug load in comparison with those of the other two studies (0.697 × 10^{-3} compared to 0.428 × 10^{-3} g/m^2 and 0.174 × 10^{-3}/m^2).

In general, the drug loading process for mesoporous particles is usually established on a case-by-case basis due to the differences in loading solvent, solubility, targeted drug load, surface area, pore size, and pore volume of the mesoporous materials being used. The use of co-spray drying in the current study as a solvent-based technique produced comparable or slightly higher drug loads to those reported in the literature using traditional solvent-evaporation. However, the real added advantage of using co-spray drying lies in the better drug loading efficiency and shorter processing time. During spray drying, there is a possibility that solute/drug diffusion towards the particle centre might facilitate drug entrapment inside the pores. The findings of higher drug loading efficiency via spray drying are also in agreement with previous studies [26,27].

3.2. In Vitro Drug Release from Mesoporous Silica

Results showed that after 30 min the dissolution of all Felodipine-Syloid or Furosemide-Syloid samples were much higher than that of the physical mixtures, both raw and spray-dried materials ($p < 0.05$). Felodipine exhibited a low dissolution within 60 min with only 9.8%, 11.4%, and 15.5% dissolution for the physical mixture, raw materials, and spray-dried materials, respectively, compared to over 70% dissolution in any Felodipine-Syloid samples (Figure 6). Similarly, within the first 60 min, the percentage of Furosemide released from mesoporous silica at 100%, 200%, and 300% of theoretical surface coverage reached 69.7%, 83.4%, and 86.5%, respectively while the physical mixture, the raw Furosemide and spray-dried materials, achieved only 56.6%, 46.2%, and 50.6% dissolution, respectively (Figure 7).

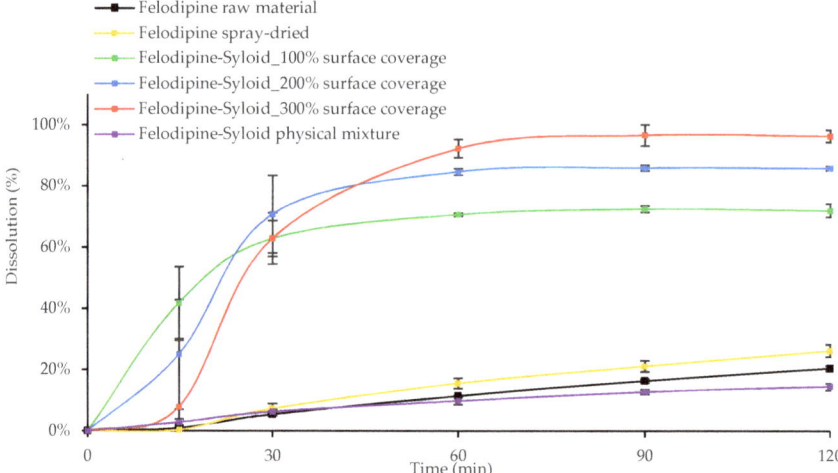

Figure 6. Release profiles of Felodipine raw and spray-dried materials, Felodipine-Syloid formulations at various drug loads, Felodipine-Syloid physical mixture. Testing conditions: phosphate buffer pH 6.5 + 0.25% SLS, 500 mL, USP apparatus 1, 50 rpm. The medium was adapted from a USP 36 monograph with SLS used to maintain a sink condition at the minimum possible amount.

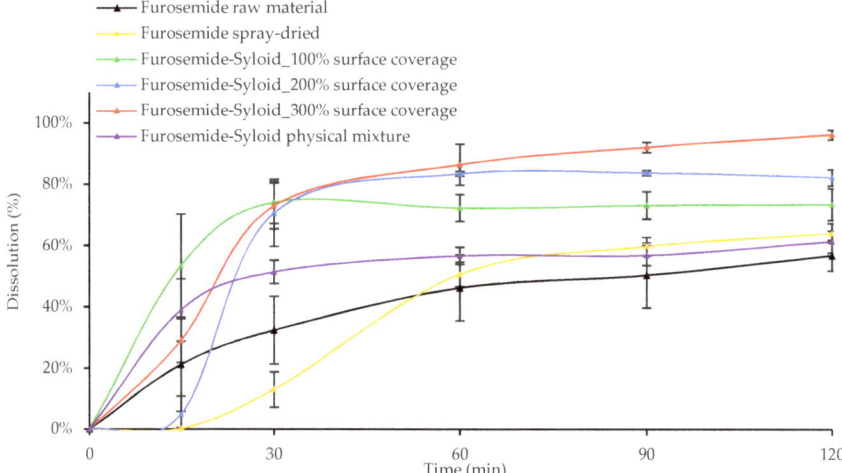

Figure 7. Release profiles of Furosemide raw and spray-dried materials, Furosemide-Syloid formulations at various drug loads, Furosemide-Syloid physical mixture. Testing conditions: HCL-NaCl medium pH 3 + 0.25% sodium lauryl sulfate (SLS), 900 mL, USP apparatus 1, 100 rpm. Medium was adapted from a study carried out by Ambrogi et al. [17] with the addition of SLS to maintain sink condition.

Although mesoporous silica is capable of enhancing the dissolution of Felodipine and Furosemide, a reversible adsorption between drug molecules and silica surface might happen. This result is in agreement with a previous study by Dening and Taylor [9], which investigated the release of ritonavir from mesoporous silica. The authors also found that drug release of ritonavir-loaded-mesoporous silica was incomplete. They suggested that there is a dynamic adsorption equilibrium between the drug adsorbed to mesoporous silica and the free drug dissolved in a dissolution medium. This dynamic equilibrium caused an incomplete drug release in dissolution testing, especially at a low drug load of 100% surface coverage (72.0% and 73.5% dissolution after 2 h for Felodipine and Furosemide, respectively). At this point, drug release from the drug-mesoporous silica samples reached a plateau, indicating that an adsorption equilibrium had been formed. However, as can be seen in our study, this dynamic equilibrium can be shifted in favour of more drug release via increasing the degree of drug load, i.e., silica surface coverage, possibly because the amount of drug adsorbed to the silica particles is negligible to that of the dissolved drug. Generally, the dissolution increased for both drugs, as the percentage of silica surface coverage increased. This is expected, as the higher amount of drug loaded inside the mesoporous silica released more drugs to the dissolution medium before reaching an adsorption equilibrium.

3.3. Influence of the Addition of HPMCAS to Mesoporous Silica on Drug Loading and Dissolution

In an attempt to increase loading efficiency and overcome the incomplete amorphisation at the highest drug load of 300% (in Section 3.1), HPMCAS was added to the drug-mesoporous silica to create a ternary system. The addition of HPMCAS to the Syloid resulted in increasing the Felodipine loading efficiency compared to that of the Syloid alone ($p < 0.05$). A weak basic drug Felodipine in a loading solution can partially dissociate to form a negatively charged anion, which might be converted into a conjugation with an acidic polymer HPMCAS. Such a conjugation between Felodipine and HPCMAS could enhance the loading efficiency as there could be an extra amount of Felodipine staying within the polymer matrix. However, there were no significant improvements in loading efficiency for Furosemide with the presence of HPMCAS ($p > 0.05$). This result could be explained because Furosemide is an acidic drug, and, therefore, no conjugation was formed between Furosemide and HPMCAS in comparison to the synergistic effect in drug loading between Felodipine and HPMCAS (Table 2).

Table 2. Drug load and loading efficiency of Felodipine and Furosimide in mesoporous silica Syloid XDP 3050 with the incorporation of HPMCAS.

% of Theoretical Surface Coverage	Felodipine-Syloid-HPMCAS			Furosemide-Syloid-HPMCAS		
	Theoretical Drug Load (%, g/g)	Actual Drug Load (%, g/g)	Loading Efficiency (%)	Theoretical Drug Load (%, g/g)	Actual Drug Load (%, g/g)	Loading Efficiency (%)
100%	12.6	13.1 ± 0.2	104.0 ± 1.6	10.8	10.7 ± 0.6	99.1 ± 5.5
200%	25.2	20.3 ± 0.5	80.6 ± 1.9	21.6	14.7 ± 1.3	68.1 ± 6.0
300%	37.8	30.3 ± 0.2	80.2 ± 0.5	32.4	23.8 ± 0.6	73.5 ± 1.9

DSC results suggested that the addition of HPMCAS to the mesoporous silica did not result in the promotion of drug amorphisation. A small endothermic peak at $139.9 \pm 0.2\ °C$ was detected in the DSC thermogram of ternary Felodipine-Syloid-HPMCAS, at a drug load of 300% surface coverage (Figure 8). This indicated a small amount of crystalline Felodipine in the ternary system, although Felodipine-Syloid stayed amorphous in the absence of HPMCAS. The negative impact on amorphisation was clearly evidenced by the DSC data of the ternary Furosemide-Syloid-HPMCAS. Furosemide-Syloid at drug load of 100% or 200% surface coverage existed in an amorphous state (Section 3.1). Contrastingly, ternary Furosemide-Syloid-HPMCAS exhibited an incomplete amorphisation at all drug loads which was confirmed by broad endothermic peaks (Figure 9).

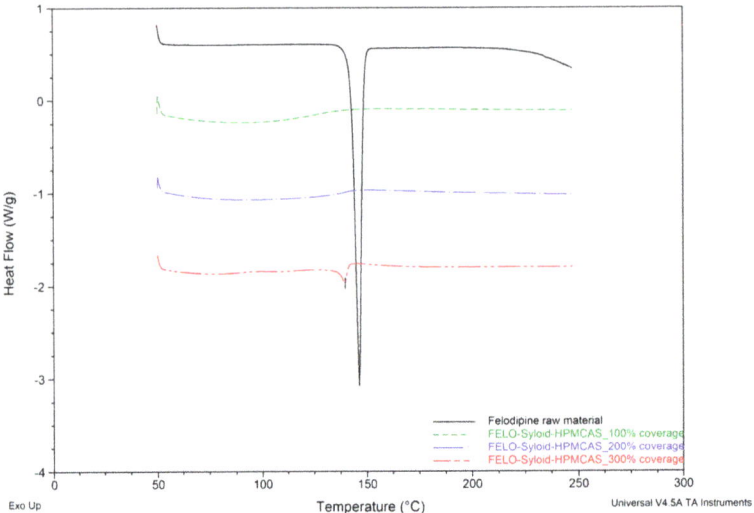

Figure 8. DSC thermograms of raw Felodipine material and FELO-Syloid-HPMCAS at various drug loads. Scanning rate: 10 °C/min. Scanning range: 50–250 °C.

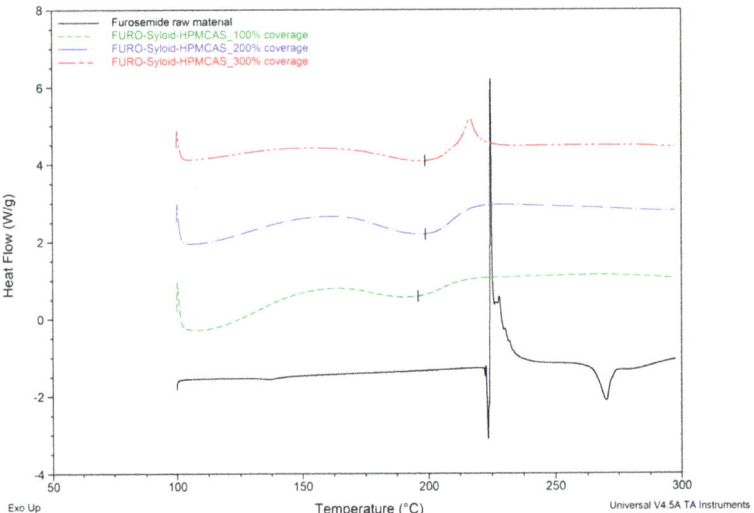

Figure 9. DSC thermograms of the raw Furosemide material and FURO-Syloid-HPMCAS at various drug loads. Scanning rate: 10 °C/min. Scanning range: 100–300 °C.

SEM images showed, to a certain extent, that HPMCAS might work as a coating polymer outside silica particles. In fact, a partial coating effect of HPMCAS as new surface ruggedness or layers could be seen in both Felodipine-Syloid-HPMCAS (Figure 4d) or Furosemide-Syloid-HPMCAS (Figure 4g), compared to the original surface of Syloid, which was relatively smoother (Figure 4a). The partial coating effect of HPMCAS was more apparent on the Furosemide-Syloid samples. There was a great number of Furosemide crystals on the external surface of the Furosemide-Syloid particles (Figure 4c), which was verified with the DSC result. Ternary Furosemide-Syloid-HPMCAS also contained crystalline Furosemide, as confirmed by DSC (Figure 9). However, there was no notable difference in the particle surface between Furosemide-Syloid and ternary Furosemide-Syloid-HPMCAS. Owing to the addition of HPMCAS, no surface crystals could be observed on the exterior of silica particles, possibly due to the layers of HPMCAS covering them (Figure 4g). SEM images, together with DSC data, suggested that HPMCAS might hinder the migration of drug molecules into the internal mesopores due to their coating effect, leading to more drug staying externally in silica particles.

The inclusion of HPMCAS with Syloid did not help in enhancing the release rate or overall dissolution of Felodipine (Figure 10). In fact, the dissolution of Felodipine-Syloid-HPMCAS was significantly lower than that of Felodipine-Syloid within the first 90 min ($p < 0.05$). This polymer is only soluble at pH 5.5 or above, hence the partial coating of HPMCAS on the mesoporous silica was not dissolved at pH 3 acidic environment (optimal pH for Furosemide absorption) and resulted in an adverse effect on the drug release of Furosemide (Figure 11).

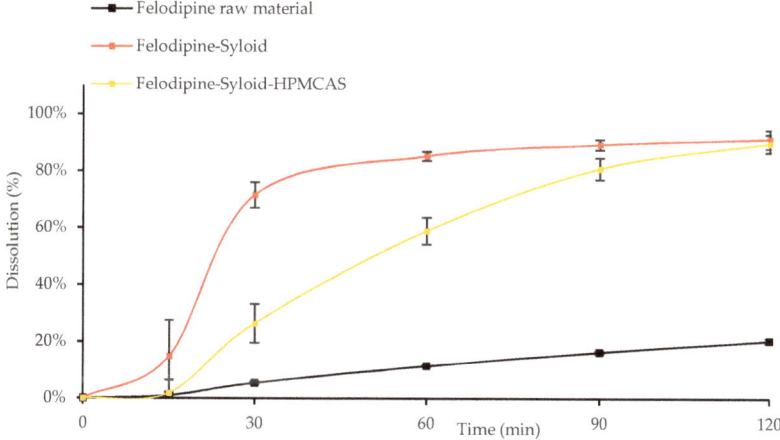

Figure 10. Release profiles of the raw Felodipine material, Felodipine-Syloid, and Felodipine-Syloid-HPMCAS at drug load of 300% monolayer coverage. Testing conditions: phosphate buffer pH 6.5 + 0.25% SLS, 500 mL, USP apparatus 1, 50 rpm.

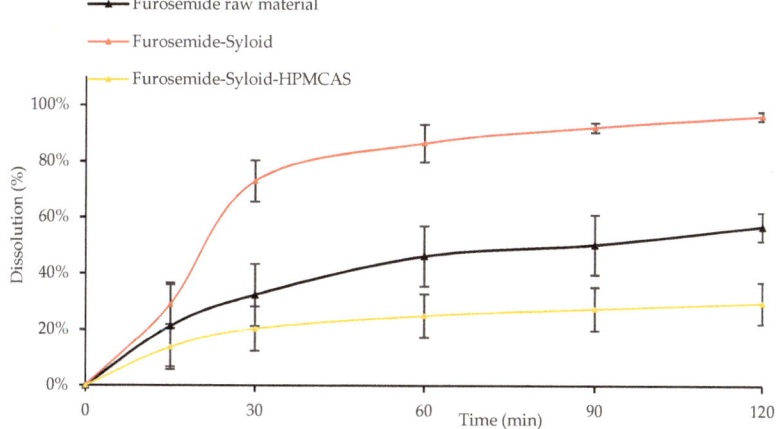

Figure 11. Release profiles of Furosemide raw material, Furosemide-Syloid, Furosemide-Syloid-HPMCAS at drug load of 300% surface coverage. Testing conditions: medium pH 3 + 0.25% SLS, 900 mL, USP apparatus 1, 100 rpm.

Furosemide-Syloid-HPMCAS still obtained 29% dissolution after 120 min as the drug was possibly released from silica particles through incomplete coating layers. A previous study [10] showed a more significant effect of HPMCAS on the dissolution of the poorly soluble drug celecoxib when loaded into the mesoporous silica Parteck (approximately a 5-fold solubility increase). The difference in the observation could be due to the different conditions, e.g., the dissolution medium of Tris buffered water pH 7.4 and a fasted state simulated intestinal fluid with pH 7.0 and the model drug Celecoxib used in their study. It is noticeable that the ratio of HPMCAS to Parteck silica was calculated as 14.7%, which was much higher than the 6.1% ratio of HPMCAS to Syloid silica in our study (calculated for Felodipine-Syloid-HPMCAS at a drug load of 300% surface coverage). Generally, a higher HPMCAS-to-silica particle ratio led to a thicker coating layer, if formed, and produced a more pronounced gastro-resistant effect, which could result in a delayed release effect in oral administration.

4. Conclusions

Syloid enhanced the dissolution of both model poorly soluble drugs, Felodipine and Furosemide, due to amorphisation within its mesoporous network. Increasing the drug load or percentage of theoretical surface coverage increased the maximum attained dissolution. However, overloading could lead to the formation of surface nanocrystals, as observed in the thermal and morphological studies. Incomplete drug release happened at all drug loads, which could be due to the reversible adsorption phenomenon mentioned in previous reports. This was predominant at the lower drug loads as less drug is available to dissolve before reaching an adsorption equilibrium. In addition, overloading resulted in a decrease of loading efficiency for both tested model drugs. A new drug load based on the drug amount to specific surface area of materials could be used in order to enable the comparison between various types of mesoporous materials. The addition of HPMCAS at a low concentration prevented the complete amorphisation of the drugs. Furthermore, it also slowed down the drug release due to partial coating, which formed on the exterior surface of mesoporous silica particles. In the future, this concept of combining mesoporous silica and polymers in hybrid structures will be further investigated for potential application in sustained release drug delivery systems.

Author Contributions: Conceptualization, T.-T.L. and A.A.-K.; Data curation, T.-T.L. and A.K.E.E.; Formal analysis, T.-T.L. and A.K.E.E.; Methodology, T.-T.L.; Project administration, A.A.-K.; Supervision, A.R.M. and A.A.-K.; Writing—original draft, T.-T.L.; Writing—review and editing, A.R.M. and A.A.-K.

Funding: This research received no external funding.

Acknowledgments: The authors would like to thank Aston University for providing Tuan Tu Le with an Overseas Bursary Scholarship towards his PhD and Cambridge Crystallographic Data Centre for providing a structural database of the model drugs.

Conflicts of Interest: The authors declare no conflict of interest.

References

1. Ting, J.M.; Porter, W.W.; Mecca, J.M.; Bates, F.S.; Reineke, T.M. Advances in Polymer Design for Enhancing Oral Drug Solubility and Delivery. *Bioconj. Chem.* **2018**, *29*, 939–952. [CrossRef] [PubMed]
2. Bosselmann, S.; Williams, R.O. Route-Specific Challenges in the Delivery of Poorly Water-Soluble Drugs. In *Formulating Poorly Soluble Drugs*; Williams, R.O., Watts, A.B., Miller, D.A., Eds.; Springer: New York, NY, USA, 2012; pp. 1–26.
3. Zografi, G.; Newman, A. Introduction to amorphous solid dispersions. In *Pharmaceutical Amorphous Solid Dispersions*; Newman, A., Ed.; Wiley: Hoboken, NJ, USA, 2015; pp. 1–41.
4. Garcia-Bennett, A.; Feiler, A. Mesoporous ASD: Fundamentals. In *Amorphous Solid Dispersions—Theory and Practice*; Shah, N., Sandhu, H., Choi, D.S., Chokshi, H., Malick, A.W., Eds.; Springer: New York, NY, USA, 2014; pp. 637–663.
5. Shen, S.C.; Dong, Y.C.; Letchmanan, K.; Ng, W.K. Mesoporous materials and technologies for development of oral medicine. In *Nanostructures for Oral Medicine*; Andronescu, E., Grumezescu, A.M., Eds.; Elsevier: Amsterdam, The Netherlands, 2017; pp. 699–749.

6. Choudhari, Y.; Hoefer, H.; Libanati, C.; Monsuur, F.; McCarthy, W. Mesoporous Silica Drug Delivery Systems. In *Amorphous Solid Dispersions—Theory and Practice*; Shah, N., Sandhu, H., Choi, D.S., Chokshi, H., Malick, A.W., Eds.; Springer: New York, NY, USA, 2014; pp. 665–693.
7. Riikonen, J.; Xu, W.; Lehto, V.P. Mesoporous systems for poorly soluble drugs—Recent trends. *Int. J. Pharm.* **2018**, *536*, 178–186. [CrossRef] [PubMed]
8. Xu, W.; Riikonen, J.; Lehto, V.P. Mesoporous systems for poorly soluble drugs. *Int. J. Pharm.* **2013**, *453*, 181–197. [CrossRef] [PubMed]
9. Dening, T.J.; Taylor, L.S. Supersaturation Potential of Ordered Mesoporous Silica Delivery Systems. Part 1: Dissolution Performance and Drug Membrane Transport Rates. *Mol. Pharm.* **2018**, *15*, 3489–3501. [CrossRef] [PubMed]
10. Lainé, A.-L.; Price, D.; Davis, J.; Roberts, D.; Hudson, R.; Back, K.; Bungay, P.; Flanagan, N. Enhanced Oral Delivery of Celecoxib via the Development of a Supersaturable Amorphous Formulation Utilising Mesoporous Silica and Co-Loaded HPMCAS. *Int. J. Pharm.* **2016**, *512*, 118–125. [CrossRef] [PubMed]
11. Edgar, B.; Lundborg, P.; Regårdh, C.G. Clinical Pharmacokinetics of Felodipine: A Summary. *Drugs* **1987**, *34*, 16–27. [CrossRef] [PubMed]
12. Chungi, V.S.; Dittert, L.W.; Smith, R.B. Gastrointestinal Sites of Furosemide Absorption in Rats. *Int. J. Pharm.* **1979**, *4*, 27–38. [CrossRef]
13. Felodipine–PubChem. Available online: https://pubchem.ncbi.nlm.nih.gov/compound/3333 (accessed on 20 March 2019).
14. Furosemide–PubChem. Available online: https://pubchem.ncbi.nlm.nih.gov/compound/3440 (accessed on 20 March 2019).
15. Shen, S.; Ng, W.; Chia, L.; Hu, J.; Tan, R. Physical state and dissolution of ibuprofen formulated by co-spray drying with mesoporous silica: Effect of pore and particle size. *Int. J. Pharm.* **2011**, *410*, 188–195. [CrossRef] [PubMed]
16. Ambrogi, V.; Perioli, L.; Pagano, C.; Latterini, L.; Marmottini, F.; Ricci, M.; Rossi, C. MCM-41 for Furosemide Dissolution Improvement. *Microporous Mesoporous Mater.* **2012**, *147*, 343–349. [CrossRef]
17. Ambrogi, V.; Perioli, L.; Pagano, C.; Marmottini, F.; Ricci, M.; Sagnella, A.; Rossi, C. Use of SBA-15 for Furosemide Oral Delivery Enhancement. *Eur. J. Pharm. Sci.* **2012**, *46*, 43–48. [CrossRef] [PubMed]
18. The United States Pharmacopoeia (USP 36). *United States Pharmacopoeia Convention*; United States Pharmacopoeia: Rockville, MD, USA, 2013.
19. HPLC Analysis of Furosemide (Lasix) in Horse Serum on Discovery® C18 after SPE Using Discovery® DSC-18. Available online: https://www.sigmaaldrich.com/technical-documents/articles/analytical-applications/hplc/hplc-analysis-of-furosemide-lasix-in-horse-serum-g003764.html (accessed on 20 March 2019).
20. Mahmah, O.; Tabbakh, R.; Kelly, A.; Paradkar, A. A Comparative Study of the Effect of Spray Drying and Hot-Melt Extrusion on the Properties of Amorphous Solid Dispersions Containing Felodipine. *J. Pharm. Pharmacol.* **2014**, *66*, 275–284. [CrossRef] [PubMed]
21. Matsuda, Y.; Tatsumi, E. Physicochemical characterization of furosemide modifications. *Int. J. Pharm.* **1990**, *60*, 11–26. [CrossRef]
22. De Cássia Da Silva, R.; Semaan, F.S.; Novák, C.; Cavalheiro, E.T.G. Thermal Behavior of Furosemide. *J. Therm. Anal. Calorim.* **2013**, *111*, 1933–1937. [CrossRef]
23. Yang, C.; Guo, W.; Lin, Y.; Lin, Q.; Wang, J.; Wang, J.; Zeng, Y. Experimental and DFT Simulation Study of a Novel Felodipine Cocrystal: Characterization, Dissolving Properties and Thermal Decomposition Kinetics. *J. Pharm. Biomed. Anal.* **2018**, *154*, 198–206. [CrossRef] [PubMed]
24. Wu, C.; Zhao, Z.; Zhao, Y.; Hao, Y.; Liu, Y.; Liu, C. Preparation of a push–pull osmotic pump of felodipine solubilized by mesoporous silica nanoparticles with a core-shell structure. *Int. J. Pharm.* **2014**, *475*, 298–305. [CrossRef] [PubMed]
25. Hu, L.; Sun, H.; Zhao, Q.; Han, N.; Bai, L.; Wang, Y.; Jiang, T.; Wang, S. Multilayer encapsulated mesoporous silica nanospheres as an oral sustained drug delivery system for the poorly water-soluble drug felodipine. *Mater. Sci. Eng. C* **2015**, *47*, 313–324. [CrossRef] [PubMed]

26. Takeuchi, H.; Nagira, S.; Yamamoto, H.; Kawashima, Y. Solid dispersion particles of tolbutamide prepared with fine silica particles by the spray-drying method. *Powder Technol.* **2004**, *141*, 187–195. [CrossRef]
27. Hong, S.; Shen, S.; Tan, D.C.T.; Ng, W.K.; Liu, X.; Chia, L.S.O.; Irwan, A.W.; Tan, R.; Nowak, S.A.; Marsh, K.; et al. High Drug Load, Stable, Manufacturable and Bioavailable Fenofibrate Formulations in Mesoporous Silica: A Comparison of Spray Drying versus Solvent Impregnation Methods. *Drug Deliv.* **2016**, *23*, 316–327. [CrossRef] [PubMed]

© 2019 by the authors. Licensee MDPI, Basel, Switzerland. This article is an open access article distributed under the terms and conditions of the Creative Commons Attribution (CC BY) license (http://creativecommons.org/licenses/by/4.0/).

Article

Investigation of Dissolution Mechanism and Release Kinetics of Poorly Water-Soluble Tadalafil from Amorphous Solid Dispersions Prepared by Various Methods

Tereza Školáková [1,*], Michaela Slámová [1], Andrea Školáková [2], Alena Kadeřábková [3], Jan Patera [1] and Petr Zámostný [1]

[1] Department of Organic Technology, University of Chemistry and Technology, Prague, Technická 5, 166 28 Prague 6, Czech Republic
[2] Department of Metals and Corrosion Engineering, University of Chemistry and Technology, Prague, Technická 5, 166 28 Prague 6, Czech Republic
[3] Department of Polymers, University of Chemistry and Technology, Prague, Technická 5, 166 28 Prague 6, Czech Republic
* Correspondence: skolakot@vscht.cz; Tel.: +420-220-444-300

Received: 2 July 2019; Accepted: 30 July 2019; Published: 2 August 2019

Abstract: The aims of this study were to investigate how the release of tadalafil is influenced by two grades of polyvinylpyrrolidone (Kollidon® 12 PF and Kollidon® VA 64) and various methods of preparing solid dispersions (solvent evaporation, spray drying and hot-melt extrusion). Tadalafil is poorly water-soluble and its high melting point makes it very sensitive to the solid dispersion preparation method. Therefore, the objectives were to make a comparative evaluation among different solid dispersions and to assess the effect of the physicochemical nature of solid dispersions on the drug release profile with respect to the erosion-diffusion mechanism. The solid dispersions were evaluated for dissolution profiles, XRD, SEM, FT-IR, DSC, and solubility or stability studies. It was found that tadalafil release was influenced by polymer molecular weight. Therefore, solid dispersions containing Kollidon® 12 PF showed a faster dissolution rate compared to Kollidon® VA 64. Tadalafil was released from solid dispersions containing Kollidon® 12 PF because of the combination of erosion and diffusion mechanisms. The diffusion mechanisms were predominant in the initial phase of the experiment and the slow erosion was dissolution-controlling at the second stage of the dissolution. On the contrary, the tadalafil release rate from solid dispersions containing Kollidon® VA 64 was controlled solely by the erosion mechanism.

Keywords: solid dispersion; tadalafil; Wood's apparatus; intrinsic dissolution rate; Weibull dissolution model; dissolution rate

1. Introduction

Poorly water-soluble drugs showing pharmacological activity are a common and ongoing issue for the pharmaceutical industry and associated with the complexity of the drug development [1–3]. It has been reported that approximately 40% of marketed drugs and 75% of active pharmaceutical ingredients under development are classified as practically insoluble in water [3].

Solid dispersions of such drugs in hydrophilic carriers have provided a promising possibility of improving their dissolution rate, and thereby absorption [4]. Basically, the solid dispersions are two-component systems which can improve drug wettability and bioavailability significantly by reducing the effective drug particle size to the absolute minimum, increasing the drug surface area, reducing its crystallinity, and increasing wettability by surrounding hydrophilic carriers due to their unique morphology [5–8]. They are very attractive for formulators due to well-known preparation processes and devices, high effectiveness, flexibility in designing the composition, or low batch-to-batch variability [6,9]. However, the ideal state of molecular dispersion may not be always achieved, and the solid dispersions may only approach the molecular level of an ideal solid solution. That is the case especially in formulations where the melting point of the drug exceeds the maximum processing temperature of the polymer, so that the dispersion is formed effectively by dissolving the drug in molten polymer or by introducing additional solvent, rather than by co-melting of the drug-polymer mixture. In such systems, the preparation method is likely to have strong impact on the apparent solubility and drug release. While different preparation techniques were reported in the literature for many systems, the link between the preparation method and the dissolution properties has not been clearly established, and therefore it requires further study [5].

The objective of this study was to investigate and to provide insights into how the drug release is influenced by various polymer carriers and methods of preparation. Tadalafil (TAD) solid dispersions formulated with two grades of polyvinylpyrrolidone (Kollidon® 12 PF and Kollidon® VA 64) were used as model system for this study. TAD was selected as a poorly soluble, hydrophobic drug, with a melting point around 300 °C, which is well above the processing temperature of common polymers used for the solid dispersion formulations. Therefore, it provides a very sensitive system for studying the effects of preparation methods on the dispersion properties and the subsequent dissolution behavior. At the same time, it represents a compound which was the subject of many formulation studies recently (see next section for the details) and thus this choice is relevant from the application point of view. The polymers were selected due to their widespread use in solid dispersion formulations. They are hydrophilic in nature, which have potential to change the crystalline drug to amorphous by solid dispersion technique, enhancing TAD wettability.

The specific goals to disclose within this paper include the comprehensive analysis of TAD solid dispersions, analysis of the dissolution profiles of different TAD formulations with respect to the erosion-diffusion mechanism, and a comparison of the dissolution behavior of TAD from physical mixtures and corresponding solid dispersions, as studied by different dissolution techniques like apparent intrinsic dissolution and flow-through cell apparatus. Those results are then linked to the preparation methods of solvent evaporation, spray drying, and hot-melt extrusion.

1.1. Theoretical Background

This section provides detailed background information summarizing the reported techniques used for improving the dissolution properties of TAD and dissolution phenomena in polymer matrix systems, which were separated from the general introduction for the sake of clarity, but which can help in understanding the methods and the discussed results, and provide necessary reference for state of the art.

1.1.1. Techniques for Improving Dissolution Rate of TAD

Based upon aqueous solubility and dissolution parameters, the efficacy TAD used for the treatment of erectile dysfunction, pulmonary arterial hypertension, and now also for therapy in pyelonephritis can be limited by its highly hydrophobic particles [10,11]. The dissolution is considered as the first step in the absorption process and therefore, a critical disadvantage of TAD is its poor water solubility [12,13]. Since the pK*a* value of TAD is 16.68, it is a non-ionizable drug via the range of physiological pH. Therefore, neither ionization nor salt formation can be applied for the enhancement of its aqueous solubility and dissolution rate and thus, it was undertaken to develop effective methods for improvement of these properties [5,14]. The transformation of pure crystalline TAD into the amorphous form is considered to be troublesome [15]. Wlodarski et al. obtained the amorphous forms of TAD by amorphization methods and without excipients, i.e., by cryogenic grinding, ball milling, spray drying, freeze-drying and antisolvent precipitation. On the other hand, they also reveal that vitrification was an inappropriate method to convert to amorphous counterpart of crystalline

TAD because of its decomposition at the melting point temperature. Their study also revealed that the techniques influence the apparent water TAD solubility insubstantially; however, disk intrinsic dissolution rate tests of amorphous TAD obtained by ball milling and spray drying did not improve in the rate of its dissolution [16]. Therefore, several technologies have been used to enhance the dissolution rate and solubility of TAD, including solid dispersion systems, e.g., in [3,6,17–20], the formulation of self-microemulsifying composition (SMEC) [11], self-nanoemulsifying drug delivery system (SNEDDS) [21], nanostructured lipid carriers (NLCs) [22], nanoparticles [13,23], cyclodextrin complexation [24], incorporation in microporous silica [25] or microemulsion system [17]. Of these formulations, however, solid dispersion is currently favored in the pharmaceutical industry [19].

Solid dispersion of TAD has been prepared using different methods, these include solvent evaporation method [8,10,12,18,19], spray drying [6,16,26], melting method [5,17], hot-melt extrusion [15], supercritical anti-solvent process [14,27], ball milling [26–28] or freeze-drying [3].

1.1.2. Dissolution Phenomena in Polymer Matrix Systems

The polymeric matrix systems, such as hydrogel-based dosage forms, are commonly used for manufacturing sustained release drug delivery systems [29,30]. Solid dispersions can also be used for controlling drug release [31]. However, the drug release from such systems involves many phenomena which can contribute to the final progress of dissolution and therefore, the complete description is very complex and not completely understood [32–34]. The critical factors in the release of drugs from matrix system are clearly summarized and reported in [35]. Briefly, when the solid dosage forms, based on a hydrophilic polymer matrix, are immersed in a dissolution medium, the surface of the matrix is wetted, the water penetrates into the systems and polymer surface swells to form a gel leading to the polymer chain relaxation, drug dissolution, drug diffusion through the hydrated polymeric network, chain disentanglement of polymer, matrix erosion and moving boundaries [36,37]. Since the polymer undergoes a relaxation leading to its erosion, the diffusion is generally not the main mechanisms that control the drug release [38]. Moreover, three fronts, and of course corresponding boundaries, can be identified in the polymer-based drug delivery, i.e., polymer glassy-rubbery transition boundary (swelling front), solid drug-drug solution boundary (diffusion front) and swollen matrix-solvent boundary (erosion front) [39,40]. Therefore, the dissolution rate of drug is affected by the movement of these fronts and the mechanisms of release can operate simultaneously [38,39]. In other words, the rate of water uptake is widely associated with the position of the swelling front [41]. However, it can be expected that erosion front movement determines the kinetics whilst the diffusion front movement determines the rate of drug release [39]. Of course, some of the phenomena occurring during water uptake can also be observed in the matrix size (its size enlargement or reduction), e.g., the swelling causes its increase and the erosion, which can only take place after complete hydration of outer layer, causes its decrease [29,42]. However, there is no universal drug release mechanism that can be valid for all systems containing polymer [43].

The mobility of the polymer chains is affected by the dissolution medium composition which should be thermodynamically compatible with the matrix [44,45]. The polymer can then undergo the relaxation process due to the decrease in the glass transition temperature, its chains become more flexible giving volume expansion which is followed by swelling of the system [44–46]. It can be expected that hydrophilic polymer may lead to improving of wetting properties leading to the enhancing the diffusion of dissolution medium to the solid powder. However, this assumption can be also greatly influenced by the intermolecular interactions and therefore, also by the arrangement of drug-polymeric carrier structures in the solid dispersions resulting in the changes of functional groups orientation. For this reason, some polar functional groups can be then available or vice versa unavailable for the interaction with dissolution medium during absorption [47].

2. Materials and Methods

2.1. Materials

Tadalafil (TAD) was obtained from Zentiva Group, a.s. (Prague, Czech Republic). Kollidon® 12 PF (K12) and Kollidon® VA 64 (K64) were purchased from BASF Pharma (Ludwigshafen, Germany). Methanol and acetonitrile were of LC/MS grades and were obtained from Fisher Scientific Ltd. (Pardubice, Czech Republic). Methanol for solvent evaporation method, hydrochloric acid, potassium dihydrogenphosphate (KDP) and sodium hydroxide were purchased from Penta s.r.o. (Prague, Czech Republic). Sodium dodecyl sulfate (SDS) was obtained from Fluka (Buchs, Germany).

2.2. Preparation of Physical Mixtures

Physical mixtures (PMs) were prepared by mixing TAD and K12 or K64 (5% of TAD in PM) with a conventional tumbling Turbula mixer (T2F model, W.A. Bachofen, Basel, Switzerland) for 1 h at 50 rpm.

2.3. Preparation of Solid Dispersions

Solid dispersions (SDs) were prepared by various methods. The methods are described in more detail below.

2.3.1. Solvent Evaporation Method (SE)

To prepare for amorphous SDs of TAD in one of the polymeric carriers at 5% drug loading, 2 g of PM was dissolved in 50 or 100 mL of methanol. The drug/polymer solution was then evaporated at 35–40 °C and at 150 rpm under vacuum (150 mbar) with rotary vacuum evaporator LABOROTA 4000 Heidolph (Maneko, Prague, Czech Republic). The thin cast film was collected in a flask and subsequently dried at 40 °C in an oven at least 24 h. The resulting SD was ground using mortar and pestle.

2.3.2. Spray Drying Technique (SPD)

PM (2 g) containing TAD and K12 or K64 was dissolved in methanol (50 mL) to form the solution which was mixed using sonification. After complete dissolution in methanol, the solution was spray dried using the Mini Spray Dryer B-290 (Büchi, Flawil, Switzerland) with an inert nitrogen loop. The aspirator flow was set at 90%, the inlet temperature was kept at 90 °C and the outlet temperature was set at 50 °C. The atomization gas flow was 35% and the pump speed was 3 mL/min.

2.3.3. Hot-Melt Extrusion (HME)

Hot-melt extrusion was performed using universal single screw (L/D ratio 19/25D) PLASTI-CORDER Lab-Station extruder (Brabender Technologie GmbH & Co. KG, Duisburg, Germany) equipped with the heating barrel which is divided into three temperatures zones and a homogenization zone at the end of the screw. The following extrusion temperatures were used for TAD-K12 PM: 140, 150, and 155 °C, and for TAD-K64 PM: 160, 165, and 160 °C. The extruder was manually fed with the 50 g of each PMs. The screw speed was set to 15 rpm for both PMs. The extrudates were then cooled to laboratory temperature on a conveyor belt and milled to fine powder using hammer mill Polymix™ PX-MFC 90 D (Kinematica™, Loughborough, UK) equipped with sieve (mesh size 0.2 mm). The rotational speed was 5000 rpm. The obtained powder was subsequently sieved using a vibratory sieve (AS 200 basic, Retsch, Haan, Germany) for 10 min (amplitude of vibration was 50 mm). The particle size fractions of 125–250 μm (HME125) and 250–425 μm (HME250) were used for further analysis.

2.4. X-Ray Powder Diffraction (XRD)

X-ray powder diffraction patterns were determined with D2 PHASER diffractometer (Bruker, Billerica, MA, USA) configured with 1D SSD detector in Bragg-Brentano parafocussing geometry. A

Cu Kα radiation was used (wavelength = 1.5416 Å, voltage = 30 kV, current = 10 mA). The samples were analyzed at room temperature over the range of 5°–80° (2θ). The time per step was 0.3 s and the increment was 0.020° (2θ). The obtained data were evaluated using X´Pert High Score Plus program with the PDF2 database (Malvern Panalytical Ltd., Royston, UK).

2.5. Scanning Electron Microscopy (SEM)

To observe the surface morphology of all components, their PMs and SDs, scanning electron microscopy studies were performed using TESCAN VEGA 3 LMU (Tescan, Brno, Czech Republic). A graphite double-sided adhesive tape was covered by samples and then the samples were made conductive by sputter-coating with 5 nm of gold using rotary-pumped sputter coater (Quorum Q150R ES, Quorum Technologies Ltd., Laughton, UK). The images were captured at magnifications factors of 500.

2.6. Fourier Transform-Infrared (FT-IR) Spectroscopy

FT-IR spectra were recorded with the NICOLET iS FT-IR spectrometer (Thermo Fisher Scientific Waltham, MA, USA). ATR module was used to measure FT-IR spectra. Each spectrum was measured using spectral resolution of 2 cm^{-1} in 4000–400 cm^{-1} range.

2.7. Stability Studies

SDs were exposed to the temperature of 40 °C and to the relative humidity (RH) of 75% in the humidity chamber, model HCP 108 (Verkon s.r.o., Prague, Czech Republic) for twelve months. The samples were stored in powder form and in the case of SDs prepared by hot-melt extrusion also in extrudate form.

2.8. Tadalafil Solubility Test

The solubility test was performed in a paddle dissolution apparatus Sotax AT7 Smart (USP 2; Sotax, Basel, Switzerland). Solubility testing was measured for powder TAD (2.5 mg) in various dissolution media (1000 mL), i.e., phosphate buffer (pH 7.2 and containing 6.8 g KDP and 0.9 g NaOH), phosphate buffer + SDS (pH 7.2 and containing 6.8 g KDP, 0.9 g NaOH and 5 g SDS), 0.1 M hydrochloric acid, and 0.1 M hydrochloric acid + SDS (containing 5 g SDS), which were heated to 37 °C. The rotational speed was set to 150 rpm. The samples of 5 mL were taken in following times: 2, 5, 10, 20, 30, and 45 min, and the concentration of TAD was measured using HPLC. Each experiment was performed twice and the mean values of TAD concentration with their standard deviations were calculated.

2.9. Dissolution Studies

The release of TAD from PMs and corresponding SDs was studied by different dissolution techniques which are described in more detail below. Each dissolution experiment was performed twice and the mean values of TAD release amount with their standard deviations were calculated.

2.9.1. Flow-Through Cell Method

The dissolution studies for SDs and PMs were carried out in the USP 4 compliant flow-through cell apparatus Sotax CE1 (Sotax, Basel, Switzerland) with piston pump Sotax CY1 (Sotax, Basel, Switzerland). The dissolution flow-through cell for powders and granules having the diameter of 12 mm and a height of 32 mm was employed to study the different samples in all experiments. Each experiment was conducted using the cell in an open-loop system with fresh dissolution medium from the reservoir continuously passing through the cell. The open-loop system was selected for samples due to the low solubility of TAD and requirement of high volume of solvent. The dissolution medium was pumped through the cell with a piston pump. One glass bead of 5 mm diameter was placed in the apex of the cone to protect the inlet tube according to the manufacturer´s instruction. The conical part of the cell was filled with glass beads of 1 mm diameter to ensure laminar flow profile. Two sieves and the amount of powder (approximately 0.02 g) to be studied were placed on the top of the layer of small beads. One sieve was also placed above the powder. The cell was closed with the filter assembly (containing glass microfiber filter GF/D, Whatman®, Fisher Scientific Ltd., Pardubice, Czech Republic) to prevent undissolved material from escaping.

The dissolution medium and the apparatus were placed into the water bath and heated to 37 °C. 0.1M hydrochloric acid with 0.5% SDS (w/v) was used as dissolution medium. The dissolution medium was then degassed. The flow rate of dissolution medium through the cell was set to 23 mL/min. The samples were collected into the beakers for the following times: 10, 20, 30, 40, 50, 60, 70, 80, 90, 100, 110, 120, 150, 180, 240, 360, 480, 600, and 900 s. Aliquots of 1 mL were then withdrawn for the HPLC analysis.

The TAD concentration in the dissolution medium was calculated using a calibration curve which was obtained from the samples of known concentration and the dissolution profiles were normalized to 100% TAD released at the end of the experiment.

The obtained concentration c was used for calculation of the dissolution rate $r(t)$ according to Equation (1):

$$r(t) = \frac{c(t) \cdot Q}{V_C}, \qquad (1)$$

where Q is the flow rate of dissolution medium (mL/min) and V_C is cell volume (dm³).

The dissolution rate can be then used for the calculation of the amount of dissolved TAD Δm (Equation (2)):

$$\Delta m = Q \int_0^t c(t) dt, \qquad (2)$$

2.9.2. Apparent Intrinsic Dissolution Rate (Wood´s Apparatus)

The apparent intrinsic dissolution rate (AIDR) technique was introduced in our previous study [48] as an extension to intrinsic dissolution rate (IDR) to measure matrix related dissolution effects of drug formulation rather than pure drug substance. Dissolution testing of compressed formulations was performed in dissolution apparatus Sotax AT7 Smart (Sotax, Basel, Switzerland) equipped by a rotating disc device. For rotating disk dissolution rate, 200 mg of the PMs or SDs were compressed by the compaction force of about 2.5 tons and using standard die of 10 mm diameter for 1 min using laboratory manual hydraulic press (Specac, Orpington, UK). The die was then attached to the rotor shaft and was immersed into 1000 mL of the dissolution medium. The rotational speed was 150 rpm. The tablets were dissolved in 0.1M hydrochloric acid with 0.5% SDS (w/v) which was maintained at 37 °C. The samples of 1 mL were taken in following times: 2, 4, 5, 6, 8, 10, 15, 20, 25, and 30 min. The concentration of TAD was measured by HPLC. The apparent intrinsic dissolution rate can be then calculated according to Equation (3):

$$IDR = \frac{V}{A} \frac{dc}{dt}, \qquad (3)$$

where V is volume of dissolution medium (L) and A is area of exposed tablet surface (cm²), c is concentration of dissolved TAD (mg/L) and t is time (s). Unlike IDR measurement, the dissolution profile can be non-linear and thus AIDR is time-sensitive. For example, IDR at $t = 0$ can be used as a measure of drug dissolution unhindered by formed polymer layer.

2.10. HPLC

The concentration of TAD in the samples was determined by LC Prominence system (Shimadzu, Kyoto, Japan) equipped with a PDA detector without further dilution and a Kinetex®, 5 µm, C18, 100 Å column (Phenomenex®, Prague, Czech Republic). Separation was performed with 20 µL injection volume. The flow rate of mobile phase was 1 mL/min and oven temperature was 30 °C. Tadalafil was monitored at 284 nm.

A solution of methanol, acetonitrile and distilled water in ratio 45:40:15 (*v*/*v*/*v*) was used as the mobile phase. The mobile phase was then degassed prior to analysis.

2.11. Differential Scanning Calorimetry (DSC)

DSC was performed on the pure TAD, polymers, PMs and SDs. DSC analyses were examined using a differential scanning calorimeter DSC 131 (Setaram, Caluire, France) under nitrogen atmosphere in the temperature range of 25–350 °C. The temperature program was set to a linear increase of temperature at a heating rate of 10 °C/min.

3. Results and Discussion

3.1. Characterization of Solid Dispersions

It has been documented that the physicochemical properties of amorphous forms can be largely dependent on methods for their production, as well [16].

3.1.1. SEM

For morphological characterization, SEM analysis was performed on the pure drug and both polymer, their PMs and SDs. The SEM images for pure drug, polymers, PMs and SDs of both polymers are shown in Figures 1 and 2. Pure drug images (Figure 1a or Figure 2a) showed fine crystalline powder of irregular shapes and uniform sizes, whereas the images of polymers (Figure 1b for K12 and Figure 2b for K64) showed a spherical particle shape with holes. From the SEM of both pure polymers, it is clear that the K64 contained larger particles in comparison to K12. In the PMs (Figures 1c and 2c), the small particle size TAD was adsorbed on the surface of polymers. The images of SDs of TAD with K12 or K64 did not show any crystalline materials (Figures 1d–g and 2d–g) and the SDs appeared as particles of irregular shape. As can be seen, the type of carrier and different methods strongly affected the morphology of the SDs. Further, only spray drying led to the formation of round and small particles.

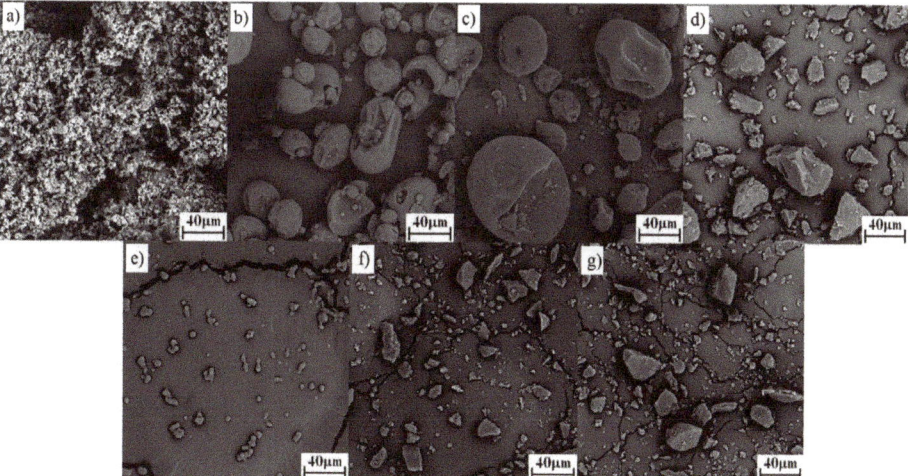

Figure 1. Morphology images of TAD, K12, their PM and SDs: (**a**) pure TAD, (**b**) pure K12, (**c**) TAD-K12 PM, (**d**) TAD-K12 SD (SE), (**e**) TAD-K12 SD (SPD), (**f**) TAD-K12 SD (HME125) and (**g**) TAD-K12 SD (HME250).

Figure 2. Morphology images of TAD, K64, their PM and SDs: (**a**) pure TAD, (**b**) pure K64, (**c**) TAD-K64 PM, (**d**) TAD-K64 SD (SE), (**e**) TAD-K64 SD (SPD), (**f**) TAD-K64 SD (HME125) and (**g**) TAD-K64 SD (HME250).

3.1.2. FT-IR

FT-IR spectra were measured to study TAD-polymer matrix interactions. In the Figures 3a and 4a, the FT-IR spectra of pure TAD, K12 and K64, their PMs and SDs at 500–4000 cm^{-1} are shown. The Figures 3b and 4b also show the detail spectra at 1600–1750 cm^{-1} or 1600–1800 cm^{-1}, respectively.

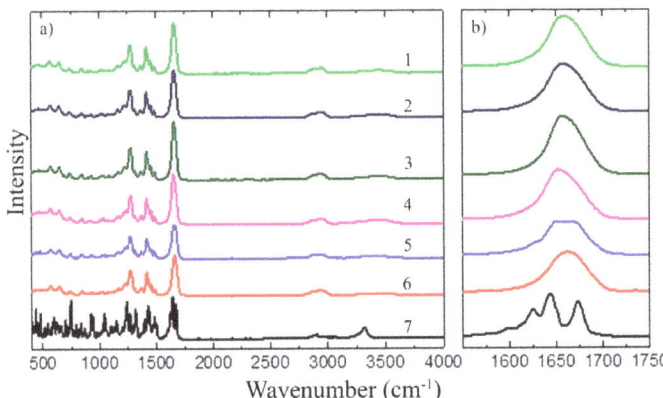

Figure 3. FT-IR spectra of pure TAD, K12, their PM and SDs: (**a**) **1**—TAD-K12 SD (HME250), **2**—TAD-K12 SD (HME125), **3**—TAD-K12 SD (SPD), **4**—TAD-K12 SD (SE), **5**—TAD-K12 PM, **6**—pure K12, **7**—pure TAD and (**b**) their detail spectra at 1600–1750 cm^{-1}.

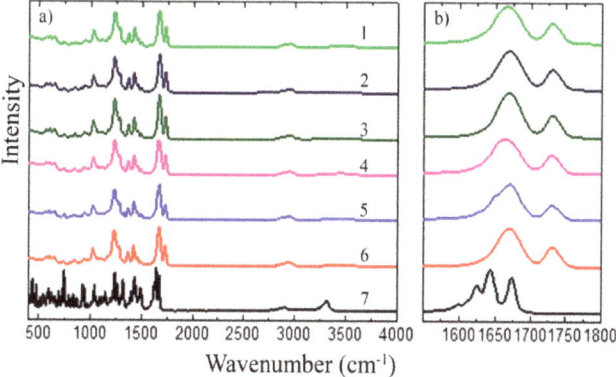

Figure 4. FT-IR spectra of pure TAD, K64, their PM and SDs: (**a**) **1**—TAD-K64 SD (HME250), **2**—TAD-K64 SD (HME125), **3**—TAD-K64 SD (SPD), **4**—TAD-K64 SD (SE), **5**—TAD-K64 PM, **6**—pure K64, **7**—pure TAD and (**b**) their detail spectra at 1600–1800 cm^{-1}.

TAD showed the signal of the stretching vibration of secondary amine group (3321 cm^{-1}), the signal of aliphatic methyl group (2904 cm^{-1}), the dual signal of carbonyl groups (1673 cm^{-1}), the signal of aromatic C=C bending (1649 cm^{-1}) and the signal of tertiary amine group (1270–1285 cm^{-1}). K12 had the carbonyl stretching band that is at 1661 cm^{-1}. The FT-IR spectra of pure K64 had aliphatic alkyl C–H stretch that is in mutual association at 2937 cm^{-1}, the signal of vinyl acetate (1730 cm^{-1}) and the signal of carbonyl group (1669 cm^{-1}). The spectra of all physical mixtures displayed tadalafil and polymer peaks with decreased peak intensity but with no shifting of the peaks. FT-IR spectra of all SDs showed the presence of some TAD peaks with decreased intensity as was described e.g., in [8]. Shifts of characteristic bands are not visible in the spectra, but only changes in their intensity or changes in the sub-band intensity of multi-component bands, as confirmed by a detailed analysis of the second spectra derivatives. All other tadalafil peaks were smoothened, indicating a strong physical intermolecular interaction of TAD with polymers. The decreasing intensity of the carbonyl group was observed for TAD-K12 PM and its SDs suggesting the presence of hydrogen bonds between the TAD secondary amine group and K12 carbonyl groups. Molecular interaction with polymer, predominantly through hydrogen bonding which was also described in [19], leads to better miscibility of TAD in the polymer matrix and the formation of amorphous SDs, as confirmed by XRD results. The interactions were stronger in the TAD-K12 SDs in comparison to TAD-K12 PM. Stronger drug-polymer interactions could be generated for example due to the melt extrusion [2].

The changes in the intensity of the carbonyl group were also observed in the TAD-K64 PM and its SDs. Even the pure polymer has other sub peaks visible on the deconvolution by curve fitting of its characteristic absorbing bands for carbonyl (1662 cm^{-1}) and for vinyl acetate (1730 cm^{-1}). These are vibrations of different types of groups, in which the carbonyl may be influenced by either two other carbonyls, optionally one or two vinyl acetate group in the neighboring position, or it may be a terminal group where the bond energy and its vibration depends on the neighboring monomer unit. After deconvolution of main bands, changes in the intensity of some peaks can be observed. In the case of a TAD-K64 PM, the vinylacetate group is also weakened. As a result of the different preparation of SDs, different relaxation of the polymer and its recombination occurs and leads to the change in the structural arrangement exhibiting different carbonyl group orientations. This is reflected in the change in the intensity ratio between the deconvoluted peaks, which is most noticeable between the TAD-K64 SD (SE) and TAD-K64 SD (HME125) spectra. Each of these mixtures was prepared at a completely different temperature and manifested in, for example, different dissolution behavior (see below). However, the interaction between the polymer and TAD can be observed with other vibrations than with the NH-vibration. Several different types of structures can be observed relative to each other for the vibrations of –CH– in the region 3000–2800 cm^{-1} range, or 2925 cm^{-1} corresponding to –CH$_2$ groups. CH deformation oscillations in the area of 1460 cm^{-1} again show the differences between the SDs and the PM when the PM exhibits more pronounced peaks, whereas, for example, in the TAD-K64 SD (SE), the absorption bands are more diffuse and overlap each other. However, no additional peak was observed in any binary system indicating absence of any another chemical interaction between TAD and polymer [5,49].

3.1.3. XRD

The powder X-ray diffraction patterns of pure TAD, K12 and K64, their PMs and SDs prepared by various method are shown in Figures 5 and 6, respectively. The diffraction pattern of the TAD had sharp intensive peaks throughout its pattern suggesting that it is crystalline in nature [15]. On the contrary, the polymeric carriers showed broad amorphous bands and no sharp diffraction peaks suggesting that the Kollidons were in amorphous state. The diffraction patterns of PMs showed sharp peaks corresponding to crystalline TAD and therefore, we can assume that crystalline TAD can be detected in its low concentration in all binary mixtures. The TAD peaks then disappeared in all SDs loaded with 5% TAD and therefore, their amorphous forms were confirmed. All extrudates were transparent, that means the dissolution of TAD in amorphous polymers.

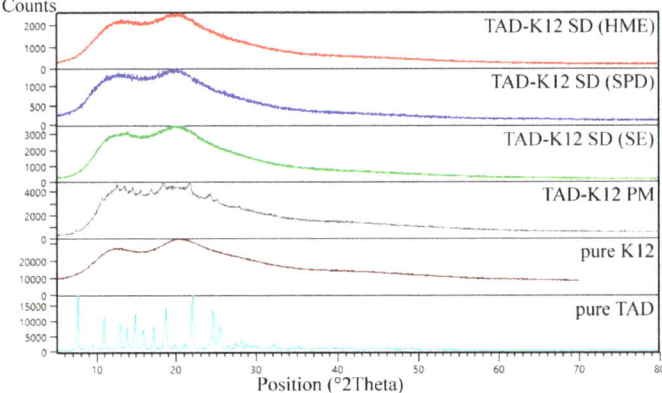

Figure 5. X-ray diffraction patterns of TAD, K12, their PM and SDs.

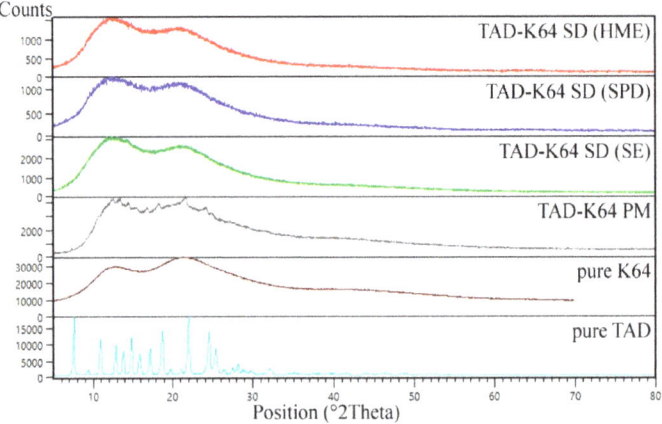

Figure 6. X-ray diffraction patterns of TAD, K64, their PM and SDs.

3.1.4. DSC Analysis

Figures A1 and A2 (see Appendix A) present DSC thermograms for crystalline TAD, amorphous K12 or K64, their PMs and SDs prepared by HME as an example. DSC thermograms of both SDs reveal the absence of any melting peak which means that these results suggest amorphous characteristics of SDs. In other words, it indicates the absence of crystalline trace of TAD in SDs. Besides, a broad endotherm ranging from 25 to 100 °C was observed in thermograms of pure K12, K64, both PMs and SDs which can be attributed to the water loss from the hygroscopic polymer upon heating. Figure A1 or Figure A2 also show a sharp melting endotherm at 302 °C corresponding to crystalline TAD. Moreover, no significant glass transition temperature was seen in thermograms. In order to detect the T_g, it is necessary to decrease heating rate. However, both Kollidons tend to degrade at lower heating rates. Figure A2 also show broad endotherms ranging from 300 to 350 °C. The degradation temperature of K64 is 230 °C. Based on this, broad endotherms can be attributed to some thermal event corresponding to the decomposed polymer.

3.2. Physical Stability of Solid Dispersions

The samples were stored in powder form and in the case of SDs prepared by HME also in extruded form. Stability testing at 40 °C and 75% RH for 12 months revealed that extrudates of both Kollidons to be amorphous for nine months and their powder forms to be amorphous for 6 months which was confirmed using XRD analysis as shown Figure A3a,b in Appendix A. Figure A3 shows results from XRD only for extrudates because X-ray diffraction patterns were identical also for powder forms prepared by various methods. Therefore, no crystalline form of the TAD was noticed after this period. The presence of hydrogen bonds between the drug and the polymer can lead to their better miscibility and effectively improves the physical stability of the SDs and therefore, they are essential [2,50]. Hydrogen bonds stiffen the structure and thus, hinder the diffusion of molecules [3]. In our case, the hydrogen bonds between the components were observed. SDs of both polymers in powder form changed their character after 6 months and for this reason, the sticky viscous solution was formed. Since the glass transition temperature is dependent on relative humidity [51], its decline can be expected leading to the glass solutions. In the case of SDs in extruded form, this phenomenon was observed after 9 months. However, all these viscous solutions were still transparent which could mean that TAD maintain its amorphous form.

3.3. Dissolution Tests of TAD Solid Dispersions

It has been reported that amorphous solid dispersion formulations require the use of many excipients for the optimal design, especially to maintain supersaturation and improve physical stability or to ensure shelf-life stability and better absorption during intestinal transit [52]. Choi et al. prepared tadalafil solid dispersion coupled with the incorporation of an acidifier and solubilizer. They found that both tartaric acid increasing wettability and Soluplus® improving solubility contribute to the dissolution rate. Their optimal formulation also contained Aerosil 200 to ensure better flow properties and drug stability [12]. In another study, Choi et al. used malic acid and meglumine at lower contents in the preparation of tadalafil solid dispersion to improve drug solubility and dissolution rate, and Aerosil 200 to improve drug dispersibility [19]. Choi et al. also investigated the effect of various weak acids and bases on tadalafil solid dispersion formulation. Their results indicate that only meglumine significantly improve the apparent drug solubility and dissolution [20]. Obeidat et al. evaluated a mixture of surfactants (Tween 80 and Span) as stabilizers [13]. However, we focused primarily on the effect of different structural polymers and methods of solid dispersion preparation on the dissolution mechanism and release kinetics of tadalafil from amorphous solid dispersions. These properties strongly depend upon the nature of all components, but the dissolution rate of the drug is mainly affected by the aforementioned factors (see Introduction) which are associated with the polymer carrier. Therefore, the drug to polymer carrier ratio was fixed at 5% drug loading in order to maximize the effect of polymer solubilisation and probability complete amorphous form. The pure TAD drug and PMs thereof with the carrier materials were used as references.

3.3.1. Solubility Testing

Prior to the dissolution tests with the polymer, the solubility studies of pure TAD (as was received) were compared in various dissolution media, as shown in Figure 7.

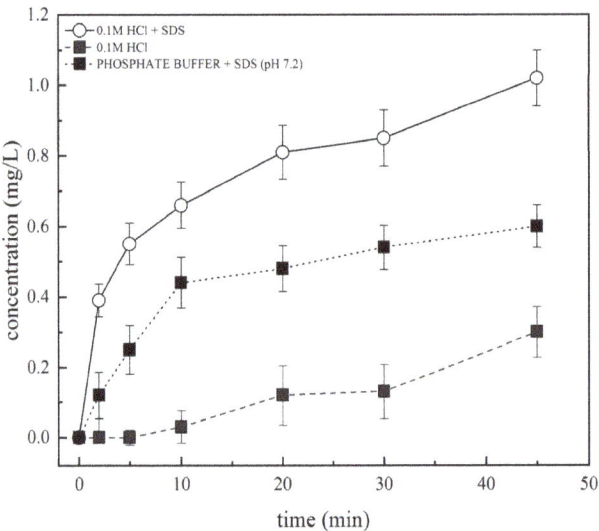

Figure 7. Solubility testing of TAD in various dissolution media.

TAD is classified as class IV in the Biopharmaceutical Classification System and is practically insoluble in water (2 µg/mL) [53]. It was found that the solubility of TAD was maximum in 0.1 M hydrochloric acid + SDS. On the other hand, the solubility of TAD was minimum in 0.1 M hydrochloric acid. Moreover, the results of TAD solubility in phosphate buffer (pH 7.2) + SDS were affected by the ion exchange between phosphate buffer and SDS. For this reason, these data were not included in the evaluation. Dissolution tests were then performed in 0.1M hydrochloric acid + SDS. The addition of SDS to the dissolution medium can have an impact on the magnitude of drug concentrations, but the relative order of the release profiles should not be changed, particularly at low drug concentration [16].

3.3.2. Assessment of the Erosion-Diffusion Mechanism of TAD Release Using Wood's Apparatus (AIDR Measurements)

A general mathematical model describing the dissolution behavior was proposed e.g., in [48,54,55]. Moreover, AIDR was described in [48]. However, there is still no model currently available to describe the solubility and dissolution behavior of amorphous SDs formulated with hydrophilic polymers [56]. Drug release from hydrophilic matrices is generally considered to be governed by the complex interactions between dissolution, diffusion and erosion mechanisms [57]. Pharmacopoeia defines the intrinsic dissolution rate for pure API (active pharmaceutical ingredient). Therefore, if some excipient is added to this drug, we can then observe whether the release rate is accelerated or retarded (see Figure 8).

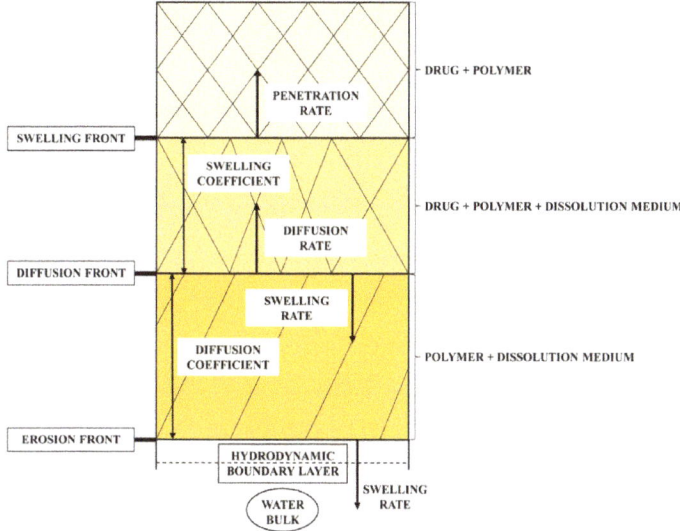

Figure 8. Scheme of front position during dissolution.

Such dissolution tests can be performed using Wood´s apparatus [1]. The transport phenomena can then only be allowed in axial direction. It gives the advantages of exposing a constant surface area of the formulation to the dissolution medium, and therefore it eliminates the influence of different surface area of powders by their compaction [26].

Prior to dissolution tests with the polymer, the release rates of pure TAD were performed as reference. The release profiles of TAD from PMs and SDs with various polymers prepared by different methods are shown in Figure 9.

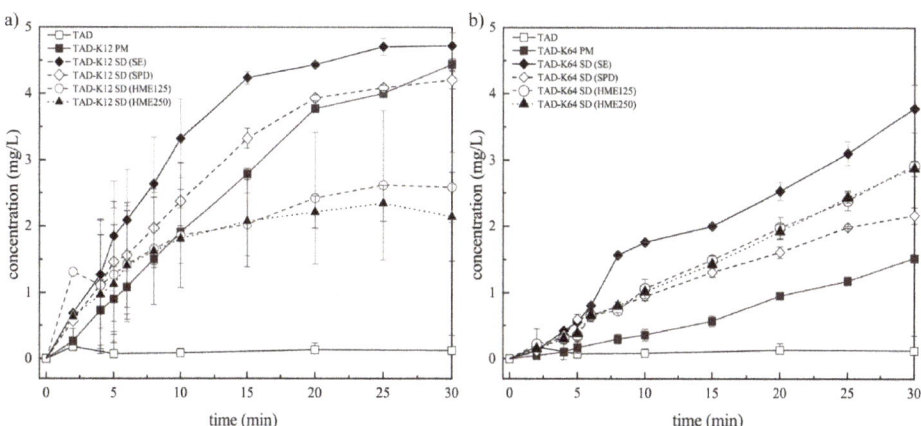

Figure 9. Release profiles of TAD from PMs and SDs prepared by different methods measured as apparent *IDR* using Wood´s apparatus: (**a**) K12 groups and (**b**) K64 groups.

Tables 1 and 2 summarize the values of AIDR for all tested samples. The slope of the initial linear part of a release profile was used as AIDR for samples which exhibited curved profiles (namely up to the 10th minute for PM and SDs containing K12).

Table 1. Apparent intrinsic dissolution rate of TAD (K12 polymer matrix).

Sample	AIDR (mg·min^{-1}·cm^{-2})
pure TAD	0.002
TAD-K12 PM	0.380
TAD-K12 SD (SE)	0.662
TAD-K12 SD (SPD)	0.433
TAD-K12 SD (HME125)	0.245
TAD-K12 SD (HME250)	0.279

Table 2. Apparent intrinsic dissolution rate of TAD (K64 polymer matrix).

Sample	AIDR (mg·min^{-1}·cm^{-2})
pure TAD	0.002
TAD-K64 PM	0.081
TAD-K64 SD (SE)	0.247
TAD-K64 SD (SPD)	0.156
TAD-K64 SD (HME125)	0.245
TAD-K64 SD (HME250)	0.194

The release profile from SDs obviously varied depending on the methods of preparation but there is notable improvement in release rate and the quantity of released TAD for all SDs over pure TAD. The release rates of TAD were found to be significantly different from each specific method and polymer. Figure 9 reveals that there was noticeable influence of polymers and preparation methods on TAD dissolution rate. It is also clear that the pure TAD had the lowest dissolution rate. As can be seen in Figure 9a, the presence of hydrophilic soluble K12 caused faster medium penetration, faster release and faster polymer erosion in comparison to pure TAD. The initial TAD release rate is faster from PM and SDs than pure TAD release rate in the first 15 min followed by a second stage with a slower almost constant drug release rate. A subsequent slowdown in the dissolution was not caused by saturation of the solution with TAD but might have been the effect of progressive swelling of polymers and hindered diffusion of TAD molecules from polymer, as was described e.g., in [3]. During this diffusion period, the thickness of the viscous gel layer on the tablet surface increased over time, leading to longer diffusion path for TAD into the bulk dissolution medium. Consequently, the release of TAD was then limited by water penetration to the tablets as well as drug diffusion through the gel layer. That suggests this type of release is only due to the combination of erosion and diffusion mechanisms.

In general, up to three phases can be observed in the AIDR profiles in Figure 9. The initial AIDR increase for SDs over that of the pure TAD is caused by improving TAD wettability by surrounding hydrophilic carrier, reducing TAD particle size effectively to single molecules and reducing its crystallinity. This is also facilitated by enhanced water penetration into the matrix and the polymer swelling. As the swelling and the diffusion fronts travel deeper into the matrix, the distance between the matrix surface (erosion front) and the diffusion front increase, thus increasing the diffusion path length, leading to reduced AIDR over time, which is represented by the second (transient) phase of gradually decreasing slope of the release profile for some formulations, especially those using the K12 polymer. This AIDR reduction means also slower progress of the diffusion front. Once the diffusion front progress become equal to that of the erosion front, the thickness of the diffusion layer and length of the diffusion path of TAD approaches its steady state, which results in the third phase of steady release at slower rate than in the initial phase. This phase is displayed for the PM and SD (prepared by SE or SPD) formulations using the K12 polymer at dissolution times over approximately 20 min. Some formulations may never reach the third phase (e.g., HME formulations using K12), which means the progress of the erosion front is negligible. Yet other formulations may maintain their initial AIDR (e.g., K64 formulations) without entering the transient second phase, which means the erosion front progress at the same rate as the diffusion front from the very start of the experiment and no significant diffusion barrier is developed. The fastest TAD release, as well as its largest amount released, was observed in SD prepared by SE. The different morphology of particles (obtained SEM) resulted in differences in disk intrinsic dissolution rate or also in medium apparent solubility (see below). This is probably because SD particles contain irregular fracture edges (Figure 1d) leading to increase in surface area and therefore, to faster TAD release. Furthermore, it is shown that TAD release rate of SD prepared by SPD is almost comparable to the corresponding PM. The incorporation or encapsulation of drug into the matrices during spray drying can affect the drug release rate [2,58]. The swellability has a significant effect on the release kinetics of an incorporated drug [43]. Therefore, TAD was probably encapsulated by K12 during spray drying and TAD release was then slower. In the case of SD prepared by HME, TAD release is the slowest in comparison to other SDs or PM containing K12. TAD dissolution rates from these extrudates were considerably greater in comparison with the pure TAD and PM at the initial phase, even though they were observed to decrease over the duration of the study. Their dissolution profiles overtake that of the PM at the beginning of the experiment (approximately during 10 min). This is probably because TAD is better wetted in the both sieve fractions of extrudates. Since their dissolution profiles overlap, they were no noteworthy differences in the porosity between both extrudates as was described e.g., in [15]. Subsequently, its release is probably retarded by the gel layer and therefore, diffusion mechanism predominated.

It is clear from Figure 9b that polymer dissolution influences the TAD release profile significantly. Drug release decreased if polymer molecular weight increased [59] and for this reason, polymer erosion rate increased with the decrease of polymer molecular weight. K12 has a lower molecular weight (2500 g/mol) in comparison to K64 (45,000 g/mol), wherein greater the molecular weight, higher the viscosity and slower the release can be observed. During SDs preparation, solid dispersions containing K64 was of high viscosity, especially in HME. Consequently, the pathways to be overcome by the TAD to be released were very much shortened in the case of the systems containing K12. For this reason, the TAD releases were influenced by the polymer molecular weight because at high molecular weights, the polymer was also more entangled. The initial polymer dissolution rate is usually zero until the entanglement strength is reduced by the increased penetrant concentration and the polymer dissolves because of chain disentanglement [60,61]. Therefore, the polymeric segments of K12 were not so entangled and for this reason, the thickness of the viscous layer on the tablet surface was weaker in comparison to tablets containing K64. The gel strength corresponds actually to resistance to dissolution media penetration [62]. The tablets containing K64 displayed a totally linear release profile, whereas the tablets containing K12 showed a non-linear release (Figure 9). Therefore, the tablets containing K12 seems to reach the plateau where no more TAD is dissolved after 15 min, whereas the TAD still continues to be released from the tablets containing K64 even after 30 min of the experiment as explained by the different erosion rate above. The linear release providing that the releasing area is kept constant was also observed in [39] when the polymer was sufficiently soluble and therefore, the gel layer thickness remained constant because the fronts in the matrix moved in a synchronized way. It can be caused by the chemical structure of the Kollidons. Figure 9b shows the swelling of the tablets (upon complete hydration approximately during 5 min) and gradual erosion of K64. It can be attributed to the fact that K64 contains lipophilic vinyl acetate which is insoluble in the aqueous medium and hydrophilic vinylpyrrolidone which can form the channels in the tablets. This phenomenon was described in [29] for Kollidon® SR. Therefore, this monomer allows better capture and penetration of water by the tablets and hence a greater viscosity of the gel. Medium then penetrates into the free spaces on the surface between the polymer chains. TAD can diffuse slowly through these channels which is reflected in its slower release rates in comparison to the tablets containing K12. However, the release rate would be controlled solely by mechanisms of erosion due to higher AIDR values in comparison to pure TAD.

Dissolution from both PMs showed also improvement in the release rate of TAD. However, the fastest TAD release, as well as its largest amount released, was also observed in SD prepared by SE as compared with that of solid dispersions and pure form of TAD. This is probably because SD particles also contain irregular fracture edges (Figure 2d) leading to increase in surface area and therefore, to faster TAD release. Furthermore, it is also shown that the TAD release rate of SD prepared by SPD was retarded in comparison to other SDs. TAD was probably also encapsulated by K64 during SPD. TAD release is then slower, although, the very small SD particles were obtained by this method of preparation (Figure 2e). There was not significant difference between the release rates of SDs sieve fractions prepared by HME and their profiles were practically superimposed. Therefore, the utilization of the same compression conditions for different SDs can paradoxically lead to varying degree of compaction and is crucial in dissolution, as was described in [16]. However, best solid dispersion form of TAD was prepared by solvent evaporation using both polymers. Slower and lower TAD release from all extrudates might also be due to the property of the polymer to undergo thermoreversible gelling at high concentration after the melting process and subsequent cooling during the formulation of SDs [5].

3.3.3. Effect of the Preparation Method on the Dissolution Behavior of TAD from Solid Dispersions

Flow-through cell apparatus for powders was used to describe the dissolution behavior of SDs. The dissolution profiles from the SDs containing K12 or K64 obviously varied depending on the methods of preparation and polymer type. Figure 10 represents normalized data. The normalization was used to more easily compare data from different SDs because possibly non-homogeneity of some samples (especially in the case of TAD-K12 SD HME125 or HME250) were observed. The original data are then shown below (namely in Figure 13). Figure 10a,b shows that the powders containing K12 allow a faster gradual release of TAD from K12 in comparison to powders containing K64.

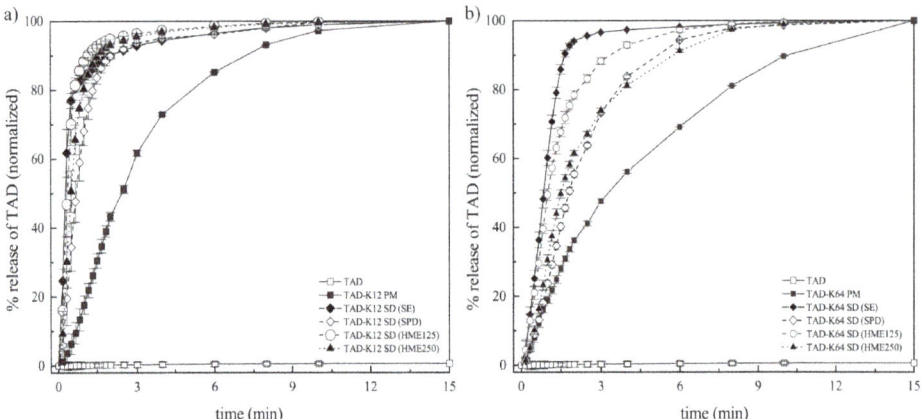

Figure 10. Release profiles of TAD from PMs and SDs prepared by different methods: (**a**) K12 groups and (**b**) K64 groups.

SDs containing K12 also show similar dissolution profiles that reach the identical asymptotic amounts of TAD in the medium. After the initial dissolution times, most of the TAD released at comparable release rates except for PM which showed the lowest release rates among all SDs. As shown in Figure 10, the TAD amounts, therefore, are very high in the first 3 min, and then the amounts become slower and are almost constant. The same trend can be observed for SDs containing K64, but the dissolution is slower compared to K12. It can be attributed to enhancement of powders wettability. Molecular dispersion of poorly water-soluble drug is the most desirable type in the theory and practice of amorphous solid dispersions because it is related to the enhancement of the solubility by the weakening of solute-solute interactions. TAD has a very high binding energy between its molecules due to its high melting point (302 °C). For this reason, the reduction in the binding energy allows for the easier passage of molecules from the amorphous solid state to the medium. It is additionally facilitated by detachment of polymer chains and subsequent increase in the contact surface with medium leading to the considerable solubility improvement [3]. Therefore, in the case of SDs containing K64, this could be attributed to the lowered mobility of TAD particles or molecules embedded within the K64 matrix system particularly in the initial phase. From the above AIDR measurements, it appears that although the SDs enhance the TAD release rate, in these dissolution experiments, there was not a significant difference between the dissolution behaviors of the SDs. It may be therefore concluded that flow-through cell data measured on powders emphasize the effects related to the prepared particles of SDs and therefore are relevant to situations corresponding to the first phase of the AIDR experiments. Since the particles are relatively small and no tablet was compressed the swelling and erosion effects are suppressed, while the particle surface effects are more pronounced. It is also obvious that the largest amount of TAD was dissolved in the case of SDs containing K12 or K64 prepared by SE. As mentioned above, their SDs contained irregular fracture edges leading to the increase in surface area (Figures 1d and 2d). On the other hand, the TAD slowest release was observed in the SDs containing K12 or K64 prepared by SPD. This was also commented above. The incorporation of TAD into the K12 or K64 matrix also affects the dissolution behavior. Therefore, SDs prepared by SE showed the highest dissolution rate among them, and both extrudates having particles of 125–250 μm showed the second highest dissolution rate among them.

The results of dissolution studies and the initial dissolution rates during the first 1.5 min (for the systems containing K12) and 4 min (for the systems containing K64) for all SDs and corresponding PMs are depicted in the Figure 11a,b and Figure 12a,b.

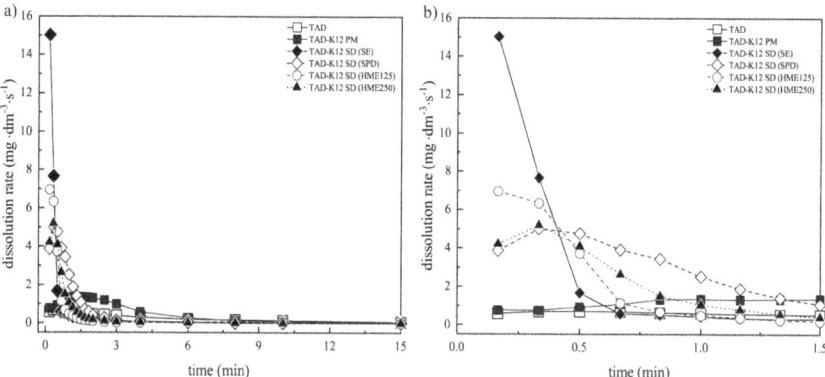

Figure 11. Dissolution rates of TAD from PM and SDs containing K12: (**a**) during 15 min and (**b**) during 1.5 min.

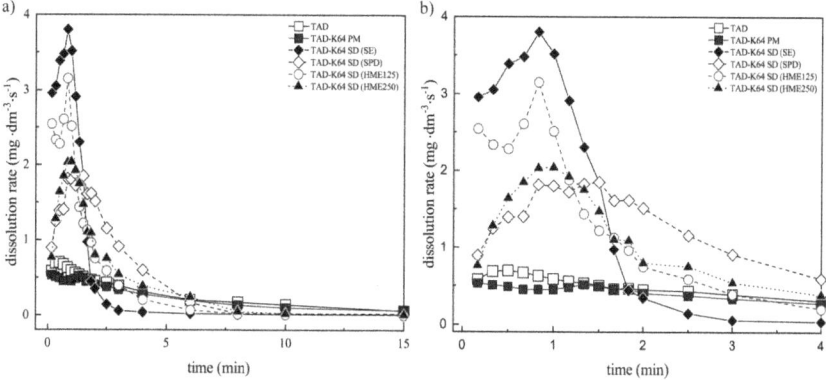

Figure 12. Dissolution rates of TAD from PM and SDs containing K64: (**a**) during 15 min and (**b**) during 4 min.

The most significant differences between the prepared samples were observed within 1.5 min. For this reason, the Figures 11b and 12b show more detail of the dissolution experiments. The results are represented as data points. From these values, dissolution rates of TAD were always higher from the SDs compared with pure TAD or the PMs. It was also found that the presence of hydrophilic K12 or K64 in the SDs caused faster dissolution rate that was in compliance with the results mentioned above. It was also observed that the SDs containing K12 showed faster dissolution rate compared to K64. The fastest TAD dissolution rate was also measured in the SDs containing K12 or K64 prepared by SE due to the presence of irregular fracture edges. The fast initial rate can be attributed to the hydrogen bonding between TAD and both polymers, which breaks relatively easily during dissolution compared with pure drug [17].

3.3.4. Weibull Dissolution Model

In order to describe the TAD release, the obtained dissolution profiles were fitted to the Weibull model which is adapted to the release process. This kinetic model can be expressed by the Equation (4):

$$A_t = A_\infty \cdot \left[1 - \exp\left(-k \cdot (t - t_0)^b\right)\right], \tag{4}$$

where A_t is the amount of drug released in time t, A_∞ is the maximum releasable amount of API, k corresponds to the reciprocal value of time scale of the release process, t_0 is the location parameter and represents the lag time before the onset of the dissolution (in most case is equal to zero) and b describes the shape of the dissolution curve [63]. When the shape parameter b is equal to one, the Weibull model corresponds to the first order kinetic model and therefore, the parameter k corresponds to the first order release rate constant. ERA software [64] was used for the fitting the Weibull equation to the experimental data. Figure 13 shows TAD release profiles fitted by the Weibull model. As can be seen in Figure 13, the model was in agreement with the dissolution profile and therefore, it was appropriately used. Therefore, the kinetic parameters of the model and their standard deviations (STDs) are summarized in Tables 3 and 4.

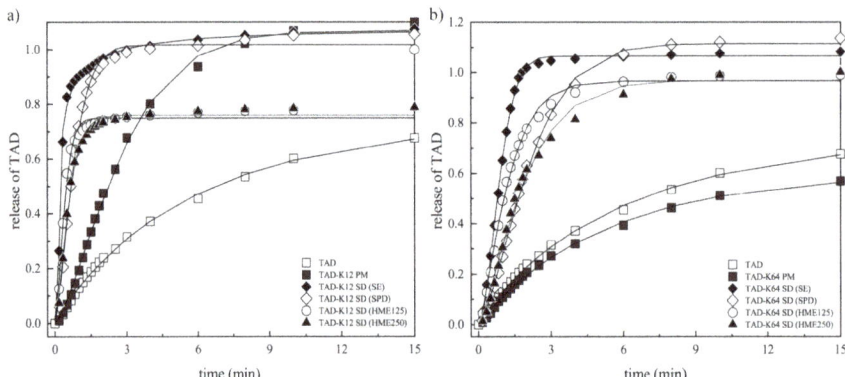

Figure 13. Dissolution profiles of TAD formulations (points) fitted by Weibull model (lines): (**a**) K12 groups and (**b**) K64 groups.

Table 3. Kinetic parameters of TAD release (K12 polymer matrix).

Sample	k (s^{-b}) ± STD	b (-) ± STD	t_0 (s) ± STD
pure TAD	0.21 ± 0.04	0.99 ± 0.10	0.00 ± 0.19
TAD-K12 PM	0.29 ± 0.04	1.13 ± 0.08	0.26 ± 0.10
TAD-K12 SD (SE)	1.92 ± 0.06	0.33 ± 0.04	0.17 ± 0.00
TAD-K12 SD (SPD)	1.27 ± 0.87	1.02 ± 1.15	0.13 ± 0.46
TAD-K12 SD (HME125)	2.30 ± 0.36	0.66 ± 0.21	0.15 ± 0.03
TAD-K12 SD (HME250)	1.76 ± 1.15	0.86 ± 1.16	0.14 ± 0.31

Table 4. Kinetic parameters of TAD release (K64 polymer matrix).

Sample	k (s^{-b}) ± STD	b (-) ± STD	t_0 (s) ± STD
pure TAD	0.21 ± 0.04	0.99 ± 0.10	0.00 ± 0.19
TAD-K64 PM	0.23 ± 0.03	0.96 ± 0.07	0.11 ± 0.12
TAD-K64 SD (SE)	0.93 ± 0.23	1.71 ± 0.29	0.00 ± 0.13
TAD-K64 SD (SPD)	0.38 ± 0.05	1.20 ± 0.10	0.21 ± 0.09
TAD-K64 SD (HME125)	0.77 ± 0.09	1.06 ± 0.14	0.13 ± 0.08
TAD-K64 SD (HME250)	0.61 ± 0.06	0.85 ± 0.10	0.41 ± 0.09

Since the parameter k corresponds primarily to the initially release rate, the Figure 14a,b represents TAD release profile with respect to this fact only up to the second minute.

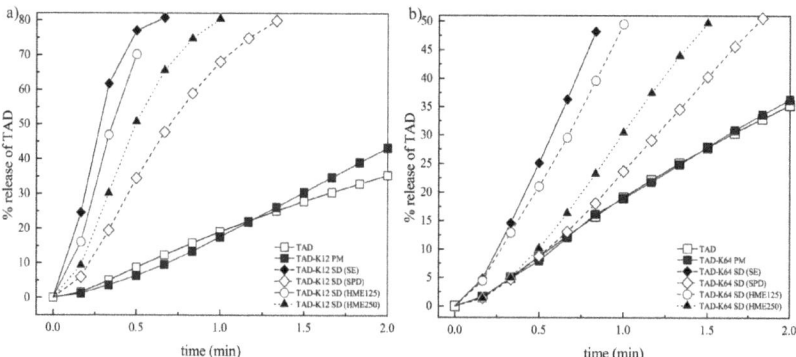

Figure 14. TAD release profile during 2 min: (a) K12 groups and (b) K64 groups.

The values of b about 1 were found for pure TAD, SD containing K12 prepared by SPD, PM containing K64 and SD containing K64 prepared by HME (specially for the sieve fraction 125–250 µm) (Tables 3 and 4). In other case, the release rate is retarded in comparison to the first order kinetic ($b < 1$), i.e., for SDs containing K12 prepared by SE and HME (both sieve fractions) and SD containing K64 prepared by HME (sieve fraction 250–425 µm). On the other hand, the release rate is accelerated in comparison to the first order kinetics ($b > 1$), i.e., for PM containing K12 and SDs containing K64 prepared by SE and SPD. These results confirm the fact that the parameter k correlate with the amount of drug released at the beginning of the dissolution experiment.

4. Conclusions

In this study, TAD amorphous solid dispersions were successfully obtained using the solvent evaporation method, spray drying technique, as well as hot-melt extrusion. Two grades of polyvinylpyrrolidone (K12 and K64) were used as a polymeric carrier. Therefore, the use of different preparation methods for SDs and polymers having different physicochemical properties can provide greater insights into the importance of various mechanisms of the dissolution process.

SEM revealed significant differences in morphology of all samples which were reflected in the dissolution tests. FT-IR spectra confirmed the interactions between the drug and both Kollidons as well as the presence of hydrogen bonds between the components. These interactions were weaker in the binary mixtures containing K64 compared to K12. The presence of hydrogen bonds between the drug and the polymers improved the physical stability of the SDs in their extruded forms. Therefore, stability studies for nine months confirmed the extrudates to be amorphous. On the contrary, the SDs of both polymers in powder form changed their character due to high relative humidity.

The Wood's apparatus was found to be suitable for determining the apparent intrinsic dissolution rate and for identification of critical factors affecting the erosion-diffusion mechanism of TAD release. TAD release from compressed SDs containing K12 was controlled by combination of erosion and diffusion mechanisms. The diffusion mechanisms were predominant in the initial phase of experiment and the slow erosion was dissolution-controlling at the second stage. TAD release rate from SDs containing K64 was controlled solely by mechanisms of erosion. The dissolution profiles obtained by Wood's apparatus were in agreement with dissolution profiles obtained using flow-through cell apparatus. The fastest TAD release, as well as its largest amount released, was observed in SD containing K12 or K64 prepared by solvent evaporation method. This is probably because SD particles contain irregular fracture edges, which was revealed by SEM, leading to increase in surface area and to faster TAD release. In the case of SDs containing K12 or K64 prepared by spray drying, TAD was probably encapsulated by K12 or K64 during spray drying. TAD release was then slower, although, the very small SD particles were obtained by this method of preparation. The effect of polymer molecular weight on the release rate was also observed. K12 has a lower molecular weight in comparison to K64. The TAD releases were then influenced by the polymer molecular weight because at high molecular weights, the polymer was more entangled. For this reason, the SDs containing K12 showed faster dissolution rate compared to K64. Weibull dissolution model was suitably used to describe the TAD release at the beginning of the dissolution experiment as well as to assess whether the release corresponds to the first order kinetics. There was noticeable influence of polymers on TAD solubility. All solid dispersions improved the TAD intrinsic dissolution rate, however, the greatest increase in the TAD dissolution rate was obtained from SD containing K12 prepared solvent evaporation. It can also be concluded that all SDs of TAD showed considerable enhancement in dissolution rate compared to both PMs and the dissolution rate of both PMs was higher compared to the pure TAD. The rapid dissolution of TAD from SDs, especially in the initial phase, may be attributed to its molecular dispersion in polymer carriers.

In order to summarize, the dissolution experiments revealed that the TAD release from the solid dispersion is controlled by different processes in combined mechanism of surface hydrophilization, dissolution, diffusion in swollen matrix, and the matrix erosion, depending not only on the polymer used, but also on the method of preparing the solid dispersion and further processing thereof. All SDs with hydrophilic polymers enhanced the initial dissolution rate of TAD in both the AIDR arrangement and the USP 4 for solid dispersion over the dissolution rate of pure TAD or TAD PMs. While the enhancement occurred for all SDs systematically, it was of different strength for both the different polymers and the different preparation methods. In general, the dissolution was more enhanced by K12 than K64 polymers and the SDs prepared by the solvent evaporation released the TAD faster than SDs prepared by other methods. Those results should be interpreted as the contribution of hydrophilization and TAD dispersion to dissolution. The AIDR experiments also showed the differences in compressed forms prepared from SD particles. Most notably, the two polymers tested exhibited entirely different behavior. While K64 SDs dissolution was not hindered by the diffusion because the polymer matrix erosion controlled the process and the AIDR proceeded at a steady rate for all samples, K12 dispersions exhibited different erosion rates for different preparation methods resulting in very different AIDR profiles and diffusion hindrance, corresponding to different diffusion layers formed by the diffusion-erosion rate equilibrium.

Apparent intrinsic dissolution rate studies and apparent solubility revealed the greatest increase in TAD solubility and significant dissolution rate enhancement for all SDs in comparison with crystalline TAD and its PMs. The proposed SDs based on K12 or K64 showed an interesting potential for improving the oral bioavailability of the poorly water-soluble tadalafil. This information can be also used either to optimize formulation to obtain the desired release profile or provide a better understanding into the mechanism of TAD release from solid dispersions.

Author Contributions: T.Š. was responsible for theoretical background, coordinating the work, analyzing data, DSC analysis and writing the article. M.S. realized the experiments (namely preparation of solid dispersions, solubility test, FT-IR spectroscopy, dissolution and stability studies). A.Š. performed SEM analysis. A.K. optimized the preparation of solid dispersions by hot-melt extrusion. J.P. contributed by FT-IR interpretation. P.Z. contributed by dissolution evaluation and DSC analysis and was responsible for final correction.

Funding: This research was financially supported by specific university research MSMT No. 21-SVV/2019. This work was realized within the Operational Programme Prague-Competitiveness (CZ.2.16/3.1.00/24501) and "National Program of Sustainability" (NPU I LO1613 MSMT-43760/2015).

Acknowledgments: The authors would like to acknowledge the BASF Pharma for providing polymers and Vojtěch Klimša for the cooperation on spray drying technique.

Conflicts of Interest: The authors declare no conflict of interest.

Appendix A

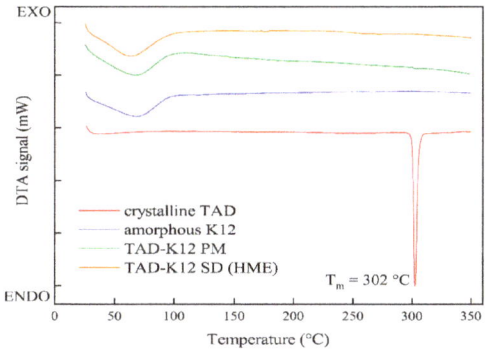

Figure A1. DSC thermograms of (from bottom to top) crystalline TAD, amorphous K12, TAD-K12 PM, and TAD-K12 SD (HME).

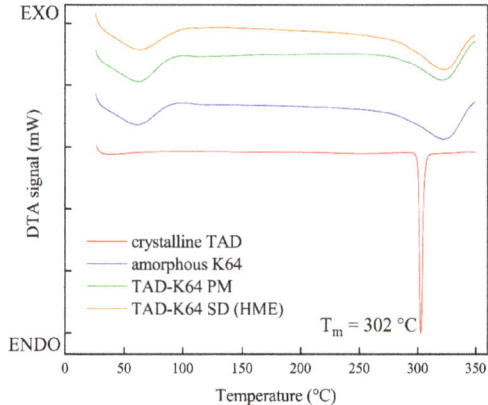

Figure A2. DSC thermograms of (from bottom to top) crystalline TAD, amorphous K64, TAD-K64 PM, and TAD-K164 SD (HME).

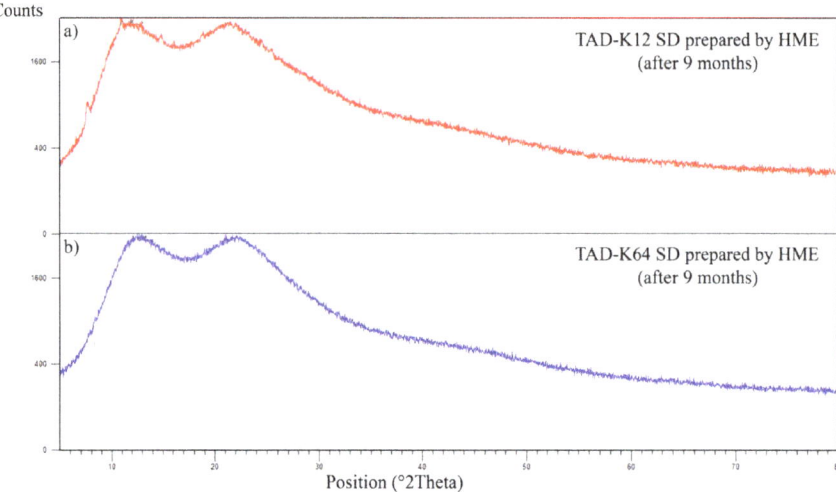

Figure A3. X-ray diffraction patterns of extruded SDs after 9 months: (**a**) TAD-K12 and (**b**) TAD-K64.

References

1. Tres, F.; Treacher, K.; Booth, J.; Hughes, L.P.; Wren, S.A.C.; Aylott, J.W.; Burley, J.C. Real time Raman imaging to understand dissolution performance of amorphous solid dispersions. *J. Control. Release* **2014**, *188*, 53–60. [CrossRef] [PubMed]
2. Davis, M.; Walker, G. Recent strategies in spray drying for the enhanced bioavailability of poorly water-soluble drugs. *J. Control. Release* **2018**, *269*, 110–127. [CrossRef] [PubMed]
3. Wlodarski, K.; Sawicki, W.; Haber, K.; Knapik, J.; Wojnarowska, Z.; Paluch, M.; Lepek, P.; Hawelek, L.; Tajber, L. Physicochemical properties of tadalafil solid dispersions—Impact of polymer on the apparent solubility and dissolution rate of tadalafil. *Eur. J. Pharm. Biopharm.* **2015**, *94*, 106–115. [CrossRef] [PubMed]
4. Craig, D.Q.M. The mechanisms of drug release from solid dispersions in water-soluble polymers. *Int. J. Pharm.* **2002**, *231*, 131–144. [CrossRef]
5. Vyas, V.; Sancheti, P.; Karekar, P.; Shah, M.; Pore, Y. Physicochemical characterization of solid dispersion systems of tadalafil with poloxamer 407. *Acta Pharm.* **2009**, *59*, 453–461. [CrossRef] [PubMed]
6. Wlodarski, K.; Sawicki, W.; Kozyra, A.; Tajber, L. Physical stability of solid dispersions with respect to thermodynamic solubility of tadalafil in PVP-VA. *Eur. J. Pharm. Biopharm.* **2015**, *96*, 237–246. [CrossRef] [PubMed]
7. Huang, Y.; Dai, W.-G. Fundamental aspects of solid dispersion technology for poorly soluble drugs. *Acta Pharm. Sin. B* **2014**, *4*, 18–25. [CrossRef] [PubMed]
8. Prasanna, T.V.; Rani, B.N.; Rao, A.S.; Murthy, T.E.G.K. Design and evaluation of solid dispersed tadalafil tablets. *Int. J. Pharm. Sci. Res.* **2012**, *3*, 4738–4744. [CrossRef]
9. Rane, Y.; Mashru, R.; Sankalia, M.; Sankalia, J. Effect of hydrophilic swellable polymers on dissolution enhancement of carbamazepine solid dispersions studied using response surface methodology. *AAPS PharmSciTech* **2007**, *8*, E1–E11. [CrossRef]
10. Sharma, P.K.; Sharma, P.; Darwhekar, G.N.; Shrivastava, B. Formulation and evaluation of solid dispersion of tadalafil. *Int. J. Drug Regul. Aff.* **2018**, *6*, 26–34. [CrossRef]
11. Mande, P.P.; Bachhav, S.S.; Devarajan, P.V. Bioenhanced advanced third generation solid dispersion of tadalafil: Repurposing with improved therapy in pyelonephritis. *Asian J. Pharm. Sci.* **2017**, *12*, 569–579. [CrossRef]
12. Choi, J.S.; Kwon, S.H.; Lee, S.E.; Jang, W.S.; Byeon, J.C.; Jeong, H.M.; Park, J.S. Use of acidifier and solubilizer in tadalafil solid dispersion to enhance the in vitro dissolution and oral bioavailability in rats. *Int. J. Pharm.* **2017**, *526*, 77–87. [CrossRef]

13. Obeidat, W.M.; Sallam, A.S. Evaluation of tadalafil nanosuspensions and their PEG solid dispersion matrices for enhancing its dissolution properties. *AAPS PharmSciTech* **2014**, *15*, 364–374. [CrossRef]
14. Park, J.; Cho, W.; Kang, H.; Lee, B.B.J.; Kim, T.S.; Hwang, S.-J. Effect of operating parameters on PVP/tadalafil solid dispersions prepared using supercritical anti-solvent process. *J. Supercrit. Fluids* **2014**, *90*, 126–133. [CrossRef]
15. Krupa, A.; Cantin, O.; Strach, B.; Wyska, E.; Tabor, Z.; Siepmann, J.; Wróbel, A.; Jachowicz, R. In vitro and in vivo behavior of ground tadalafil hot-melt extrudates: How the carrier material can effectively assure rapid or controlled drug release. *Int. J. Pharm.* **2017**, *528*, 498–510. [CrossRef]
16. Wlodarski, K.; Sawicki, W.; Paluch, K.J.; Tajber, L.; Grembecka, M.; Hawelek, L.; Wojnarowska, Z.; Grzybowska, K.; Talik, E.; Paluch, M. The influence of amorphization methods on the apparent solubility and dissolution rate of tadalafil. *Eur. J. Pharm. Sci.* **2014**, *62*, 132–140. [CrossRef]
17. Mehanna, M.M.; Motawaa, A.M.; Samaha, M.W. In sight into tadalafil-block copolymer binary solid dispersion: Mechanistic investigation of dissolution enhancement. *Int. J. Pharm.* **2010**, *402*, 78–88. [CrossRef]
18. Refaat, A.; Sokar, M.; Ismail, F.; Boraei, N. Tadalafil oral disintegrating tablets: An approach to enhance tadalafil dissolution. *J. Pharm. Investig.* **2015**, *45*, 481–491. [CrossRef]
19. Choi, J.S.; Park, J.S. Design of PVP/VA S-630 based tadalafil solid dispersion to enhance the dissolution rate. *Eur. J. Pharm. Sci.* **2017**, *97*, 269–276. [CrossRef]
20. Choi, J.S.; Lee, S.E.; Jang, W.S.; Byeon, J.C.; Park, J.S. Tadalafil solid dispersion formulations based on PVP/VA S-630: Improving oral bioavailability in rats. *Eur. J. Pharm. Sci.* **2017**, *106*, 152–158. [CrossRef]
21. El-Badry, M.; Hag, N.; Fetih, G.; Shakeel, F. Solubility and dissolution enhancement of tadalafil using self-nanoemulsifying drug delivery system. *J. Oleo Sci.* **2014**, *63*, 567–576. [CrossRef] [PubMed]
22. Baek, J.S.; Pham, C.V.; Myung, C.S.; Cho, C.W. Tadalafil-loaded nanostructured lipid carriers using permeation enhancers. *Int. J. Pharm.* **2015**, *495*, 701–709. [CrossRef]
23. Bhokare, P.L.; Kendre, P.N.; Pande, V.V. Design and characterization of nanocrystals of tadalafil for solubility and dissolution rate enhancement. *Inventi Impact Pharm. Process Dev.* **2015**, *2015*, 1–7.
24. Badr-Eldin, S.M.; Elkheshen, S.A.; Ghorab, M.M. Inclusion complexes of tadalafil with natural and chemically modified β-cyclodextrins. I: Preparation and *in-vitro* evaluation. *Eur. J. Pharm. Biopharm.* **2008**, *70*, 819–827. [CrossRef] [PubMed]
25. Mehanna, M.M.; Motawaa, A.M.; Samaha, M.W. Tadalafil inclusion in microporous silica as effective dissolution enhancer: Optimization of loading procedure and molecular state characterization. *J. Pharm. Sci.* **2010**, *100*, 1805–1818. [CrossRef] [PubMed]
26. Wlodarski, K.; Tajber, L.; Sawicki, W. Physicochemical properties of direct compression tablets with spray dried and ball milled solid dispersions of tadalafil in PVP-VA. *Eur. J. Pharm. Biopharm.* **2016**, *109*, 14–23. [CrossRef] [PubMed]
27. Krupa, A.; Descamps, M.; Willart, J.-F.; Jachowicz, R.; Danède, F. High energy ball milling and supercritical carbon dioxide impregnation as co-processing methods to improve dissolution of tadalafil. *Eur. J. Pharm. Sci.* **2016**, *95*, 130–137. [CrossRef] [PubMed]
28. Nowak, P.; Krupa, A.; Kubat, K.; Węgrzyn, A.; Harańczyk, H.; Ciułkowska, A.; Jachowicz, R. Water vapour sorption in tadalafil-Soluplus co-milled amorphous solid dispersions. *Powder Technol.* **2019**, *346*, 373–384. [CrossRef]
29. Reza, M.S.; Quadir, M.A.; Haider, S.S. Comparative evaluation of plastic, hydrophobic and hydrophilic polymers as matrices for controlled-release drug delivery. *J. Pharm. Pharm. Sci.* **2003**, *6*, 282–291.
30. Caccavo, D.; Cascone, S.; Lamberti, G.; Barba, A.A.; Larsson, A. Swellable hydrogel-based systems for controlled drug delivery. In *Smart Drug Delivery System*; Sezer, A.D., Ed.; IntechOpen: London, UK, 2016; pp. 237–303. [CrossRef]
31. Ozeki, T.; Yuasa, H.; Kanaya, Y. Controlled release from solid dispersion composed of poly (ethylene oxide) —Carbopol® interpolymer complex with various cross-linking degrees of Carbopol®. *J. Control. Release* **2000**, *63*, 287–295. [CrossRef]
32. Grassi, M.; Colombo, I.; Lapasin, R. Drug release from an ensemble of swellable crosslinked polymer particles. *J. Control. Release* **2000**, *68*, 97–113. [CrossRef]
33. Gajdošová, M.; Pěček, D.; Sarvašová, N.; Grof, Z.; Štěpánek, F. Effect of hydrophobic inclusions on polymer swelling kinetics studied by magnetic resonance imaging. *Int. J. Pharm.* **2016**, *500*, 136–143. [CrossRef] [PubMed]

34. Vueba, M.L.; Batista de Carvalho, L.A.E.; Veiga, F.; Sousa, J.J.; Pina, M.E. In vitro release of ketoprofen from hydrophilic matrix tablets containing cellulose polymer mixtures. *Drug Dev. Ind. Pharm.* **2012**, *39*, 1651–1662. [CrossRef] [PubMed]
35. Maderuelo, C.; Zarzuelo, A.; Lanao, J.M. Critical factors in the release of drugs from sustained release hydrophilic matrices. *J. Control. Release* **2011**, *154*, 2–19. [CrossRef] [PubMed]
36. Wan, L.S.C.; Heng, P.W.S.; Wong, L.F. Relationship between swelling and drug release in a hydrophilic matrix. *Drug. Dev. Ind. Pharm.* **1993**, *19*, 1201–1210. [CrossRef]
37. Lamoudi, L.; Chaumeil, J.C.; Daoud, K. Swelling, erosion and drug release characteristics of Sodium Diclofenac from heterogeneous matrix tablets. *J. Drug Deliv. Sci. Technol.* **2016**, *31*, 93–100. [CrossRef]
38. Sujja-areevath, J.; Munday, D.L.; Cox, P.J.; Khan, K.A. Relationship between swelling, erosion and drug release in hydrophilic natural gum mini-matrix formulations. *Eur. J. Pharm. Sci.* **1998**, *6*, 207–217. [CrossRef]
39. Colombo, P.; Bettini, R.; Santi, P.; De Ascentiis, A.; Peppas, N.A. Analysis of the swelling and release mechanisms from drug delivery systems with emphasis on drug solubility and water transport. *J. Control. Release* **1996**, *39*, 231–237. [CrossRef]
40. Colombo, P.; Bettini, R.; Peppas, N.A. Observation of swelling process and diffusion front position during swelling in hydroxypropyl methyl cellulose (HPMC) matrices containing a soluble drug. *J. Control. Release* **1999**, *61*, 83–91. [CrossRef]
41. Colombo, P.; Bettini, R.; Massimo, G.; Catellani, P.L.; Santi, P.; Peppas, N.A. Drug diffusion front movement is important in drug release control from swellable matrix tablets. *J. Pharm. Sci.* **1995**, *84*, 991–997. [CrossRef]
42. Lamberti, G.; Galdi, I.; Barba, A.A. Controlled release from hydrogel-based solid matrices. A model accounting for water up-take, swelling and erosion. *Int. J. Pharm.* **2011**, *407*, 78–86. [CrossRef] [PubMed]
43. Siepmann, J.; Peppas, N.A. Modeling of drug release from delivery systems based on hydroxypropyl methylcellulose (HPMC). *Adv. Drug Deliv. Rev.* **2001**, *48*, 139–157. [CrossRef]
44. Wu, N.; Wang, L.-S.; Tan, D.C.-W.; Moochhala, S.M.; Yang, Y.-Y. Mathematical modeling and in vitro study of controlled drug release via a highly swellable and dissoluble polymer matrix: Polyethylene oxide with high molecular weights. *J. Control. Release* **2005**, *102*, 569–581. [CrossRef] [PubMed]
45. Chavanpatil, M.D.; Jain, P.; Chaudhari, S.; Shear, R.; Vavia, P.R. Novel sustained release, swellable and bioadhesive gastroretentive drug delivery system for ofloxacin. *Int. J. Pharm.* **2006**, *316*, 86–92. [CrossRef] [PubMed]
46. Colombo, P.; Bettini, R.; Santi, P.; Peppas, N.A. Swellable matrices for controlled drug delivery: Gel-layer behaviour, mechanisms and optimal performance. *Pharm. Sci. Technol. Today* **2000**, *3*, 198–204. [CrossRef]
47. Punčochová, K.; Heng, J.Y.Y.; Beránek, J.; Štěpánek, F. Investigation of drug-polymer interaction in solid dispersions by vapour sorption methods. *Int. J. Pharm.* **2014**, *469*, 159–167. [CrossRef] [PubMed]
48. Zámostný, P.; Petrů, J.; Majerová, D. Effect of maize starch excipient properties on drug release rate. *Procedia Eng.* **2012**, *42*, 482–488. [CrossRef]
49. Ford, J.L. The current status of solid dispersions. *Pharm. Acta Helv.* **1986**, *61*, 69–88.
50. Vo, C.L.-N.; Park, C.; Lee, B.J. Current trends and future perspectives of solid dispersions containing poorly water-soluble drugs. *Eur. J. Pharm. Biopharm.* **2013**, *85*, 799–813. [CrossRef]
51. Thielmann, F.; Williams, D. Determination of the glass transition temperature of maltose and its dependence on relative humidity by inverse gas chromatography. *Dtsch. Lebensm.-Rundsch.* **2000**, *96*, 255–257.
52. Hurley, D.; Potter, C.B.; Walker, G.M.; Higginbotham, C.L. Investigation of ethylene oxide-co-propylene oxide for dissolution enhancement of hot-melt extruded solid dispersions. *J. Pharm. Sci.* **2018**, *107*, 1372–1382. [CrossRef] [PubMed]
53. Ghadge, O.; Samant, M.; Khale, A. Determination of solubility of tadalafil by shake flask method by employing validated HPLC analytical method. *World J. Pharm. Res.* **2015**, *4*, 1370–1382.
54. Petru, J.; Zamostny, P. Analysis of drug release from different agglomerates using a mathematical model. *Dissolut. Technol.* **2014**, *21*, 40–47. [CrossRef]
55. Petru, J.; Zamostny, P. Prediction of dissolution behavior of final dosage forms prepared by different granulation methods. *Procedia Eng.* **2012**, *42*, 1463–1473. [CrossRef]
56. Li, N.; Taylor, L.S. Tailoring supersaturation from amorphous solid dispersions. *J. Control. Release* **2018**, *279*, 114–125. [CrossRef] [PubMed]
57. Shoaib, M.H.; Tazeen, J.; Merchant, H.A.; Yousuf, R.I. Evaluation of drug release kinetics from ibuprofen matrix tablets using HPMC. *Pak. J. Pharm. Sci.* **2006**, *19*, 119–124. [PubMed]

58. Khan, N.; Craig, D.Q.M. The influence of drug incorporation on the structure and release properties of solid dispersions in lipid matrices. *J. Control. Release* **2003**, *93*, 355–368. [CrossRef] [PubMed]
59. Zuleger, S.; Lippold, B.C. Polymer particle erosion controlling drug release. I. Factors influencing drug release and characterization of the release mechanism. *Int. J. Pharm.* **2001**, *217*, 139–152. [CrossRef]
60. Borgquist, P.; Körner, A.; Piculell, L.; Larsson, A.; Axelsson, A. A model for the drug release from a polymer matrix tablet-effects of swelling and dissolution. *J. Control. Release* **2006**, *113*, 216–225. [CrossRef]
61. Katzhendler, I.; Hoffman, A.; Goldberger, A.; Friedman, M. Modeling of drug release from erodible tablets. *J. Pharm. Sci.* **1997**, *86*, 110–115. [CrossRef]
62. Chirico, S.; Dalmoro, A.; Lamberti, G.; Russo, G.; Titomanlio, G. Analysis and modeling of swelling and erosion behavior for pure HPMC tablet. *J. Control. Release* **2007**, *122*, 181–188. [CrossRef]
63. Costa, P.; Lobo, J.M.S. Modeling and comparison of dissolution profiles. *Eur. J. Pharm. Sci.* **2001**, *13*, 123–133. [CrossRef]
64. Zamostny, P.; Belohlav, Z. A software for regression analysis of kinetic data. *Comput. Chem.* **1999**, *23*, 479–485. [CrossRef]

© 2019 by the authors. Licensee MDPI, Basel, Switzerland. This article is an open access article distributed under the terms and conditions of the Creative Commons Attribution (CC BY) license (http://creativecommons.org/licenses/by/4.0/).

MDPI
St. Alban-Anlage 66
4052 Basel
Switzerland
Tel. +41 61 683 77 34
Fax +41 61 302 89 18
www.mdpi.com

Pharmaceutics Editorial Office
E-mail: pharmaceutics@mdpi.com
www.mdpi.com/journal/pharmaceutics

www.ingramcontent.com/pod-product-compliance
Lightning Source LLC
LaVergne TN
LVHW071947080526
838202LV00064B/6697